W9-CIK-130

SCIENTIFIC MAN
The Humanistic Significance of Science

INTRODUCTION TO SCIENTIFIC HUMANISM

A series of exploratory studies undertaken by the Institute for Scientific Humanism to prepare a synthesis of science and the humanities.

Initial entries:

ENRICO CANTORE, *Director*

> *Scientific Man: The Humanistic Significance of Science*

> *Science and Human Authenticity: The Philosophical Quest for Humanism in the Scientific Age* (in preparation)

> *Humanistic Approach to Science: Methodological Investigations and Reflections* (in preparation)

FREDERICK E. TRINKLEIN, *Deputy Director*

> *Ethical and Religious Implications of Science: Interviews with Internationally Leading Scientists* [a new, enlarged edition of *The God of Science*] (in preparation)

ISH PUBLICATIONS
NEW YORK

ENRICO CANTORE

DIRECTOR
INSTITUTE FOR SCIENTIFIC HUMANISM

SCIENTIFIC MAN

The Humanistic Significance of Science

No Longer Property of
Phillips Memorial Library

PHILLIPS MEMORIAL
LIBRARY
PROVIDENCE COLLEGE

ISH PUBLICATIONS

NEW YORK

Copyright © 1977 by Enrico Cantore
All rights reserved. No part of this book may be reproduced in any form
or by any means, electronic or mechanical, including photocopying, re-
cording, or by any information storage and retrieval system, without
permission in writing from the publisher.

Library of Congress Cataloging in Publication Data

Cantore, Enrico.
    Scientific man.

    (Introduction to scientific humanism)
    Bibliography: p.
    Includes indexes.
    1. Science—Philosophy. 2. Humanism. I. Title.

II. Series.
Q175.C243   501   76-28439
ISBN 0-917392-01-9

Publisher: INSTITUTE FOR SCIENTIFIC HUMANISM, Lowenstein Center
    at Fordham University, New York, N.Y. 10023, U.S.A.

Distributor: INTERNATIONAL SCHOLARLY BOOK SERVICES, Inc.,
    P.O. Box 555, Forest Grove, OR. 97116, U.S.A.

Manufactured in Italy

TYPIS PONTIFICIAE UNIVERSITATIS GREGORIANAE — ROMAE

To the revered memory of
WERNER HEISENBERG
SCIENTIST AND HUMANIST
WITH PERENNIAL GRATEFULNESS

TABLE OF CONTENTS

Chapter 2

DYNAMICAL ORDERLINESS:
THE SCIENTIFIC EXPERIENCE OF NATURE

CHAPTER 3

WONDER AND AWE: THE SCIENTIFIC EXPERIENCE OF ULTIMATES

CHAPTER 4

THE HUMANISTIC EXPERIENCE OF SCIENCE:
ACHIEVEMENT AND DRAMA

PART TWO

THE HUMANISTIC PERSPECTIVES OF SCIENCE

CHAPTER 5

EXPERIENCE AND REFLECTION:
THE EPISTEMOLOGICAL PERSPECTIVE OF SCIENTIFIC HUMANISM

CHAPTER 6

SUPEREMINENT INTELLIGIBILITY:
THE ONTOLOGICAL PERSPECTIVE OF SCIENTIFIC HUMANISM

CHAPTER 7

PERSONAL CORESPONSIBILITY:
THE ETHICAL PERSPECTIVE OF SCIENTIFIC HUMANISM

Chapter 8

THE HUMANISTIC SIGNIFICANCE OF SCIENCE: A SYNTHESIS

PREFACE

The aim of this book is to explore the significance of science for man—a topic familiar to contemporary public opinion. The widespread discussions about the so-called two cultures have focused general attention on the theoretical aspect of the subject. Recent technological developments—the taming of nuclear power, the increasing practicality of sociobiological engineering, major ecological alterations—have alerted the public to the social relevance of applied science. These days it would seem that to speak of science as significant to man amounts to a commonplace. And yet, although there is very much agreement about science as being influential for man, there is very little accord concerning the actual human meaning of science itself. Some people praise science as the source of universal salvation; others condemn it as the root of all evils; most rest content with a pragmatic attitude that ignores the basic questions involved. The aim of this book is to explore systematically the meaning of science for contemporary man.

The thesis of this book is that science constitutes an essential factor of the historical development of man as a cultural being. In this sense I speak of science as humanistically significant. To help the reader to assess the thesis as expounded in the course of the book, here is the train of thought that gave rise to the present work. When I first started to address myself to science, my concern was strictly philosophical. I desired to detect the philosophical presuppositions and implications of science as they operate in the effective practice of science itself. To this end I developed an approach, which I called inductive-genetic, and applied it to the analysis of atomic physics. The outcome of this study, which I published in book form, was the dawning conviction that science was not only a philosophically important endeavor, but also a humanistic one. [a] That is, I began to perceive that science is human not only because it is produced by man, but also because it is in itself an agent fashioning man in a culturally new way. As a consequence, I decided to pursue this insight along two converging lines. First, I discussed in a number of essays the methodology to be applied in order to bring to light the

---

[a] *Atomic Order: An Introduction to the Philosophy of Microphysics* (Cambridge, Mass.: MIT Press, 1969). See especially chapters 1 and 10.

humanistic significance of science. Second, I examined the concrete meaning of science for its practitioners by means of extensive conversations
with some leading scientific creators. The upshot was the project
embodied in this book. Here I have extended to the whole of science
the tentative humanistic interpretation of it that I had previously perceived through my several partial investigations.

My goal here has been to analyze the evidence that justifies calling
science humanistic and to explore the perspectives that arise therefrom.
I have tried to attain this goal only by way of *general introduction*. The
reasons that have moved me to this delimitation of my work can be easily
surmised, given the complexity of the issues involved. In any case they
will be discussed, along with other methodological considerations, in the
Preliminaries. At this point, therefore, it may be enough to point out
some typical features of this work regarding both its obvious limitations
and its hopefully original contributions. Concerning the latter, I wish
to draw the attention of the reader particularly to the *documentation*
I offer to support my interpretation of science. Such documentation is
embodied in numerous quotations, which should be considered an essential part of this work. My main effort in this regard has been to do justice
to the authors quoted by giving attention to their context and even their
emphasis. Concerning the limitations of this work, one stands out prominently. I do not examine the attempt made by other philosophers to
understand and evaluate the significance of science for man. The reason
for such a procedure is methodological. The writings of philosophers on
the meaning of science are so numerous and complex that I did not think
I could take them properly into account except by dedicating a whole
book to their study. Hence I decided to forgo their analysis here. My
book dealing with the historical relationships between science and philosophy is currently in preparation and should be made available to the
public in a few years. [b]

The readers I had in mind while writing this book are those included
in the three categories of people whose views greatly contributed to the
composition of the book itself. In the first place, I have been thinking
of the practicing scientists. Contrary to a widespread misconception, many
of these persons are deeply concerned with the humanistic relevance of
their work. This situation is familiar to those directly acquainted with
the mentality of scientific researches—all popular clichés notwithstanding.
To support this contention, just a piece of personal evidence may suffice.
Some time ago I published a summary of my methodological approach

---

[b] The book, scheduled for publication in the same INTRODUCTION TO SCIENTIFIC
HUMANISM series as the present one, is tentatively entitled *Science and Human Authenticity:
The Philosophical Quest for Humanism in the Scientific Age.*

to science. [c]   Although the essay appeared in a philosophical journal, the response of practicing scientists was extraordinarily encouraging.   Specialists in all branches of research wrote to me in large numbers for reprints from all over the world.   This book, therefore, is first intended to meet the humanistic expectations of such concerned scientists.

The second category of persons I had in mind while composing this book is the traditional humanists: philosophers, theologians, artists, literary writers and critics.   I know from experience that many of these persons are earnestly interested in the humanistic relevance of science. Frequently, however, they feel frustrated in their understanding by the lack of a suitable introduction to the subject.   By discussing science as a cultural factor, therefore, I hope to meet the expectations of such concerned contemporary humanists.

The third category of people I had in mind while composing this book is scholarly university students specializing in either the scientific or humanistic areas.   My experience as a teacher has disclosed to me both the seriousness of the dangers and the greatness of the opportunities to which these young people are exposed.   The key to the dangers is their instinctive tendency to take science for granted.   Coming to maturity in a time when science is a controversial component of our culture, they are inclined to adopt dogmatic positions in its regard.   Some simply dismiss science in the name of man, whereas others are willing to make man a pawn of science.   Such a partisanship is very dangerous because it perpetuates and further aggravates the crisis of understanding and orientation that afflicts contemporary man.   However, if the dangers are great, so are the opportunities.   To successfully meet these opportunities, and thus definitively conquer the dehumanizing dangers of the contemporary cultural split, the student has to act consistently with his inborn tendency toward the ideals of intellectual honesty and ethical sensitivity. To the more scholarly minded among these young people, therefore, I offer the present book as a contribution of encouragement and guidance toward fulfilling one of the most pressing and stirring tasks of our time: the restoration of authentic humanism.

It is a pleasant duty for me to thank publicly some leading scientist-humanists whose intellectual leadership and friendly support supplied me with the most convincing evidence about the humanistic significance of science.   To all of them I owe hearfelt gratitude, but above all I feel obliged to the late Professor Werner Heisenberg.   He proved decisive for all my research.   In particular, the numerous conversations I had with him as his guest at the Max Planck Institut für Physik, Munich, were

---

[c] "Humanistic Significance of Science: Some Methodological Considerations," *Philosophy of Science* **38** (1971), pp. 395-412.

essential to the interpretation of science presented here. This book is therefore dedicated to him. I tried to have it appear as a homage for his seventieth birthday. Unfortunately, publishing difficulties prevented me from succeeding. I hope it will serve to perpetuate and deepen the humanizing influence of his work.

Other outstanding scientist-humanists whose help proved particularly important are Professor Vasco Ronchi, former Director of the Italian Institute of Optics and President Emeritus of the International Union of History and Philosophy of Science, and Professor Jean Piaget, the renowed genetical psychologist and philosopher.

Still another scientist-humanist who deserves special mention here is Professor Wilhelm Magnus of the Polytechnic Institute of New York. He has been an incomparable friend and advisor for many of the trying years during which this book was developed.

Finally, I am glad to acknowledge various forms of support that made possible the preparation, composition and publication of this book. I am indebted to several institutions. The American Philosophical Society gave me a grant to meet the basic expenses of this research; the Niels Bohr Library for History and Philosophy at the American Institute of Physics supplied me with invaluable assistance through its comprehensive collection of bibliographic material and its unfailingly courteous service. But I am especially obliged to many benefactors who, individually or in groups, on both sides of the Atlantic, stood by me financially and psychologically. They contributed money and hospitality unstintingly, and some behaved literally like brothers in my regard. Their numbers and modesty enjoin me from naming them. But I hope this anonymous mention will suffice to emphasize the crucial importance of their help and to commend their encouraging and generous open-mindedness. In a world where humanism seems to be dying because of intellectual intolerance, group factionalism and greed, it is heartening to be able to testify that such authentically human beings continue to exist and operate in our midst.

New York
May 1976

ENRICO CANTORE

PRELIMINARIES

THE PROBLEM OF THE HUMANISTIC SIGNIFICANCE OF SCIENCE

A basic challenge confronts the humanistic investigator of science. He must substantiate his assertions that science may be considered humanistically relevant. In particular, he must indicate why and how a humanistic investigation of science can be feasible and profitable, for many people deny that science can be humanistically significant. Others, on the contrary, see science as so obviously significant for man that they reject as superfluous any research into the subject. The aim of this introduction is to examine this preliminary problem. Here are the steps I shall follow: first, a discussion of the terminology employed; second, an analysis of the standpoint from which science can be viewed as humanistically important; and third, an outline of the methodological guidelines to be followed in the course of this book.

1. Terminological Clarification

(a) *The Notion of Humanism*

Humanism, as is well known to anyone who has grappled with the subject, is a term that is peculiarly elusive in character. Etymologically, any doctrine that tries to understand man as such can claim to be humanistic. Thus humanism can be conceived as the most simple and comprehensive of philosophical positions. In the words of a leading proponent of this view:

Humanism is really in itself the simplest of philosophic standpoints: it is merely the perception that the philosophic problem concerns human beings striving to comprehend a world of human experience by the resources of human minds. [1]

But the very simplicity and comprehensiveness of the etymological definition tend to make the idea of humanism so vague as to be confusing.

---

[1] F. C. S. Schiller, *Studies in Humanism* (London: Macmillan, 1907), p. 12.

1

Confusion arises because the term becomes susceptible to a limitless number of interpretations. Any philosophical system, ideological position, or religious perspective can be a humanism. [2] Hence the number of humanisms easily grows out of bounds. Much worse, the term itself becomes not a rallying point but rather a bone of contention among the most disparate forms of faith and nonfaith. [3]

The difficulty when dealing with humanism consists in giving a definition that be meaningful without being dogmatic. The danger of dogmatism is always present because of the very subject matter at stake. What is at issue here is the self-understanding of man. But individuals or groups are only too inclined to assume that they understand enough about man to appropriate the term "humanism" to their own particular worldview. Since an inadequate interpretation of our term could easily doom all our subsequent research to failure, we must insist on this point. In particular, we must exclude a popular interpretation that conceives humanism as meaning that man is the measure of all things. The term is used in this sense by some current authors. [4] Also, societies often tend to monopolize the term as though it could be used exclusively to designate their own worldview. [5] The confusion that arises from such a specialized use of the term is serious because members of these societies frequently claim that their worldview is a necessary consequence of the scientific mentality. [6] In addition, people of this persuasion not too rarely use the term humanism in an exclusivistic sense, especially in regard to religion or to any doctrine of absolutes. [7] We must

---

[2] Some examples: M. Buber, *A Believing Humanism: My Testament*, trans. M. Friedman (New York: Simon and Schuster, 1967); J. Maritain, *Humanisme Intégral: Problèmes temporels et spirituels d'une nouvelle Chrétienté* (Paris: Aubier, 1936); E. Fromm, ed., *Socialist Humanism: An International Symposium* (New York: Doubleday, 1965).

[3] See, for instance, A. Etcheverry, *Le Conflit Actuel des Humanismes* (Rome: Gregorian University Press, 1964).

[4] "... man, indeed, is the 'measure of all things'." E. Fromm, *Man For Himself: An Inquiry into the Psychology of Ethics* (Greenwich, Conn.: Fawcett, 1968), p. 23.

[5] The societies referred to are mainly those organized in the International Humanist and Ethical Union. The American Humanist Association publishes an official journal significantly entitled *The Humanist*.

[6] One of the most outspoken proponents of specialized humanism as an expression of scientific thinking, especially of evolutionist biology, is Julian Huxley. See his *Essays of a Humanist* (New York: Harper & Row, 1964), especially p. 73.

[7] "It [humanism] will have nothing to do with Absolutes, including absolute truth, absolute morality, absolute perfection and absolute authority, but insists that we can find standards to which our actions and our aims can properly be related." J. Huxley, *op. cit.*, pp. 73f. For a synthesis of this way of conceiving humanism see J. Huxley, ed., *The Humanist Frame* (London: G. Allen, 1961).

reject all of these particularized interpretations as inadequate in expressing the genuine nature of humanism. [8]

To avoid dogmatism, I shall start this investigation simply with a *heuristic definition* of humanism. Our concern here is not that of determining precisely what the complex characteristics of a full-fledged humanism should be. For the purpose of our research it is sufficient to know in general what is essential to genuine humanism as such. Then the meaning of the term is unequivocal. Ever since the Renaissance, humanism has been taken to express such a comprehensive view of man that man himself may thereby recognize his proper position in the whole of reality and be helped to attain the ideal for which he was born. This therefore will be our definition of this fundamental term. To quote an acknowledged expert in the field:

Humanism is the ideal pattern supposed to reveal the true nature of man and the task for which he was born—the task of shaping himself into a true man and thereby creating a society worthy of him to be transmitted to future generations. [9]

(b) The Notion of Science

The second concept in need of preliminary clarification is that of science. Here, again, we should be particularly careful in the use of the term. In fact, everybody speaks of science, and almost everybody is inclined to think that he knows enough to pass judgment on it. But a loose usage of this term is a frequent source of misunderstandings and embarrassing equivocations. Accordingly, I wish to detail with some precision the sense in which the term is going to be employed in this book.

Science can be taken in a generic as well as a specific acceptation. I intend to use it here in its specific or strict meaning as *natural or experimental science*. Thus, when speaking of science, I always want to refer to a well-defined type of human endeavor characterized both by a typical aim and an unmistakable method. The aim is to discover the intelligible structure of observable reality or nature. The method is systematic observation, theoretical elaboration of observed data, and experimental testing of theoretical deductions. The term science, therefore, will be employed here exclusively to designate any and all of those endeavors that fit such qualifications. Science, then, will be not only physics and biology, but also psychology, sociology, and any other disciplines that meet the criteria given.

---

[8] See, for instance, H. J. Blackham, ed., *Objections to Humanism* (Baltimore: Penguin, 1965).

[9] R. N. Anshen, in her Editorial Introduction to the book series entitled *Perspectives in Humanism* (Cleveland: World, 1967 on).

Another illuminating way of defining science consists in looking at it from the historical angle. Against the widespread tendency of calling science any form of study of nature, I shall use the term according to the precise meaning that it has acquired ever since Galileo. Galileo is rightfully named the father of science because he developed the observational-theoretical approach just mentioned. The way of knowing reality that he initiated was quite novel when compared with former types of knowledge. "In a very real sense Galileo is the first of the moderns; as we read his writings, we instinctively feel at home; we know that we have reached the method of physical science which is still in use." [10]

To characterize more clearly the sense in which science will be taken in this book, a word can be added about what science should *not* be taken to mean. The popular mind ordinarily lumps together experimental science with mathematics and technology. I am not going to accept this identification. As for the reasons of this decision, they are clear to anyone acquainted with the subject matter. Mathematics, though invaluable for science proper, is itself a specific form of knowledge whose aim and method are quite different from those of natural science. As for technology, one should admit a practical and psychological connection with science proper. Historically, technology did much to bring about the rise of science. [11] However, it remains clear that technology is another type of undertaking, with specific aim and method, which do not coincide with those of science. The aim of technology is to bring about practical changes that man deems desirable; its method consists of actual skills capable of controlling the dynamism of nature. The distinction between science and technology is particularly clear from the historical perspective. Technology developed and thrived quite independently of science for many centuries, both in pre-Galilean and post-Galilean times. It is only in the comparatively recent past, beginning with the middle of the last century, that technology came to rely more and more on science. However, even now, technology is not science, despite its tendency to depend more and more on scientific information. For technology always was and still is a self-contained body of practical skills or know-how, whereas science is eminently a search for knowledge. Thus, although technology may be closely related to science, it is never the same thing. In the following discussions, therefore, I shall deal with technology only in connection with examining the social implications of the scientific mentality.

---

[10] W. C. Dampier, *A History of Science and Its Relations with Philosophy and Religion* (New York: Cambridge, 1966), p. 129.

[11] See, for instance, E. J. Dijksterhuis *The Mechanization of the World Picture,* trans. C. Dikshoorn (Oxford: Clarendon Press, 1961), pp. 241-247.

(c) Scientific Humanism: Reasons for a Synthetical Formula

To complete our terminological analysis, I wish to outline briefly the central expression of this study—*scientific humanism*—and the reasons for its adoption. The expression itself is not new, hence there should be no reason for finding it objectionable. [12]   However, the current usage of the expression is laden with all the equivocations we have mentioned when discussing the notion of humanism. It is therefore necessary here to justify my proposal to adopt this expression, and explain the sense in which I intend to use it. The motivation for adopting the expression is a semantic one. We need a synthetical formula that may facilitate communication by avoiding bothersome circumlocutions when discussing the mutual relations of science and humanism. But there is also a psychological advantage in the use of the expression proposed—provided one succeeds in not misinterpreting it. The advantage consists in stressing the importance of keeping open the channels of communication between science and humanism.

On the other hand, one should avoid ascribing to the expression a meaning that is not intended for it. When speaking of scientific humanism, I have no intention of opposing it to the literary form of humanism itself; nor do I claim that science alone suffices to establish a new or better humanism when contrasted with the prescientific forms of humanism. On the contrary, the reason I employ the term is simply to designate a hoped-for attainment with all possible clarity. Indeed, if science is really relevant for man as such, it follows immediately that a humanism must be possible that genuinely satisfies man living in the age of science. Scientific humanism, then, designates the aim to be striven after by man, who is both humanistic and scientific. That is, scientific humanism is the humanism that should result from the harmonious integration of traditional humanistic concerns and the new insights brought to light by science.

2. SCIENCE AS A HUMANISTIC FACTOR

Having settled the question of terms, we must hasten to face the substantive issue of our entire discussion. The purpose of this book is to study the humanistic significance or relevance of science. In this preliminary chapter

---

[12] Some examples: F. Burkhardt, ed., *The Cleavage in Our Culture: Studies in Scientific Humanism in Honor of Max Otto* (Boston: Beacon, 1952); D. Dubarle, *Scientific Humanism and Christian Thought,* trans. R. Trevett (New York: Philosophical Library, 1955); H. J. Muller, "The Purport of Scientific Humanism" in his *Science and Criticism: The Humanistic Tradition in Contemporary Thought* (New York: Braziller, 1956), pp. 241-298.

we must clarify how one may look at science in order to realize its humanistic relevance.

Although we have already agreed to deal only with experimental science, this term can still be taken in a variety of ways. The principal forms are: science as an accomplished result, science in the making, and science as an activity of the whole person. In other words, one can speak of science in the formal sense of the term, or in the methodological sense, or finally in the experiential sense. A brief discussion will determine how science may be recognized as humanistically relevant.

The *formal* meaning of the term "science" is the most widespread. It stands for the actual contents of scientific knowledge, the results which are made visible in research papers and school manuals. Is this type of science relevant from the humanistic standpoint? Certainly, at least in some ways. The humanistic investigator, in fact, needs to know what science says in order to understand its meaning for man. Hence one should not belittle the importance of scientific information to the humanistic understanding of science. However, the study of formal science is by itself insufficient to disclose the humanistic significance of science itself. The reason is plain. One can know an enormous amount of formal science without understanding much of what it really means for man. This is indeed a widespread complaint against the scientific specialist: that he knows very much science, but only formal science.

A second way of interpreting science is that of viewing it as a process —that is, science in the making, science as a *methodological approach*. This second acceptation of science is of great importance to the humanistic investigator, even more than the preceding one. For, of course, one cannot come to have an idea of the actual significance of science for man without perceiving the concrete attitude, the goals and the means, that man adopts in order to "do" science. However, the methodological interpretation of science is still inadequate to disclose the profound significance of science itself to man. The reason is again obvious. One can know very well the methods of science and employ them quite successfully, but nevertheless fail to realize why science itself may deserve to be called humanistic. The cause of this situation is that the humanistic study of science does not fall into the province of science proper.

The third attitude one can adopt consists in considering science as an activity of the whole person. This is what is frequently called concrete or real science. If understood this way, science is certainly humanistic, because it is a factor which molds the entire personality of scientific man. To designate this acceptation I speak of science as *lived experience*.

The word *experience* can be illuminating if it is properly understood. For experience is the most common occurrence in human life and constitutes the source from which humanistic meaning arises in any field of human endeavor. But since experience is so universal and accessible, one should strive the harder to avoid vagueness and equivocations in the use of the term. I suggest that experience be understood by means of personal reflection. On this basis, we see that experience has two fundamental senses, as witnessed by ordinary parlance. The first sense is that of an event or occurrence. The event consists of some sort of direct contact, wherein a person becomes aware of another being by realizing or experiencing the characteristic features of the latter. Thus one experiences, for instance, the color of a flower, the taste of a fruit, the bitterness of humiliation. The second sense of the word "experience" is that of a collection of past experiential events. In this sense one speaks of an experienced person, a man with much experience. However, even if these two acceptations are fundamental to an understanding of the meaning of experience, they are far from sufficient. For experience, if taken in the fullness of its meaning, implies all the complex and rich reactions that arise in man as a result of his personal contact with other beings. In other words, experience cannot be reduced to a simple contact between the subject and the object, as though the participants would be left intimately unaffected by their encounter. To experience means to become aware of another's being through one's total participation in the encounter. Hence experience can have different levels: the superficial one of sense impression and immediate reactions, the deeper one attained through continued reflection. In addition, experience is objective, in that it can come about only by meeting another being. But experience is also subjective, because the one who experiences is the subject. Furthermore, experience is dynamic. In fact, any experience brings about a change in the experiencing subject. As an overall consequence, experience is something that affects and colors the whole worldview of the person—his perception of the world external to him, and his very perception of self.

Obviously, there are good reasons for employing the notion of experience in connection with science, provided one does not misinterpret it. Misinterpretations arise whenever the meaning of the term is restricted to some specialized acceptation. Hence, on the objective side, experience should not be identified with experiment or observation. For observation is indeed experience, but of a special kind. It is a publicly testable and repeatable experience. As for experiment, it is but an instrumental and precise observation. On the subjective side, experience should not be reduced to pure feeling or emotion, because these are but accompanying manifestations of experience itself.

In the light of the preceding, the approach that promises to reveal the humanistic relevance of science should be clear. The central requirement is to realize how science, as a lived activity of the whole man, constitutes a characteristic way of experiencing reality and man himself. This is therefore the first objective of the present book: to explore science, in a systematic fashion, as an experience affecting the whole man who does science.

By developing further the line of reasoning just adopted, we can find an additional motive for hypothetically ascribing humanistic relevance to science. ⌜If science is an original experience of the whole man, it is likely that science itself will disclose new perspectives to the self-understanding of man in general.⌝ Hence the second objective of the book: to discuss in an introductory way how science touches upon the basic areas of any humanistic synthesis.

3. Methodological Guidelines

After having determined the goal of our investigation we must still discuss the question of method. How can we discover the relevance of science to man? The following methodological considerations outline the approach adopted in this book.

(a) The Inductive-Genetic Attitude

As a suitable starting point let us take a close look at science as it exists in actual fact. We notice immediately two main data. First, although it makes sense to speak of science in general, what actually exists is not one single but rather a collection of many individual sciences. Second, science is not a permanently fixed body of knowledge but a continually developing one. Accordingly, the first methodological indication to be followed may be called an inductive-genetic attitude toward science.

Inductivity is the methodological principle that originates from an obvious consideration. To speak knowledgeably about science, it is clear that one must have some direct acquaintance with it. But this is what is implied by the inductive principle proposed here. According to this principle, therefore, the humanistic investigator should not content himself with discussing science in general, but must strive to acquire a direct acquaintance with science itself. In practice, unavoidable limitations of time and energy will restrict very much the applicability of this principle. But then the spirit of inductivity requires that the humanistic student of science be at home in at least one concrete field of research. This personal acquaintance will introduce him to the spirit of science in general.

Is the inductive attitude practical? It is not without risk, because too great a familiarity with a given area of science can make a person biased when interpreting other areas. With this qualification, however, the approach is promising. For, as said, not only is it true that science exists embodied in many concrete sciences, each of which is characterized by special traits, but also it is true that there is such a thing as science in general, namely an overall mentality that applies to all fields of science. Accordingly, one can expect that, by starting from a concrete field of science, one may come to understand the nature of science as such.

Geneticity as a methodological principle is, again, the consequence of an obvious consideration. Any entity that develops in time can be understood adequately only if one studies the various stages marking its development. This is obvious, for instance, in the case of life. One cannot properly understand a living being by just studying it as an embryo or an adult. But science is characterized by typical developmental features. It starts with germinal insights and grows into full-fledged theories. As a consequence, a genetic approach is indispensable if one genuinely wants to understand science. In the words of an outstanding scientist and educator:

To understand science one must see a problem unfold from its beginnings, see progress impeded by traditional ways of thought, learn that scientists make mistakes as well as achieve successes, and observe what experiments brought illumination, and why. [13]

In practice, of course, one cannot follow in great detail the genesis of science. But the spirit of it—embodied especially in the psychology of scientific creation and the history of science—remains essential. For one can only understand the true meaning of scientific statements if he realizes what their formulators meant to say through them and what evidence was presented to justify them.

To sum up, inductivity and geneticy together furnish a solid basis for understanding science. Inductivity provides concreteness, since it makes the humanistic investigator realize what scientists are actually talking about. Geneticity supplies the key to take full advantage of inductivity. For it makes the humanist realize why scientists talk the way they do. As regards this work, inductivity and geneticity have effectively provided its foundations, as I have explained briefly in the Preface. In point of fact, the views that I am going to propose here about the general meaning of science for man emerged gradually in my mind through direct acquaintance with a concrete field of research, namely atomic physics.

(b) The Experiential-Reflective Procedure

The inductive-genetic attitude is necessary but not sufficient to provide a humanistic understanding of science. It is necessary because, as has been seen, it makes the humanist understand science objectively. It is not suf-

---

[13] B. Glass, *Science and Liberal Education* (Baton Rouge: Louisiana State University Press, 1959), p. 62.

ficient, however, because science is more than a body of objective information; it is an all-encompassing experience. Hence the inductive-genetic
attitude has to be complemented by another methodological guideline,
which can be called the experiential-reflective procedure.

Since science is an experience, it is clear that the humanistic investigator should study the experiential nature of science itself. The *experiential
procedure* advocated here is but the codification of such a requirement.
The humanist should try to relive from within the total experience that
constitutes science as such. This is an essential requirement for a variety
of reasons. Many equivocations and polemics stem from the failure of
humanists to realize the experiential nature of science. The fact is that
science is the source of very great personal involvement on the part of
its practitioners. And yet, the opposite view is widely taken for granted.

"Scientific work is no impersonal, cut-and-dried matter, yet the rationality of scientific method is frequently confused with the internal
experiences and feelings of scientists." [14] This remark by a leading psychologist of science is important, and should be heeded. But, of course,
the experiential procedure also presents other reasons for adoption. The
main one, as hinted, is that of coming to realize from within what science
actually is as an activity of the whole man. There is need of much effort
here if we want to attain a genuine understanding of science. "Great
[scientific] thinkers were neither logical machines nor magicians. We
cannot follow the working of their minds, but by studying what they did
we can free ourselves from many misconceptions and perhaps catch something of the inspiration that made them what they were." [15]

However, the experiential procedure in studying science, even when
coupled with the inductive-genetic attitude, is not yet sufficient to enable
man to understand properly the humanistic significance of science itself.
For one may well share fully the experience of science—in both the
objective and subjective meanings of the term—without thereby being
able to realize adequately the humanistic significance of science itself.
Genuine humanism, in fact, demands a critical and systematic worldview.
Hence one has no genuine scientific humanism unless one succeeds in
integrating the new insights and perspectives disclosed by science into such
a systematic structure. But this cannot be achieved by pure experience.
What has to be added is reflective interiorization of experience itself. *Re-
flectivity* is defined here as the effort to become consciously and comprehensively aware of the significance of science for man. That is to say,
the humanistic investigator of science should not be content with reliving

---

[14] B. T. Eiduson, *Scientists: Their Psychological World* (New York: Basic Books,
1962), p. 93.

[15] H. Dingle, *The Scientific Adventure: Essays in the History and Philosophy of Science*
(New York: Philosophical Library, 1953), p. 38.

from within the scientific experience. In addition he should endeavor to express explicitly and consistently the scientific experience itself. For then science reveals itself as truly humanistic when it becomes clear in what sense it affects man's conception of reality, his ethical and religious views, and the very understanding of self.

(c) The Philosophical Character of Investigation

The two guidelines we have just analyzed appear to suffice in principle for a humanistic study of science. But the question must now be asked: how can one apply them in practice? The answer is: by means of a philosophical approach.

The reason for speaking of philosophy here is the reflective nature of philosophy itself. To put it with Collingwood: "Philosophy is reflective. The philosophizing mind never simply thinks about an object; it always, while thinking about any object, thinks also about its own thought about that object." [16]   But, of course, just to mention philosophy in the present connection is not sufficiently illuminating because the term is frequently interpreted in many contrasting ways. Hence a succinct elaboration of the notion of philosophy as adopted here is in order. [17]

We are going to take philosophy as a specific form of knowing, defined by precise objective and method. The method is reflection, the objective is understanding. Man philosophizes, that is, he understands philosophically in that he reflects on his own activity as a knower. The essential point for our following consideration is that philosophy be taken strictly as an autonomous form of cognition. Philosophy is the reflective cognition of man's own knowing to be pursued in an explicit, critical, and systematic manner. But man's knowing is threefold. Hence the three main branches of philosophy itself. First, knowledge is an activity of the subject. Accordingly, the first branch of philosophy is epistemology, or the study of knowledge as an activity of the subject. Second, knowledge is information about an object. Accordingly, the second branch of philosophy is ontology or the reflective study of the objective intelligibility of reality made accessible to man through his own knowledge. Third, knowledge moves the knower to act in accordance with what he has come to know. Therefore the third branch of philosophy is ethics or the reflective study of knowledge as source of striving after goals and values.

As will become increasingly clear, the contribution of philosophy to

---

[16] R. G. Collingwood, *The Idea of History* (New York: Oxford, 1946), p. 1.

[17] I have discussed at some length the relationships of science and philosophy in two methodological articles: "Science and Philosophy: Some Reflections on Man's Unending Quest for Understanding," *Dialectica* **22** (1968), 132-166; "Science and Humanism: The Sapiential Role of Philosophy," *Dialectica* **24** (1970), 215-241.

the humanistic understanding of science is essential. We can see the main reason of this fact at once. Knowledge is the activity that characterizes man as man. And science is eminently knowledge. But, naturally, one should take knowledge, and science, in the fullness of their meaning, instead of restricting their interpretation according to any preconceived scheme. In particular, when speaking of philosophy in connection with science, one should reject the widespread view that reduces the role of philosophy to that of a logical analysis of science. This view is wrong because it fails to take into account the concretely human nature of knowledge in general and of science in particular. Knowledge is not a disembodied vision but an activity of the whole man. As a consequence, philosophy of science should not be an abstract logical analysis but rather a reflective understanding of science in all of its human richness.

In particular, the vital importance of philosophy for the humanistic understanding of science stands out plainly if we contrast its contribution to those of other disciplines that also study the nature of science. I refer especially to sociology and psychology of science. Of course, one should value these disciplines highly, and take their findings into serious consideration. However, they are not adequate by themselves in bringing to light the humanistic significance of science. The reason is that psychology and sociology are themselves sciences. As a consequence, they can indeed know science, but only as an object rather than as a personal activity of the subject. Thus, such disciplines can supply a great amount of invaluable information about science—the ways the average scientist thinks, the different phases of scientific creation, and the like. But, in regard to the overall goal of understanding science as an activity of the whole person, they remain silent. These disciplines cannot tell us why scientists behave the way they do—that is, describe the critical justification of their thinking and the ethical justification of their acting. Sociology and psychology of science cannot offer this information simply because they are not personally reflective the way philosophy is. Here lies the decisive contribution of philosophy to the humanistic understanding of science. Accordingly, this is the reason why our investigation will have an essentially philosophical character.

(d) The Multiplicity and Unity of Science

Continuing our methodological analysis, we must now face an objection that seems to jeopardize our enterprise. The objection is leveled against the assumption underlying our entire procedure. We have been speaking continually of science and scientists. But is the scientific attitude so standardized, are scientists so uniform in their mentality, that one can hope to discover *the* mentality of science? In particular, even though one could be justified in speaking of a typical scientific mentality, who can be considered the authentic representatives or spokesmen for such a mentality?

The objection is serious, but not insurmountable. Since it refers to a question of fact, the answer should be given factually. What are the factual data, as supplied by psychology and sociology of science? To begin with, it is obvious that there is an immense variety of scientific personalities.

Scientists are people of very dissimilar temperaments doing different things in very different ways. Among scientists are collectors, classifiers and compulsive tidiers-up; many are detectives by temperament and many are explorers; some are artists and others artisans. There are poet-scientists and philosopher-scientists and even a few mystics. [18]

These words, which express the experience of an outstanding scientist, find confirmation in the systematic research of the psychologist of science. "Their [scientists'] personality pictures cut across diagnostic classifications and classical personality configurations." [19]

However, as soon as we ask why all of these different people can be called scientists, and why they succeed in harmoniously doing science together, we find that scientists are not just disparate people. In fact, psychological investigation shows that, for all their dissimilarities, scientists do have something in common that would justify the use of a common name.

The results of statistical analysis of psychological test data show that the scientists are a very homogeneous group in thought and perception. They organize and systematize material very similarly, developing their intellectual resources in such a way that we can say they think about phenomena and look at them with a common orientation or 'set'—not so far as content, but so far as the kinds of stimuli they look for and in which they become interested. [20]

This basic psychological result finds, in turn, an easily verifiable expression on the social level. As is well known, C. P. Snow in his celebrated lecture did no more than utter a universal experience when he spoke of a "culture" as typical of the scientific mentality. Of course, it is true that scientists differ from each other both as human beings and as scientific workers. It is also true that scientists frequently do not understand one another's field and may even dislike each other. Nevertheless,

there are common attitudes, common standards and patterns of behavior, common approaches and assumptions. This goes surprisingly wide and

---

[18] P. B. Medawar, *The Art of the Soluble* (London: Methuen, 1967), p. 132.
[19] Eiduson (note 14), p. 113; cf. p. 192.
[20] *Ibid.*, p. 122; cf. p. 247.

deep. It cuts across other mental patterns, such as those of religion or politics or class.... In their working, and in much of their emotional life, their attitudes are closer to other scientists than to non-scientists who in religion or politics or class have the same labels as themselves. [21]

We can now give a reply to the objection proposed. Is the scientific attitude such that one can speak of *the* mentality of science despite the obvious diversity noticeable in scientists? Yes, because *science is both one and many*. Science is one in the objective sense of the term. All scientists—no matter what their individual idiosyncrasies—share the basically same outlook toward reality and knowledge. But science is many when the subjective attitudes of scientists are involved. For scientists remain human. Hence they preserve their individuality also when doing science and thinking about it.

The answer given, however, raises with new urgency the objection we are examining. Indeed, if science is one but the individual scientists are so different from one another, who are the authentic representatives or spokesmen of science? No simple answer to this question can be found. For experience shows that even the great scientist cannot be automatically taken as an adequate interpreter of science. Some humanistic investigators are very harsh in judging the situation. To quote one prominent sociologist and historian of science:

From the little that is known about the psychology and sociology of scientists one fact is outstanding—their actual motivations, reasons, and reasoning in science are nearly always quite different from what they say. [22]

This attitude of distrust is shared even by scientists. Einstein, for instance, warns that one cannot come to understand what a theoretical scientist does by listening to what he says but only by examining what he does.

If you want to find out anything from the theoretical physicists about the methods they use, I advise you to stick closely to one principle: don't listen to their words, fix your attention on their deeds. [23]

The problem may appear unsolvable. If the professional scientist is not a reliable interpreter of science, who can be? The humanistic investigator is ordinarily too little of a scientist himself to venture his own interpretation of science as such. The key to the solution can be found

---

[21] C. P. Snow, *The Two Cultures: And a Second Look* (New York: Mentor Books, 1963), p. 16.

[22] D. J. de Solla Price, "The Science of Science," in M. Goldsmith and A. Mackay, eds., *Society and Science* (New York: Simon and Schuster, 1964), p. 197.

[23] A. Einstein, *Ideas and Opinions*, trans. S. Bargmann (New York: Crown, 1954), p. 270.

by recalling some basic points we have already made. The humanistic significance of science is a fact, in that science is an experience of the whole person. But an experience can be lived in many different ways, more or less profoundly and comprehensively. In particular, science as an experience can be perceived more or less adequately by those who participate in it. The humanist who intends to understand science, therefore, will do well to be on his guard against taking naïvely what scientists say about science. However, he would be wrong were he to reject the testimony of scientists in principle. For only those who do science can be genuine witnesses to the experiential meaning of science itself. As for a proper formulation of the humanism of science, this can only come about as a result of a collaboration between scientists and humanists. The scientist provides the information, the humanist reflects philosophically on it. The solution to our problem, as a consequence, can be found if the humanist is willing to take seriously what the scientist says—but not too literally. Humanists frequently commit the error of taking the philosophical expressions of scientists too literally. But, as Bridgman points out, replying to one critic, this is unfair because the the scientist "is not completely determined by what he says and . . . what he does is often wiser than what he says." [24]

To sum up, this is going to be our methodological procedure in the light of the objection considered. We start our research with the assumption that there is such a thing as a universal structure of science and that it is significant for man as such. Such a structure can be called the implicit philosophy of science. This, with Eddington, can be described briefly in experiential terms.

It is the philosophy to which those who follow the accepted practice of science stand committed by their *practice*. It is implicit in the methods by which they advance science, sometimes without fully understanding why they employ them, and in the procedure which they accept as giving assurance of truth, often without examining what kind of assurance it can give. [25]

Our goal, in the course of this book, will be to make systematically explicit what constitutes the implicit philosophy of science and to explore the implications that derive from it concerning the self-understanding of man.

---

[24] P. W. Bridgman, Introduction to J. B. Stallo, *The Concepts and Theories of Modern Physics* (1881) (Cambridge, Mass.: Harvard University Press, 1960), p. xxiii.

[25] A. Eddington, *The Philosophy of Physical Science* (Ann Arbor: University of Michigan Press, 1958), p. vii.

(e) A Contribution to the Integrity of Science

One brief consideration may serve to clarify further the methodology of this book. Our investigation aims to offer a contribution to the integrity of science.

In recent years there has been a mounting concern about science as a man-affecting factor. In particular, scientists have begun to speak of the integrity of science as a value to be recognized and fostered for the good of man and of science itself. An authoritative panel has explored the issue and brought it to the attention of the public. The integrity of science is a reality in that science, as an overall endeavor, presents a "unified internal structure" and is oriented to a specific goal. The goal, "its guiding imperative," is "the search for objective knowledge." [26] But the integrity of science is currently exposed to a growing danger of erosion both inside the scientific community and inside society at large.

The threat to integrity inside the scientific community arises from the very growth of science and the pressures that result from it. Science is proliferating. Hence new generations of scientists tend to be trained in a very specialized way, which makes them lose sight of the complex totality of the scientific phenomenon. Science is accelerating its productivity. Hence scientific workers feel compelled to dedicate their entire energies to research and publication, without any leisure for reflection and interiorization. Science depends increasingly on economic, political, and military sources for the funding of its projects. Hence scientists are exposed to the temptation of justifying their work by gratifying the expectations of those who finance them, rather than being chiefly concerned with the humanization of man.

The threat to the integrity of science inside society originates from the preponderance of the scientific phenomenon and the manifold reactions it arouses. Science has become an essential part of our civilization. Hence no one can ignore its presence or remain indifferent to it. Everybody speaks of science and feels entitled to pass judgment on it. But science is an extremely complex phenomenon. Thus the judgments passed on science frequently lack sufficient profundity and balance to contribute to a better understanding of science itself. It follows that science becomes a source of growing polarization among contemporary people. The more science becomes visible, the more people are split, for or against science itself.

The crisis arising from the erosion of scientific integrity needs no

---

[26] Committee on Science in the Promotion of Human Welfare of the American Association for the Advancement of Science, under the chairmanship of Barry Commoner, "The Integrity of Science," in G. Holton, ed., *Science and Culture: A Study of Cohesive and Disjunctive Forces* (Boston: Beacon, 1967), pp. 291-332, especially pp. 291f.

stressing.   Modern man can greatly damage himself, and impair his science, if he does not succeed in situating the integrity of science within an overall understanding of himself and of his role in the world.   The aim of this book is to contribute to the integrity of science taken in this humanistic sense.   The contribution is to be philosophical in nature.   The integrity of science is presupposed as the starting point and the final goal.   But since this integrity can be blurred by many factors, this book aims to explore it and make it explicit by means of a systematic philosophical reflection.   Needless to add, the contribution intended here can only be an introductory one.   Science affects man in his entirety.   Hence the problems it raises are too profound and far-reaching to permit their solution within the compass of one book.   But, within the limits of an introduction, our work should be a profitable one, both negatively and positively. Negatively speaking, we shall attempt to remove various widespread misunderstandings about the nature of science and its significance for man. Positively speaking, we shall try to see how science can be integrated into a comprehensive humanism that truly meets the expectations of people living in the scientific age.

PART ONE

THE HUMANISTIC EXPERIENCE OF SCIENCE

CHAPTER 1

CREATIVE DEPENDENCY:
THE SCIENTIFIC EXPERIENCE OF KNOWLEDGE

In our attempt to understand science as an experience of the whole man we may begin from the obvious. Science is a cognitive undertaking. Thus we may start our exploration by studying the characteristic way scientists experience knowledge. These are therefore the topics to be discussed in this first chapter. ⌜Science is a form of cognition marked by a creative attitude on the part of the knower.⌝ Our first subject will be the creativity of the scientific approach. Science is a form of cognition that strives to be faithful to objective reality. Our second subject will be the dependency of the scientific approach. Finally, science is a form of knowing which involves the whole personality of the knower. Our third subject will be the features that distinguish the scientific way of experiencing knowledge.

1. The Creativity of the Scientific Approach

Creativity means, in general, the ability to bring about something hitherto nonexistent. But science is widely acknowledged as a creative undertaking. Thus our initial concern will be to explore in what way scientific cognition deserves to be called creative. To find an answer to this question we shall analyze the basic structure of scientific research. Research begins with observation, progresses to theorization, and attains its aim through experimentation. Accordingly, we begin by examining these three stages of scientific research.

(a) The Activity of Observation

Experimental science begins with observation. This is an obvious fact that is rooted in the psychological and even physiological nature of man. To cite an acknowledged master:

Experimental ideas are by no means innate. They do not arise spontaneously; they must have an outer occasion or stimulant, as is the

21

case in all physiological functions. To have our first idea of things, we must see those things; to have an idea about a natural phenomenon, we must, first of all, observe it. [1]

But, if observation is so fundamental to science, the question arises immediately: what is the role of man in observation? It may appear at first sight that observation is essentially passive. Man has nothing to do but keep his eyes open, trying to take in every detail of the objects with which he deals. Then, all of a sudden, discovery will strike him: luminous, precise, unchangeable. This is the uncanny gift which people not directly acquainted with science frequently ascribe to the scientific mind. But such an apparently obvious interpretation is fallacious. Scientific man becomes a discoverer only by means of a very active attitude as a patient and persevering seeker.

Historical information sheds convincing light on this subject. Frequently, people think that the reason science originated in post-Renaissance times was the availability of sophisticated apparatus that made precise observation possible. But the historical evidence shows that the situation was far more complex. Fundamentally, the great breakthroughs in science took place without the help of any sophisticated instruments. Examples include Galileo's observations of the motion of falling bodies, Vesalius' studies of the structure of the human body, and Harvey's investigations of blood circulation. Not even that most exact observer, Tycho Brahe, had much more to assist him in his astronomical observations than his bare eyes supplemented by rather primitive contraptions consisting of rods and wheels. Thus, when sophisticated instruments of observation were invented, it was not the case of straight dependency of science on technology. Rather, it was a question of active minds who knew how to take advantage of new technological possibilities. A persuasive example is Galileo's use of the telescope. He was not the inventor of the instrument. Contemporary people amused themselves with that funny gadget which enlarged and distorted the images of things. But Galileo, precisely because he had a personal sense of the activity of observation, realized immediately, upon hearing of the gadget, that it was a device "of inestimable usefulness." [2] At once, he set out to develop better models of the telescope itself and obtained his famous astronomical discoveries. Clearly, scientific observation is not merely a matter of passive receptivity on the part of man; it implies a highly complex activity. We must now detail the principal aspects of such an activity.

---

[1] C. Bernard, *An Introduction to the Study of Experimental Medicine,* trans. H. C. Green (New York: Dover, 1957), pp. 32f.

[2] Quoted in V. Ronchi, *Storia del Cannocchiale* (Vatican City: Pontifical Academy of Sciences, 1964), p. 789.

(aa) Courageous Open-Mindedness

Mach, discussing Galileo's work on the inclined plane, remarks that he manifested "such scientific greatness" in that "he had the intellectual audacity to see, in a subject long before investigated, more than his predecessors had seen, and to trust to his own perceptions." [3] This observation offers the key to understanding the scientific mind. A courageous determination to break away from the apparently obvious in order to see better and deeper is the basic requirement of science as such. In other words, fundamentally science consists in asking questions whose answers convention makes appear self-evident. But such an enterprising attitude can only be the expression of a quite active personality. For "It requires a very unusual mind to undertake the analysis of the obvious." [4]

The essential reason scientific observation demands activity on the part of man is the psychological tendency of human nature to take appearance for truth. It seems foolish and fruitless to question the obvious. Thus the first effort the scientist has to make is that of striving to be open-minded. That is to say, scientific creativity is not primarily detection of new phenomena, but perception of familiar events in a new light. *Open-mindedness* as typical of the scientists means refusing to take things for granted—not in order to be skeptical, but to keep oneself fully open to the intelligibility of the things themselves. In other words, to be open-minded means to make oneself sensitive to the actual features of nature without allowing accepted views to prejudge their interpretation. Science was born when great persons arose who were able to take this exacting attitude. Science progresses through the ever-renewed adoption of such an attitude. Of course, one should not misunderstand the nature of open-mindedness. In particular, one should not equate it with the lack of information about past opinions on the subject. For an observer is ordinarily the more successful the more comprehensive is his information about his subject matter. But what is demanded is the determination not to become a prisoner of assumed knowledge while striving for authentic knowledge. "Only we must keep our freedom of mind . . . and must believe that in nature what is absurd, according to our theories, is not always impossible." [5]

---

[3] E. Mach, *The Science of Mechanics: A Critical and Historical Account of Its Development,* trans. T. J. McCormack (La Salle, Ill.: Open Court, 1960), p. 63.

[4] A. N. Whitehead, *Science and the Modern World* (New York: Mentor Books, 1948), p. 5.

[5] Bernard (note 1), p. 38.

(ab) Industrious Familiarization

A second aspect of scientific observation further clarifies the effort toward open-mindedness that characterizes science. It can be termed industrious familiarization.

Familiarization is in general the process through which a person makes himself at home with someone or something by means of direct and prolonged contact. But science is founded on such an attitude. No scientist has made any significant discoveries, however theoretical, without some sort of direct contact with the material objects of his research. A moving example in this connection is a confession by Einstein. He appears to be the very personification of the theorist. Yet he confides in his autobiography that, despite his excellent mathematical professors at the Polytechnic Institute of Zürich, he failed to get a sound mathematical education there. The reason was, as he tells us, "I worked most of the time in the physical laboratory, fascinated by the direct contact with experience." [6] Familiarization is a slow process, and a difficult one. Frequently it comes to fill the entire life of the person who has decided to open new fields of research. For instance, Tycho Brahe—the great astronomer whose data paved the way for the work of Kepler and Newton—devoted most of the nights of the 20 years he spent at Hveen to a systematic exploration of the skies. Leeuwenhoek spent literally all the time he could spare from his business to attain his masterful expertise with the microscope.

In order to realize how man can actually attain successful scientific observation, one important aspect of the familiarization process needs to be emphasized. Scientific observation does not ordinarily come about just as a result of long and sympathetic acquaintance with a given subject matter. To this one must add *industriousness*. That is, the aspiring observer must sharpen his wits and engage all his resources in the enterprise. Frequently it is even a matter of working with one's own hands, by collecting specimens or constructing instruments. Galileo, for instance, succeeded in transforming the telescope from a toy into an instrument "of inestimable usefulness" because of his industriousness. As he tells the reader in his *Starry Messenger*, the first report he heard about the telescope

...caused me to apply myself wholeheartedly to inquire into the means by which I might arrive at the invention of a similar instrument.... Finally, sparing neither labor nor expense, I succeeded in constructing for myself so excellent an instrument...[7]

---

[6] In P. A. Schilpp, ed., *Albert Einstein, Philosopher-Scientist* (New York: Harper Torchbooks, 1959), p. 15.

[7] In S. Drake, ed. and trans., *Discoveries and Opinions of Galileo* (New York: Doubleday Anchor Books, 1957), p. 29.

Other great observers manifested the same active involvement. William Herschel and Christian Huygens spent countless hours grinding lenses to improve their telescopes. Leeuwenhoek, too, became an outstanding microscopist through his hard-won skill in lens grinding. In another area of science, Darwin is famous for his relentless dedication to collecting evidence. As he himself puts it:

I think that I am superior to the common run of men in noticing things which easily escape attention, and in observing them carefully. My industry has been nearly as great as it could have been in observation and collection of facts. [8]

The outcome of the industrious familiarization described constitutes scientific observation at its best—the ability to see ordinary things in a really new way. As Galileo, for instance, speaks of his astronomical vision of the skies: ". . . that which presents itself to mere sight is as nothing in comparison with the high marvels that the ingenuity of learned men discovers in the heavens by long and accurate observation." [9] The observational apparatus, when employed, becomes a virtual extension of the body of the researcher—a companion or cooperator that requires attention and respect but, in return, provides invaluable information. Scientific observation, in short, is essentially a matter of dedicated activity. As Leeuwenhoek put it succinctly: "Through labor and diligence we can discover matters which we had thought inscrutable before." [10] The testimony of this incomparable observer is particularly appropriate. Although simply a self-taught amateur, Leeuwenhoek became such an outstanding microscopist that even experts had difficulty in checking his discoveries. The secret of his achievements? Martin Folkes, vice president of the Royal Society, commented suitably upon receiving the microscopes that Leeuwenhoek himself had bequeathed to the Society:

His own great judgment, and experience in the manner of using them [microscopes], together with the continual application he gave to that business, and the indefatigable industry . . . cannot but have enabled him to form better judgments of the nature of his objects, and see farther into their constitution . . . [11]

---

[8] In N. Barlow, ed., *The Autobiography of Charles Darwin* (*with original omissions restored*) (New York: Norton, 1969), pp. 140f.

[9] "Letter to Madame Christina," in Drake (note 7), p. 197.

[10] Quoted in A. Schierbeek, *Measuring the Invisible World: The Life and Works of Antoni van Leeuwenhoek* (New York: Abelard-Schuman, 1959), p. 202.

[11] Quoted in C. Dobell, *A. van Leeuwenhoek and His "Little Animals"* (New York: Russell and Russell, 1958), pp. 104f.

(ac) Self-Availability for Discovery

The activity demanded by scientific observation presents still one more important aspect. Scientific observation is not merely the perception of some data that other people have hitherto failed to notice. Above everything else, observation is scientific in that it contains the germinal realization of a profound, previously unsuspected intelligibility of nature. That is to say, observation acquires full scientific status when the person becomes aware that the phenomenon he is studying is but an individual case of an overall regularity hitherto undetected. A classic example is offered by the tale of Newton coming to realize the existence of universal gravitation through observation of a falling apple. But, if observation is significant especially because of the insight it leads to, it is clear that scientific observation itself requires a special kind of activity on the part of the person who pursues it. For insight cannot be attained through mere striving. This third aspect of the activity characterizing scientific observation can be called *self-availability for discovery*.

The importance, as well as the difficulty, of the kind of activity we are discussing should be properly stressed. For the attitude in question involves the whole personality of the scientist, including particularly what is most hard to give: a courageously persistent effort in the search for a solution, even when results seem to completely elude the search itself. The investigator must be enthusiastic and laborious, but he must also be unconquerably steadfast and practice unbroken concentration. He must bring all his powers to bear on the object of his research, while patiently waiting for the insight to surface. This was, for instance, the opinion of Newton in his famous words to Conduitt that truth was "the offspring of silence and unbroken meditation." [12] Newton was speaking out of experience and personal practice. When asked how he had come to make his discoveries, he replied: "I keep the subject constantly before me, and wait till the first dawning open slowly by little and little into the full and clear light." [13]

In other terms, the basic activity required of the searching scientist is that of holding himself in a *state of readiness*. The scientist has to do whatever possible on his part to sensitize himself to the manifestations of intelligibility presented by his object. Then, as soon as a lead appears, he has to follow it to its very end. This essential aspect of scientific research is particularly clear in the so-called chance discoveries. These are events in which a researcher, suddenly confronted with an unexpected phenomenon, rapidly succeeds in realizing a general regularity of

---

[12] Quoted in F. E. Manuel, *A Portrait of Isaac Newton* (Cambridge, Mass.: Harvard University Press, 1969), p. 86.
[13] *Ibid.*

nature as embodied in the phenomenon itself. A famous example is the discovery of X rays. Several prominent physicists had already noticed strange effects on photographic plates which had been exposed to electric discharges in vacuum tubes. William Crookes had been annoyed and had had the plates sent back to the manufacturer as defective. Goodspeed of Philadelphia had even made an X-ray photograph (February 22, 1890), but without realizing the significance of what he had done. Then, in December 1895, Röntgen also happened to notice the same phenomenon. The difference was that he paid close attention to the data and investigated them for more than one month of hectic work. His conclusion: the strange spots on the plates were in no way accidental or meaningless appearances; rather, they manifested a well-defined property of matter, hitherto unheard of. Hence the very name of X —that is, "unheard-of"—rays. [14] This example shows how sense perception becomes genuinely scientific observation through the state of active readiness of the researcher.

In general, it seems even questionable to distinguish between chance discoveries and other discoveries. For in all discoveries there is not only an element attributable to chance, but also an element having to do with the purposeful activity of the researcher. The element of chance consists of the fact that, had not a given person been exposed to certain information, he would never had had sufficient evidence to make the discovery in question. But the role played by chance is never all-determinant. For even the discoveries most clearly due to chance are the result of the preparedness of the mind. Pasteur summarized the situation in the famous dictum: "In the field of observation, chance favours only the prepared mind." This paradoxical expression contains a profound truth. Chance offers the challenge. Only the mind that has patiently tried to gather all the clues needed for the solution is able to meet the challenge successfully and thus achieve discovery. A great neurophysiologist wrote in this connection: "The accidental discovery is a prize, going to the persevering investigator." [15] These words are apt because scientific discovery is essentially a breakthrough into the unknown, starting from sense experience. The breakthrough comes about only as a result of the active attitude on the part of the researcher—through his unflagging persistence and indomitable perseverance. In other terms, the researcher is a dedicated seeker, and discovery his reward. But scientific observation is the persistent striving after this reward. As Konrad Lorenz, the master observer, put it movingly:

---

[14] Details in O. Glasser, *Wilhelm Conrad Roentgen* (Springfield, Ill.: Charles C. Thomas, 1958), especially pp. 84f.

[15] S. Ramòn y Cajal, quoted in E. H. Craigie and W. C. Gibson, *The World of Ramòn y Cajal with Selections from His Nonscientific Writings* (Springfield, Ill.: Charles C. Thomas, 1968), p. 197.

"What an observer fails to notice at all on an object for the first twenty times, he will finally perceive clearly the two hundredth time." [16]

(b) The Originality of Theorization

So far we have only attained the threshold of science. Observation is the gate to discovery. But discovery itself is a highly complex affair, which reaches deeply into the intelligibility of nature and affects scientific man in a pervasive way. To continue our investigation we must now examine discovery itself. In particular, we must analyze the creative activity that enables the researcher to perceive nature in a new way. I call this activity *theorization*. Theory, etymologically, means vision. Theorization, then, is the process through which the mind comes to see nature in the way that is characteristic of science. In accordance with our guidelines, we shall explore the role played by theorization in science by examining the data of the scientific experience.

(ba) The Search for Patterns

As has been seen, science arises as a result of open-minded self-availability on the part of the scientist. But the question presents itself: is science just unstructured open-mindedness or is it an investigation aimed at a precise goal? The interpretation of scientific research as an unstructured undertaking appears to be widely accepted by the nonscientific public. The scientist is seen as a sort of superman. Like the lynx-eyed detective of fiction, he is supposed to be guided by no hypothesis. Rather, he is ascribed the uncanny ability to survey all details and them come up automatically with the right solution. Such an interpretation is, of course, no better than a caricature. For the first step toward scientific understanding consists precisely in striving after a goal. The goal is to perceive an overall intelligible pattern as embodied in the observational data of one's own field of research.

The situation in which a researcher finds himself while striving for understanding is illustrated by the testimony of Heisenberg. He compares his own research in quantum mechanics to the experience of mountain climbing. The climber, immersed in fog, has only a vague idea of the location and conditions of the peak he intends to ascend. Determined not to give in to difficulties, he keeps going, one step at a time, but he does not really know whether he is moving toward the peak or not. Then, all of a sudden, the fog clears away for a moment. The climber recognizes the goal of his striving as well as the surrounding situation. He sees a

---

[16] Translated from K. Lorenz, *Gestaltwahrnehmung als Quelle wissenschaftlicher Erkenntnis* (Darmstadt: Wissenschaftliche Buchgesellschaft, 1964), p. 45.

pattern. "In the very moment that you have seen that, then the whole picture changes completely. . . ." [17] This is the first step of theorization. The researcher looks for a pattern that may guide him in the understanding of nature. Of course, perceiving a pattern is just the beginning, and many difficulties remain to be overcome. In the analogy, the climber—once he has seen the peak—may still not know how to reach it. However, the perception of the goal of one's striving is what makes all the difference. The investigator now sees an overall connection, a unitary picture of the whole field. "I think it makes such an enormous difference, at least from my own impression, whether I only see details, or whether I see the picture." [18] Heisenberg is firm on this point. He considers it an essential manifestation of the experience of scientific research. In particular, he speaks of physics. "What quite frequently happens in physics is that, from seeing some part of the experimental situation, you get a feeling of how the general experimental situation is. That is, you get some kind of picture. . . . This 'picture' allows you to guess how other experiments might come out." [19] Thus, although perceiving a general pattern is not sufficient to achieve scientific understanding, Heisenberg insists that this is the first decisive step. For now nature has begun to disclose the intelligibility of its interconnections: ". . . nature itself has such a close connection in itself that one can guess it. One sees how things are connected." [20]

In practice, however, the researcher can achieve scientific understanding only by solving one difficulty at a time. Hence it is to be expected that other scientists may take a position that seems to contradict the one just examined. Dirac, for instance, discussing his methodology with Heisenberg, insisted that his work was simply a sequence of individual steps, aimed at solving individual difficulties as they arose. Heisenberg admits that this is a practical necessity. However, he refuses to accept it as an adequate interpretation. The reason is that, to him, one can come to solve individual difficulties satisfactorily only when one has realized the general structure of the problem. For otherwise—to return to his analogy—how could one really progress, albeit one step at a time, when one does not know where the peak is or whether the peak can be climbed at all? Thus, to the basic objection that one can never solve more than one difficulty at a time, he replies that such a piecemeal approach does not, by itself, lead to any solution at all:

[17] W. Heisenberg, in T. S. Kuhn, J. L. Heilbron, P. Forman, L. Allen, eds., *Sources for History of Quantum Physics* (copyright, American Philosophical Society); interview February 25, 1963, p. 4.

[18] *Ibid.*

[19] *Ibid.,* interview February 11, 1963; pp. 16-18.

[20] *Ibid.*

One can never solve only one single difficulty, but one is always obliged to solve many of them at once.... The genuine solution of a difficulty consists in the fact that one, through it, comes in contact with the simple overall interconnections of nature (*die einfachen grossen Zusammenhänge*). And thereby other difficulties are automatically removed which one had not even considered in the first place. [21]

The controversy may appear unsolvable. But this is not the case. Dirac himself, when questioned about his contributions to quantum mechanics, hinted at the solution. Although he had been one of the most original and powerful developers of the field, he modestly admitted that he had needed the inspiration of others in order to attain his successes. In particular, he referred to the leadership of Heisenberg. In his own words:

I know I was very much impressed by action and angle variables. Far too much of the scope of my work was really there; it was much too limited. I see now that it was a mistake; just thinking of action and angle variables one would never have gotten on to the new mechanics. So without Heisenberg and Schrödinger I should never have done it by myself. [22]

That the creativity of the scientific mind in its search for understanding consists basically of trying to perceive unitary patterns is confirmed by psychological investigation. Max Wertheimer, particularly, studied systematically what he calls "productive thinking," namely the process through which the mind comes to achieve understanding. His conclusion is that the mind must be active in many ways. Understanding is achieved only when the person is able to perceive a whole—that is, a unitary pattern as embodied in many details or parts that hitherto appeared to form a meaningless disconnected collection of individual cases. He discusses in particular the understanding of geometric figures.

There is grouping, reorganization, structurization, operations of dividing into sub-wholes and still seeing these sub-wholes together, with clear reference to the whole figure and in view of the specific problem at issue. [23]

Wertheimer applies his result to Galileo's discovery of the principle of inertia, as far as is ascertainable from the extant literature. We cannot dwell at length on his interpretation. [24] But it is fascinating to notice

---

[21] Translated from W. Heisenberg, *Der Teil und das Ganze: Gespräche im Umkreis der Atomphysik* (Munich: Piper, 1969), p. 143.

[22] In Kuhn *et al.* (note 17); interview May 7, 1963; p. 6.

[23] M. Wertheimer, *Productive Thinking* (London: Social Science Paperbacks, 1966), p. 41.

[24] *Ibid.*, pp. 205-212.

how the origin of this fundamental theoretical principle can be traced to the search for a unitary pattern of behavior of moving bodies, when all external causes of both acceleration and deceleration are thought away. Of course, given the paucity of historical material on this subject, we cannot be entirely sure that Wertheimer's interpretation fits the case under discussion. However, after having studied Wertheimer's work, we have the impression that when we read Galileo we finally understand him. I refer in particular to a famous boast of Salviati in the *Dialogue*. This spokesman of the new science asserts with a straight face that he is able to make Simplicio agree with his own inertial interpretation of the motion of a stone falling from the mast of a moving ship. And this he promises to do "without experiment."[25] Clearly Galileo must have been thinking here of the creative power of the mind as being able to overcome, through its own activity, even the most deeply ingrained instinctive opinions of people. In general, it is obvious that Galileo's own success in giving rise to science was due to the creative attitude of his mind. It was this attitude that enabled him to seek and find regular patterns of behavior where to all appearances only meaningless chaos was present. To illustrate the power of such creativity it suffices to recall Galileo's achievements in the study of falling bodies and projectile motion.

In the light of the preceding, we begin to realize the cognitive nature of science. Science is not just a matter of attentive observing; above all it is an active mental search of an intelligibility that is present but hidden in the objects of observation. We can summarize the situation with Bronowski: "All science is the search for unity in hidden likenesses. . . . For order does not display itself of itself; if it can be said to be there at all, it is not there for the mere looking."[26] In a true sense, then, as the same writer goes on to say, one can speak of scientific creativity: ". . . order must be discovered and, in a deep sense, it must be created. What we see, as we see it, is mere disorder."

(bb) The Decisive Role of Theory

To understand the nature of scientific knowing, we must now ask the central question: what is the decisive factor in the making of science? Many people accept the Baconian view, which ascribes prominence to facts. Scientific facts are taken to be well-defined events, publicly observable and measurable. Science is therefore seen as an immediate and absolute form of knowing: nothing but factuality.

---

[25] Galileo Galilei, *Dialogue concerning the Two Chief World Systems, Ptolemaic and Copernican,* trans. S. Drake (Berkeley: University of California Press, 1962), pp. 144-147, especially p. 145.

[26] J. Bronowski, *Science and Human Values* (New York: Harper Torchbooks, 1965), pp. 13f.

The inadequacy of the Baconian interpretation comes to light if we raise a simple critical point. Is the distinction between theory and fact, which is so widely taken for granted, an obvious one? Experience shows that this is not normally the case. Facts are often views that come to be accepted as evident through a long cultural process. William Whewell speaks rightly in this connection of a veritable "education of the senses." [27] He is right because scientific fact is mostly not what a person perceives through the senses, but what his mind tells him he should consider an objective state of affairs. This is what distinguishes fact from mere appearance. The foundation of fact, therefore, is some sort of theoretical interpretation. An obvious, everyday example is the assurance, shared universally by contemporary Western people, that the earth is moving around the sun. Do people take this as a fact because of what they see with their eyes or because of what they think with their minds? Clearly, there is no justification of principle to predicate science as pure factuality in opposition to theory. We must now try to determine with some precision what is the decisive role played by theory in the making of science.

We consider the experience of the scientists. Starting with what they say, we notice that they often speak of facts, and claim the possibility of detecting them. However, practicing scientists never identify science with a collection of facts. On the contrary, they frequently go so far as to warn against too great a reliance on facts themselves. Huygens, for instance, while discussing his research on the laws of impact, states explicitly: "Don't believe that I follow facts (*experientias*). I know indeed how misleading they are...." [28]

Claude Bernard, the great experimentalist, sternly reminds aspiring scientists: "I think that blind belief in fact, which dares to silence reason, is as dangerous to the experimental sciences as the beliefs of feeling or of faith which also force silence on reason." [29] T. H. Huxley, another prominent experimentalist, states bluntly that "those who refuse to go beyond fact rarely get as far as fact." [30]

What are the reasons given by creative scientists to motivate their wariness against factuality in science? One is determinant. *Facts do not suffice by themselves to make science.*

---

[27] W. Whewell, *The Philosophy of the Inductive Sciences Founded upon Their History* (London: Parker, 1840), vol. II, pp. 502-504.

[28] Quoted in R. Dugas, *Mechanics in the Seventeenth Century: From the Scholastic Antecedents to Classical Thought,* trans. F. Jacquot (Neuchâtel: Griffon, 1958), p. 282.

[29] Bernard (note 1), p. 53.

[30] In C. Bibby, ed., *The Essence of T. H. Huxley: Selections from His Writings* (New York: St. Martin's, 1967), p. 43.

By simply noting facts, we can never succeed in establishing a science. Pile up facts or observations as we may, we shall be none the wiser. To learn, we must necessarily reason about what we have observed, compare the facts and judge them by other facts used as controls. [31]

In Poincaré's famous comparison: "Science is built up of facts, as a house is built of stones; but an accumulation of facts is no more a science than a heap of stones is a house." [32]   In other words, facts remain sterile for science until they are evaluated in the light of a theory.   "Facts . . . are never duly appreciated till, in the hand of some skilful observer, they are made the foundation of a theory. . . ." [33]   This declaration by Dalton, the father of atomic theory, is strikingly appropriate in expressing the scientific attitude.

If facts are not supreme in scientific research, it is obvious that the decisive role is played by another factor.   This is variously referred to as hypothesis, theory, or simply idea.   The essential point is the intellectual character of such a factor.   To cite Poincaré again:

The isolated fact attracts the attention of all, of the layman as well as the scientist. But what the true scientist alone can see is the link that unites several facts which have a deep but hidden analogy.... Facts would be barren if there were not minds capable of selecting between them and distinguishing those which have something hidden behind them and recognizing what is hidden—minds which, behind the bare fact, can detect the soul of the fact. [34]

This conviction is universal among scientists.   In particular, Bernard the experimentalist is tireless in propounding it.   To him discovery is essentially a matter of idea:

We usually give the name of discovery to recognition of a new fact; but I think that the idea connected with the discovered fact is what really constitutes the discovery.... A great discovery is a fact whose appearance in science gives rise to shining ideas.... [35]

If ideas, not facts, make science, it follows that the speculative activity of the mind should not be banned from science, but rather should be recognized as central to it.   With Bernard one must admit that the first step of science is to form a hypothesis, as novel and comprehensive as the observational evidence allows.   Of course, one should take care not to

---

[31] Bernard (note 1), p. 16.

[32] H. Poincaré, *Science and Hypothesis*, trans. W. J. G. (New York: Dover, 1952), p. 141.

[33] J. Dalton, quoted in L. K. Nash, *The Atomic-Molecular Theory*, Harvard Case Histories of Experimental Science (Cambridge, Mass.: Harvard University Press, 1950), p. 32.

[34] H. Poincaré, *Science and Method*, trans. F. Maitland (New York: Dover), pp. 27f.

[35] Bernard (note 1), p. 34.

be carried away by the hypothesis advanced. "We must be enemies of hypothesis as a conclusion, but must always begin from it. . . ." [36] However, hypothesis is vital. The main reason is the very nature of hypothesis. A tentative interpretation of observational data is the only way open to man to achieve understanding. But without understanding, there is no genuine science, no matter how abundant observational data may be. A second reason for the importance of hypothesis is psychological. Frequently, the lack of a hypothesis makes it impossible to gather relevant observational data in the first place. Indeed, to collect data, one must know what they are relevant to. But light on this issue can only be shed by the theorizing effort of the mind. Thus theorization appears decisive in the making of science, since "Facts cannot be observed as facts except in virtue of the conceptions which the observer himself unconsciously supplies." [37] The situation can be summed up with Bernard: "Facts are necessary materials; but their working up by experimental reasoning, i.e. by theory, is what establishes and really builds up science. Ideas, given form by facts, embody science." [38]

If we want to substantiate the scientists' words about theorization with their deeds, the evidence is so abundant that the only embarrassment is that of choice. To begin from the very origin of science, how did Galileo succeed in bringing about the new scientific view of phenomena, including such apparently absurd doctrines as the principle of inertia and the law of uniform velocity of falling bodies? The answer is clear. It was the theorizing power of his mind that enabled him to create in thought an ideal world of pure vacuum with total absence of disturbing factors. Through this theorization, what up to then had appeared a mere figment of the mind turned out to be a fact, whereas what everybody had hitherto accepted as fact disclosed itself as mere appearance.

Another example of the vital role of theory in science making is presented by the work of Darwin. Since he lived in the heyday of positivism and since his activity consisted largely in amassing an enormous amount of evidence, Darwin has been frequently presented as the very type of the Baconian collector of facts. But Darwin himself never admitted this view, and all his work gives the lie to such an interpretation. In his late twenties, he was already speaking of "his theory" about biological species and their mutual relationships. His theoretical bent is obvious from what he wrote at that time (1837):

---

[36] Translated from C. Bernard, *Philosophie: Manuscrit inédit* (Paris: Boivin, 1937), p. 19.

[37] W. Whewell, quoted in P. B. Medawar, *The Art of the Soluble* (London: Methuen, 1967), p. 149.

[38] Bernard (note 1), p. 26.

The line of argument often pursued throughout my theory is to establish a point as a probability by induction and to apply it as hypothesis to other points and see whether it will solve them. [39]

When the *Origin of Species* came out, Darwin's detractors circulated the remark that he was just a good observer, but had no power of reasoning. He rebutted this criticism by stressing that *Origin* "is one long argument from the beginning to the end ... No one could have written it without having some power of reasoning." [40]   In general, Darwin insisted frequently that one could not be a good observer without being an active theorizer. "How odd it is that anyone should not see that all observation must be for or against some view if it is to be of any service." [41] "I have an old belief that a good observer really means a good theorist." [42]

An outstanding example which documents how theory effectively leads man to a satisfactory scientific interpretation of observational data is offered by Heisenberg, who documents the case carefully. It regards his own discovery of the uncertainty relations, a momentous breakthrough in the understanding of atomic matter. The genesis of his success is to be traced to a disconcerting remark by Einstein. Early in 1926 the young Heisenberg had visited the famous master in Berlin and discussed methodological questions with him. In the conversation he had expressed, with great conviction, the view that physics had to be built exclusively on observational data. His conviction in uttering this view had been great, particularly because he shared the common opinion that saw Einstein's relativity as a confirmation of such a methological principle. To his interlocutor's surprise, Einstein rejected roundly the validity of Heisenberg's position. In his own words, as recollected by Heisenberg:

Possibly I have used this kind of philosophy, but it is nonetheless nonsense. Or I can put it more cautiously: it may be heuristically advantageous to keep in mind what one has really observed. But, from the point of view of principle, it is entirely false to try to found a theory only on observable magnitudes. The theory is the first to decide what one can observe. [43]

---

[39] Quoted in G. de Beer, *Charles Darwin: A Scientific Biography* (New York: Doubleday Anchor Books, 1965), p. 95.

[40] In Barlow (note 8), p. 140.

[41] Letter to Fawcett: Sept. 18, 1861. Quoted in P. B. Medawar, *Induction and Intuition in Scientific Thought* (Philadelphia: American Philosophical Society, 1969), p. 11, note.

[42] Letter to Bates: Nov. 22, 1860. Quoted in Medawar, *loc. cit.* See also explicit testimony of his son Francis, himself a distinguished biologist, in Barlow, ed. (note 8), pp. 162f. For a general discussion of Darwin's methodological approach valuable considerations are to be found in M. T. Ghiselin, *The Triumph of the Darwinian Method* (Berkeley: University of California Press, 1969).

[43] Translated from Heisenberg (note 21), p. 92.

Probably the young scientist did not know at the time what sense to make of these words. But the following year the great crisis came. It was now a question of reconciling the corpuscular properties of electrons with their wavelike features. This reconciliation was essential in order to develop an atomic physics that really made sense of observable data. But precisely these data seemed to defy any meaningful solution. Electrons, in fact, as shown by their tracks in the cloud chamber, seemed to have only pointlike, corpuscular features. But then, in other experiments, they also manifested wavelike properties. How was it possible to reconcile these two apparently contradictory characteristics? Heisenberg and Bohr searched and discussed for many months without achieving any result. Then one night, when walking in a Copenhagen park, Heisenberg recalled his conversation with Einstein. Was Einstein perhaps not right when asserting that it was the theory, not the facts, that decided what could be observed? If so, Heisenberg realized immediately the source of his and Bohr's error. They had only too easily taken for granted that observation really manifested a pointlike trajectory of the electron in the cloud chamber. What was actually observed—he argued to himself— was probably less, namely only a sequence of unprecisely determined positions of the electrons. But if this was the case, the reconciliation of the corpuscular and wavelike properties of electrons did not present any difficulty of principle at all. In fact, within a few days Heisenberg was able to complete his basic paper on the uncertainty relations. Through them, quantum theory finally came of age.[44]

We can now sum up our discussion of the importance of theory in science. Theory is essential first for observing, and then for interpreting the observed data. In Heisenberg's own summary, when dealing with atomic physics:

Only when you have the complete theory can you say what can be observed.... So long as you have no laws in physics you don't observe anything. Well, you have impressions and you have something on your photographic plate, but you have no way of going from the plate to the atoms. [45]

The decisive role of theory in the making of science could not be expressed more convincingly.

(bc) The Theoretical Essence of Science

We have seen that theory is essential to the achievement of scientific discovery. We must now ask: what is the significance of theory in rela-

---

[44] *Ibid.*, pp. 110-112.
[45] Heisenberg, in Kuhn *et al.* (note 17); interview February 25, 1963, p. 19.

tion to science as such? Must theory be thrown out as a useless tool once discovery has been achieved? This view is frequently propounded by pragmatic and positivist philosophers. They speak of theory as of crutches to be discarded after use. The reason for such a stand is their interpretation of science. Science for them is essentially the ability to calculate with precision the occurrence of future events. As against any such view, the experience of the creative scientist is that theory is far from a mere tool for discovery. Rather, theory *is* discovery and, as such, is the very essence of science. The experiential standpoint in regard to this issue is paramount. Theory, as we have already seen, means vision. But for the creative scientist, science is precisely vision—a characteristic new way of seeing with the mind the intelligibility that is present in nature but which is overlooked by those who are not scientists.

Who is right in defining the essence of science? To decide this question let us recall what is the radically new element that discriminates the scientific from the prescientific approach to nature. Restricting our consideration to astronomy, it is clear that the novelty of the Galilean and Newtonian position was not the ability to predict events in a better way than was possible in the Ptolemaic system. This system, in fact, was based on accurate observations, and was excellently suited for practical calculations and predictions. In the medieval phrase, it was quite adequate "to save the appearances." The originality of Galileo consisted in refusing such a "saving" as worthy of the scientist. He rejected it as a merely pragmatic skill that gave no insight into the intimate structure of reality, hence his opposition to those whom he calls "the mathematical astronomers" and his siding with "the philosophical astronomers." The latter alone are, in his view, the genuine scientists, because they "seek to investigate the true constitution of the universe—the most important and most admirable problem that there is." [46]

Often the question of the essence of science is belittled. It does not matter—so the widespread contention goes—whether science is essentially theory or just a calculating device. What matters is that science works. On the contrary, history shows that a proper conception of science makes the entire difference for the very existence and thriving of science itself. As an example we revert to Galileo. He could have avoided all trouble had he simply followed Bellarmine's suggestion of speaking only "hypothetically"—that is, had he defended the Copernican system as merely a better calculating device. [47] Why did Galileo not accept this premise? Because to do so would have been a betrayal of science as such. No better proof can be had for the importance of the conviction

[46] "*On Sunspots,*" in Drake (note 7), p. 97.

[47] See Bellarmine's text in G. de Santillana, *The Crime of Galileo* (Chicago: University of Chicago Press, 1955), p. 99.

about the theoretical essence of science. The scientist suffers and struggles, animated by only one desire: to come to perceive with his mind the objective intelligibility of reality. As Einstein put it, "Behind the tireless efforts of the investigator there lurks a stronger, more mysterious drive: it is existence and reality that one wishes to comprehend." [48]

We can conclude the discussion of this all-important point by quoting a classical study of de Broglie on "The Grandeur and Moral Value of Science." To him there is no doubt that the goal of science consists

...in penetrating further into the knowledge of natural harmonies, to come to have a glimpse of a reflection of the order which rules in the universe, some portions of the deep and hidden realities which constitute it. [49]

As to the objection that science can be interpreted as pure pragmatism, he replies by referring to Pierre Duhem, himself a leading pragmatist. This thinker had to recognize that theories "establish between the phenomena a 'natural classification,' allowing us to sense the existence of 'an ontological order' which is beyond us." [50] As a consequence, de Broglie reiterates his conviction:

...all scientists, when they are sincere, recognize that the search for truth is the real reason that justifies the efforts of pure science and constitutes its nobility. Moreover, on this important question of the goal of disinterested science, all true scientists, in spite of the differences of opinion which can separate them, are without doubt nearer to being in agreement than they themselves often imagine. [51]

(bd) The Ancillary Function of Mathematics

A final consideration may serve to complete our exploration of the originality of scientific knowing. This consideration is aimed at removing a common misunderstanding. The layman sees science as essentially mathematical and feels discouraged in his desire to understand science by the abstractness and complexity of the mathematical formulation. However, this view misrepresents the essence of science.

The function of mathematics in science is essentially ancillary. The first reason for this assertion is the datum that science is fundamentally experiential. But experience is something that is lived through rather

---

[48] A. Einstein, *The World as I See It,* trans. A. Harris (New York: Covici Friede, 1934), pp. 137f.

[49] L. de Broglie, *Physics and Microphysics,* trans. M. Davidson (New York: Grosset & Dunlap, 1966), p. 207.

[50] For Duhem's views referred to here see P. Duhem, *The Aim and Structure of Physical Theory,* trans. P. P. Wiener (New York: Atheneum, 1962), pp. 26f.

[51] L. de Broglie (note 49), p. 208.

than formulated in a language, let alone the abstract and logically rigorous language of mathematics. At the request of a friend, Einstein tried to express the experience of scientific creation as lived by him. He began by declaring that words as such were absent.

The words or the language, as they are written or spoken, do not seem to play any role in my mechanism of thought. The psychical entities which seem to serve as elements in thought are certain signs and more or less clear images which can be "voluntarily" reproduced and combined. [52]

Einstein insists on the experiential nature of such elements which, according to him, are essentially "of visual and some of muscular type." [53] To him scientific creation is basically an effort of combination of the elements mentioned.

...taken from a psychological viewpoint, this combinatory play seems to be the essential feature in productive thought—before there is any connection with logical construction in words or other kinds of signs which can be communicated to others. [54]

From these texts we can derive a basic reason for denying the identification of science and mathematics. The reason, as we have seen, is the experiential character of science as such.

The second reason for denying the identification of science with mathematics is the goal of science itself. Science aims at understanding. But understanding is not a question of mathematics—of abstract and logically rigorous formulation. Rather, understanding consists in becoming aware of what is there, in nature. An illuminating testimony in this connection is that of Heisenberg. He, as a creator of quantum mechanics, came to use quite abstract mathematical formalisms. Nevertheless, in his reminiscences he repeatedly insists on the need of not exaggerating the importance of mathematics. In his experience, science begins with a sense of uneasiness ("a trouble") concerning the intelligibility of a certain area of nature. But, he remarks, "a trouble is something that is not in mathematics, but in physics." [55] Why is mathematics of secondary importance in science? Because it is a rationalization. "...rationalization, as everybody knows, is always a later stage and not the first stage." [56] In other terms, in Heisenberg's view, two steps are necessary to create science. First, one has "to cover the experimental situation by means of concepts

---

[52] In J. Hadamard, *An Essay on the Psychology of Invention in the Mathematical Field* (New York: Dover, 1954), p. 142.

[53] *Ibid.*, p. 143.

[54] *Ibid.*, p. 142.

[55] In Kuhn *et al.* (note 17); interview February 28, 1963, p. 30.

[56] *Ibid.*; interview February 11, 1963, p. 17.

that fit.   Later on you have to put the concepts into mathematical forms, but that is then more or less a trivial process which has to be solved." [57] From these words we can see the primacy of understanding over rigorous mathematical formulation.   Heisenberg is insistent on this point because of its importance.   One can never hope to solve a problem if, instead of concentrating on the real issue, understanding, one shifts his attention to a question of mathematical rigor.

One first tries to see how things are connected—what they really mean. ...one tries to make it more difficult by forgetting about mathematical schemes and at the same time one comes to a kind of substance of things which one is inclined to forget if one works in the mathematics alone. [58]

What, then, is the significance of mathematics in science? It is that of a language that the creative scientist learns to handle, with greater or smaller skill, but never mistakes for science as such.   As examples we can mention two of the most seminal scientists who ever existed, Newton and Bohr.   Newton was acknowledgedly a mathematical genius. And yet, history shows that he never made mathematics the central element of scientific research, but constantly subordinated mathematics to physics proper.[59]   In the case of Bohr, his limitations in the mathematical field are amply documented, beginning with his own cheerful admission of being "really an amateur." [60]   And yet we know that practically all the major contributors to modern quantum physics acknowledge Bohr as the source of their inspiration and success.   Some of these contributors were far superior to him in mathematical ability—as, for instance, Dirac and Heisenberg.   What did they find in Bohr?   Heisenberg provides the clue by speaking of his own experience.   Although he had already studied quantum physics under a highly competent teacher, Sommerfeld, he asserts that his "true scientific development first began" with an afternoon walk that he took with Bohr the first day he met the great master (Göttingen, 1922).[61]   The fundamental factor that influenced the young scientist was Bohr's experiential way of doing science, something quite distinct from mathematical ability.

One of the things that impressed me most was this kind of intuition of Bohr—Bohr knew the whole Periodic System.   At the same time, one could

---

[57] *Ibid.*, interview November 30, 1962, p. 14.

[58] *Ibid.*, interview February 28, 1963, p. 30.

[59] "Newton's Conception of the Significance of Mathematical Demonstrations in Physics," in Dugas (note 28), pp. 416f.

[60] "You know, I am really an amateur. And if they go really into high mathematics I can't follow."   Testimony reported by J. Franck in Kuhn *et al.* (note 17); interview July 11, 1962, p. 5.

[61] Heisenberg (note 21), pp. 59-63.

easily learn from the way he talked about it that he had not proved anything mathematically, that he just knew that this was more or less the connection.... The intensity of imagination which he had—that made an enormous impression. [62]

As Heisenberg goes on to say, this way of doing science with the whole being was the decisive factor for inspiring a young scientist.

A young man learned that one can do a lot of work oneself—it's not finished at all. At the same time, it is almost clear that the whole picture [of atomic physics] cannot be much different from what Bohr says, but still, all the details have to be filled out.... [63]

To conclude, the essence of science, even in its most abstract part —theory—is experiential, not mathematical. Mathematics enters it only indirectly, being essentially a language. However, by the same token, we can easily see why many tend to identify mathematics with science. The error arises from the instinctive tendency to identify the content with the language used to express it. Unfortunately, such an erroneous interpretation is quite widespread, not only among laymen. The situation is deplored by many experts. For instance Eddington, himself a distinguished mathematical physicist, complains about the tendency even among physicists "to treat the mathematical development of a theory as the only part which deserves serious attention." [64] Against this view he insists:

But in physics everything depends on the insight with which the ideas are handled before they reach the mathematical stage. [65]

Mathematics is a useful vehicle for expression and manipulation; but the heart of the theory is elsewhere... [66]

(c) The Diligence of Experimentation

Trying to understand the activity of man in the making of science, we have dwelt at some length on the two first stages of the scientific approach. We have seen the role played by the mind in observation and theorization. We must now reflect on the third stage, that of experimentation. Experimentation complements the scientific approach in that it tests the validity of the theoretical interpretation proposed about the data supplied by observation. The question is now: is the experimenter as such creative and, if so, what does his creativity consist of? Before discussing this

---

[62] In Kuhn et al. (note 17); interview February 13, 1963, p. 7.

[63] Ibid.

[64] A. Eddington, The Philosophy of Physical Science (Ann Arbor: University of Michigan Press, 1958), p. 55.

[65] Ibid.

[66] Ibid., p. 74.

question, let us introduce a word of clarification about the distinction between observation and experiment.

Frequently, in the ordinary and also scientific parlance, observation and experiment are two terms that are used interchangeably. There are good reasons for this usage. In fact, both the observer and the experimenter are persons who try to see and record events as they can be detected through inspection of nature. However, there are also grounds for a distinction. The reason is that observation and experiment, respectively, belong to different stages of the development of science. The observer is a man who gathers evidence as made available to him by the spontaneous course of nature. The experimenter, on the other hand, is the person who intervenes actively in the course of nature. In short ". . . observation is investigation of a natural phenomenon, and experiment is investigation of a phenomenon altered by the investigator." [67]

The general motive for calling experimentation creative is obvious. Experiment is undertaken to test a theory; in turn, theory applies to data supplied by observation. The mental activity of experimentation, therefore, is basically an extension of that of theorization. Science starts from observable evidence and ends up with observable evidence. But the mind of man must provide the connection between the two sets of evidence. Theorization consists in the attempt to reconstruct the data of original observation according to an intelligible scheme. The thinking that must precede experimentation amounts to devising a way to put the theory to the test of controlled observation. This is a difficult task that demands much ingenuity—authentic creativity. For the connection between what can be checked by experiment and the contents of a theory are far from easy to determine. As Einstein put it:

Here, too, the observed fact is undoubtedly the supreme arbiter; but it cannot pronounce sentence until the wide chasm separating the axioms [of the theory] from their verifiable consequences has been bridged by much intense, hard thinking. [68]

The activity demanded by experimentation is both logical and inventive. The scientist must—as Galileo put it—"develop and resolve" the theory "into its principles," then conceive an instrumental structure to determine whether the consequences of the theory agree with reality as observable.[69] In both phases, it is plain that the mental activity of man, or idea, is of paramount importance. Experimentalists reflecting upon their work like to insist on this point. As an example Bernard, the great physiologist, writes:

---

[67] Bernard (note 1), p. 15.

[68] A. Einstein, *Ideas and Opinions*, trans. S. Bargmann (New York: Crown, 1954), p. 282.

[69] Translated from A. Carugo and L. Geymonat, eds., G. Galilei, *Discorsi e Dimostrazioni Matematiche intorno a Due Nuove Scienze* (Turin: Boringhieri, 1958), p. 414.

I consider it... an absolute principle that experiments must always be devised in view of a preconceived idea, no matter if the idea be not very clear nor very well defined. [70]

The role of experimentation is frequently expressed in terms of a question put to nature. The simile is only vaguely appropriate. For the experimenter does indeed put a question, but at the same time he also tries to find what the answer to the question must be. Bernard prefers to compare the experimenter to the watcher of a dumb show. Such a watcher tries to learn the intentions of the actors by merely viewing their external behavior. Thus he is far more than a spectator or even a questioner. He is an interpreter. Likewise the experimenter is a continually active interpreter.

The creativity of experimentation becomes most clear when one considers the execution of experiment and its interpretation. Experiment is a creation in that it is pure as an event. Every event that takes place spontaneously in nature is very complex. Many factors are simultaneously at work on it. Experiment, on the contrary, is a pure event because the experimenter excludes from it, or at least minimizes, all factors except for the ones he directly studies. With Bernard one can say that, through experimentation, "man becomes an inventor of phenomena, a real foreman of creation." [71]

Interpretation of an experiment demands creativity because the data furnished by the experiment become meaningful only through much thinking. Usually researchers express the data of experiments by means of points on graph paper. All the experiment supplies directly is a limited number of disconnected points. But disconnected points make no sense by themselves. They cannot manifest a universal regularity of nature, much less disclose the reason for such regularity. The experimenter, therefore, has to exert his mind if he is to detect meaningful information as embodied in the experimental data.

Examples that illustrate the creativity of experimentation are naturally countless. Every branch of science has been successfully initiated and brought to maturity by the creative activity of experimenters. But abundance of documentation is not needed to establish our point. I shall make only a few remarks about Galileo's experimental study of falling bodies. The discussion of Galileo's attitude toward experiment is illuminating because it shows simultaneously how much inventiveness is needed for experimentation and how easily the work of an experimenter can be misunderstood by the nonscientific mind. Beginning with the latter point, we see at once why Galileo did not carry out the experiment on the

---

[70] Bernard (note 1), p. 23.
[71] *Ibid.*, p. 18.

uniform velocity of falling bodies which his enthusiastic admirers ascribed
to him.   Despite the explicit testimony of his late disciple Vincenzo Viviani, modern historians of science have proved satisfactorily that Galileo
never dropped bodies from the top of the Leaning Tower of Pisa to prove
the validity of his famous law. [72]   Why did Galileo not perform such a
spectacular act of showmanship, which would have made a tremendous
impression on his contemporaries?   The theoretical reason is clear.   Such
an attempt would have been meaningless from the scientific standpoint,
failing the necessary condition of vacuum.   In point of fact we have at
least indirect evidence that Galileo condemned the attempts of contemporary scientists to execute the experiment. [73]   But what sort of experiment did Galileo carry out concerning falling bodies?   As far as we can
ascertain historically, he tested his other famous discovery—the law of
constant acceleration—by means of the experiment that he describes with
so much love of detail in the Third Day of his *Discorsi* and which he
declares typical of the true scientific spirit. [74]   This experiment involved
measuring the acceleration of bronze balls sliding down a wooden groove
on an inclined plane.   Various critics have denied the historicity of such
an experiment on the strength of the many sources of inexactitude inherent
in Galileo's primitive instrumentation.   And yet we have no objective
reason for not taking Galileo's own words seriously.   For the experiment,
even when done with Galileo's simple instruments, proves to be substantially precise and convincing. [75]   To conclude our analysis of Galileo's
work in this area, crucial for the beginning of science, why did Galileo
move so confidently in the experimental realm to the discomfiture of
admirer and critic alike?   The reason can only be the outstandingly creative
power of his mind.   Theoretical creativity made all the difference.   The
philosopher Hanson put it aptly:

Galileo's theoretical vision made him a better empiricist than his contemporaries. . . .   It enabled him to see more of the world than they could. [76]

---

[72] Viviani was a very late disciple of Galileo's, and full of hero worship for the famous
master.   He is not always historically reliable.   See an accurate discussion of the issue in
Carugo and Geymonat (note 69), pp. 689-692.

[73] Attempts to prove the law experimentally were actually carried out by Giovan Bat-
tista Baliani and Vincenzo Rinieri.   Galileo seems to have been very critical of them.   See
Carugo and Geymonat, *op. cit.*, pp. 691f.

[74] In Carugo and Geymonat, *op. cit.*, pp. 198-200.

[75] The experiment has been carried out recently, with the instruments and the theory
described by Galileo himself.   The experimenter was an engineer and historian of science,
T. B. Settle.   See his "An Experiment in the History of Science" in *Science* 133 (1961),
19-23.

[76] N. R. Hanson, in M. F. Kaplon, ed., *Homage to Galileo: Papers Presented at the
Galileo Quadricentennial University of Rochester* (Cambridge, Mass.: MIT Press, 1965),
pp. 48f.

2. The Dependency of the Scientific Approach

If activity of man is the most prominent feature of the scientific approach, it does not in itself suffice to characterize the scientific approach. For man, in doing science, is not a totally autarkic performer. He is rather a dependent, though active, partner. Thus, to continue our analysis of the scientific experience of knowledge, we must now study the dependency of scientific cognition. To this end, I shall divide the investigation into three main points. First, I shall examine the responsive character of research. Second, I shall discuss the docility of science, especially as expressed in the concern for precision and accuracy. Third, I shall say a word about discovery as submission to objective reality.

(a) The Responsiveness of Research

What is the reason for saying that science, for all its creativity, is essentially a response? The fundamental reason is psychological in nature. We can become aware of it if we survey briefly the process through which science comes into being.

When relating his first impressions as a geologist, Darwin wrote significantly: "On first examining a new district nothing can appear more hopeless than the chaos of rocks." [77] This sense of hopelessness must be experienced by any scientific beginner. Nothing, in fact, can appear more discouraging than the welter of possibly meaningful events among which the scientist has to pick his way in order to come to discovery. This is the reason why scientists sometimes say that the most difficult step in science is to find the problem. Indeed, the structure of the problem is something highly puzzling, as philosophers have repeatedly underscored, from Plato to Polanyi. For a problem consists, by definition, in the fact that many elements—which appear totally disconnected—actually form an intelligible whole. But, then, how can man determine what elements, among the countless ones he perceives, truly belong together as to form an objective, unitary whole? "How can we tell what things not yet understood are capable of being understood?" [78] Obviously, the nature of the problem is paradoxical. Both horns of the dilemma appear inescapable. Either you do not know what you are looking for (and then you do not look for anything) or you know already what you are looking for (and then you cannot look for it any more).

---

[77] In Barlow (note 8), p. 77.
[78] M. Polanyi, *Science, Faith and Society* (Chicago: University of Chicago Press, 1964), p. 14.

As against the puzzling nature of the problem, experience shows that the researching scientist never proceeds by trial and error, but aims straight at his goal—albeit in a slow and tentative way. This is the unanimous testimony of the psychologists of science and the creative scientists themselves. Wertheimer, for instance, investigated—in conference with Einstein—the long work that led the youthful researcher to the formulation of the theory of relativity. His conclusion is the unequivocal rejection of any blind trial-and-error procedure.

Scrutiny of Einstein's thought always showed that when a step was taken this happened because it was required.... Quite generally... one knows that any blind and fortuitous procedure was foreign to his mind... there was no mathematical guesswork in it. [79]

But this feature of scientific thought does not appear only in the case of exceptionally gifted minds like Einstein's. Eiduson comes to the same conclusion after having studied a number of variously gifted scientists.

...the scientist goes into his work with a certain scientific eye... he is selective, discriminative, and quickly recognizes what might or might not be appropriate. [80]

This therefore seems to be truly an essential feature of scientific procedure: to move almost instinctively toward the sought-after solution. In this context Poincaré speaks of discovery as being a kind of selection. "Discovery is discernment, selection." [81] But then he goes on to explain that the selection is practically automatic, since only plausible and useful ideas are retained by the mind of the researcher. In his own words: "Unfruitful combinations do not so much as present themselves to the mind of the discoverer."

The question that arises by contrasting the nature of the problem and the actual behavior of the scientific researcher can only be solved by supposing that the researcher himself is in some way guided in his research by the intelligibility of the reality he has set out to discover. According to Polanyi:

A potential discovery may be thought to attract the mind which will reveal it—inflaming the scientist with creative desire and imparting to him intimations that guide him from clue to clue and from surmise to surmise. [82]

---

[79] Wertheimer (note 23), "Einstein: The Thinking that Led to the Theory of Relativity," pp. 213-233, especially pp. 232f.

[80] B. T. Eiduson, *Scientists: Their Psychological World* (New York: Basic Books, 1962), p. 124.

[81] Poincaré (note 34), pp. 51f.

[82] Polanyi (note 78), p. 14.

Here, of course, as Polanyi himself hints in this context, there is no need to speak of extrasensory perception.   It is enough to speak of docile attention and consequent heightened sensitivity on the part of the researcher.   A passage by Duhem reflecting on his experience as a physicist is illuminating.   The scientist does not choose his way in the search for truth any more than a flower chooses the grain of pollen which will fertilize it.   In his own words:

... the flower contents itself with keeping its corolla wide open to the breeze or the insect carrying the generative dust of the fruit; in like manner, the physicist is limited to opening his thought through attention and reflection to the idea which is to take seed in him without him. [83]

We can begin to understand the dependency experienced by the working scientist.   To do science is fundamentally to respond.   As Galileo used to say, "Nature leads one by the hand."   Of course, scientific research is not a passive response, but a very active one.   Nevertheless it is an essentially responsive process.   The process begins with a sort of vague stirring—interest mixed with uneasiness.   Man gets the impression that there must be something in a given area of nature that is not yet understood but that is interesting and worth investigating.   As already mentioned, Heisenberg speaks in this connection of a sense of "trouble." [84]   If man does not divert his attention but keeps reflecting and seeking, little by little he feels attracted toward a plausible solution.   In a simile, Lorenz speaks of "a good hunting dog, which begins to pull on the line in a certain direction." [85]   But the solution does not come immediately.   Responsiveness then takes the form of patience and perseverance.   The researcher must adopt the attitude of the person who looks for a pattern in a jigsaw puzzle.   He knows that the pattern is there, but he cannot force the pattern itself to appear before his eyes.   He must rather seek attentively and wait patiently.   Then, suddenly, the pattern becomes visible.   At this moment the researcher reaches the goal of his investigation: insight or discovery.   Discovery, thus, arises from responsiveness.   In other words, discovery is indeed a manifestation of creativity, but the creativity involved is of a dependent kind.

(b) The Docility of Precision

The active dependency of science, which is clear in the process leading to the insight of discovery, becomes even clearer in the subsequent critical step.   Although an insight may be attractive, to be scientifically con-

---

[83] Duhem (note 50), p. 256.
[84] See note 55.
[85] Lorenz (note 16), p. 45.

vincing it must be put to the test and found valid. This is the rigorous aspect of the scientific procedure emphasized, for instance, by Whewell:

To form hypotheses, and then to employ much labour and skill in refuting, if they do not succeed in establishing them, is a part of the usual process of inventive minds. Such a proceeding belongs to the rule of the genius of discovery.... [86]

The most visible manifestation of the spirit of dependency that animates science is *experiment*. Experiment is a profession of dependency for, through it, man submits his views to the testing of nature. But this is why scientists are unanimous in their praise of experiment. Already Galileo characterized "the true scientist" as the person who "confirms with sensible experiences (*sensate esperienze*) the [theoretical] principles." [87] Other researchers praise experiment because of its rigor. As the famous biologist Richet put it, one must be "as rigorous in experiment as audacious in hypothesis." [88] All scientists concur in extolling experiment because of its objectivity. In Bernard's classic formulation:

...we must never make experiments to confirm our ideas, but simply to control them; which means, in other terms, that one must accept the results of experiments as they come, with all their unexpectedness and irregularity.... In a word, we must alter theory to adapt it to nature, but not nature to adapt it to theory. [89]

If consideration of experiment is important to understand the spirit of science, equally important is reflection on the language which alone makes experiment meaningful. Experiment, we saw (Chapter 1, Section 1c), stands in need of planning and interpretation. But in both such phases of science it is possible for man to incur the danger of subjectivism unless he can count on an instrument of complete objectivity. The central reason, then, that *mathematics* is so indispensable in science is the inherent objectivity of the mathematical language. As Einstein stressed it:

We do not ask how *we* can describe nature by mathematical schemes, but we say that *nature* always works so that the mathematical scheme can be fitted to it. [90]

The objectivity of mathematics is vital because it does not tolerate any subjective penumbra, but compels the mind to take a completely unequivocal stand. However, scientists prize mathematics not as an end in

---

[86] Whewell (note 27), vol. II, p. 221.
[87] Translated from Carugo and Geymonat (note 69), pp. 198f.
[88] C. Richet, *Natural History of a Savant*, trans. O. Lodge (London: Dent, 1927), pp. 47 and 123.
[89] Bernard (note 1), pp. 38f.
[90] Quoted by Heisenberg in Kuhn *et al.* (note 17); interview February 25, 1963, p. 16.

itself but rather as an instrument for the discovery of truth (see Chapter 1, Section 1bd). That is, they are concerned not with mathematics as such but with the intelligibility of nature that the application of mathematics brings to light. Mathematics, as Poincaré put it, reveals "the hidden harmony of things." [91] Accordingly, one can realize why scientists are at times so enthusiastic about mathematics. A much-quoted passage of Galileo's deserves to be mentioned here:

Philosophy is written in this grand book, the universe, which stands continually open to our gaze. But the book cannot be understood unless one first learns to comprehend the language and read the letters in which it is composed. It is written in the language of mathematics.... [92]

To confirm the importance of mathematics as a revealer of objective truth, we can listen to Heisenberg speaking out of his own experience:

I would say, I had never much fun in mathematics where you have to prove something. But I had much fun in mathematics where you have to find out how things are. [93]

How is the scientific insistence on objectivity and precision to be assessed in human terms? Nonscientists frequently dismiss it as a kind of pedantry. The word normally used is objectivism—as though the scientist would stress unduly the submission of man to material reality. As experienced by the scientist, however, the search for objectivity is rather a form of duty. It manifests an attitude of active docility that results from a sense of *respect for reality* as such. An example of such respect, and of its far-reaching consequences for man, is offered by the history of astronomy. One of the decisive pieces of evidence that helped bring about the downfall of the Aristotelian-Ptolemaic cosmology was the accurate observation of celestial phenomena. New stars and comets had been observed from remote antiquity, but no one had realized that these phenomena were incompatible with the Aristotelian doctrine of the immutability of the heavens. When new stars were sighted, in 1572 and 1577, quantitative observations—made especially by Tycho Brahe—proved beyond doubt that the new stars were really a change in the heavens, since they were more removed from the earth than the moon itself. [94] Tycho, impressed by this discovery, decided to explore the skies system-

---

[91] H. Poincaré, *The Value of Science,* trans. G. B. Halsted (New York: Dover, 1958), p. 79.

[92] In Drake (note 7), pp. 237f.

[93] In Kuhn *et al.* (note 17); interview November 30, 1962, p. 9.

[94] See C. D. Hellman, "The Role of Measurement in the Downfall of a System: Some Examples from Sixteenth Century Comet and Nova Observations," in A. Beer, ed., *Vistas in Astronomy: New Aspects in the History and Philosophy of Astronomy* (New York: Pergamon, 1967), pp. 43-52.

atically in order, as he himself said, "to lay the foundations of the revival of astronomy." [95]   He accomplished his self-appointed task admirably. However, his data needed theoretical elaboration so as to provide understanding of the motion of the planets.   This was the task that Kepler took upon himself.   He labored for many years to determine the orbit of Mars, which he took as a test case.   His calculations seemed to prove that the orbit was circular.   Then he noticed a discrepancy between the conclusions issuing from his theory and Tycho's data.   If his theory was right, Tycho had been wrong by 8 minutes of arc in his determination of the positions of Mars.   At this point Kepler had to choose between his own hard-won results and the precision of Tycho's data.   Had he been less respectful of precision, he could easily have explained the discrepancy as due to the unreliability of Tycho's data.   Kepler chose the alternate solution.   With genuine scientific spirit, he rejected his own interpretation and started his theoretical calculations all over again.   After much effort, the outcome was the discovery of the elliptical orbit of Mars and of the other planets. This discovery, as is well known, was one of the main foundations of the Newtonian synthesis and, thus, of modern science as a whole.   With Kepler himself one can conclude: "These eight minutes showed the way to a renovation of the whole of astronomy." [96]   And we can add that they contributed substantially to renovating the entire knowledge of reality by making science firmly established—striking vindication for the scientists' conviction about the importance of precision.

What we have seen in an example, history shows to have general validity.   Speculation and phenomenism cease and science begins when man learns to respect nature by being precise in observing and interpreting it.   Thus, for instance, optics was born when Newton decided to observe experimentally and quantitatively the behavior of light rays.   Chemistry was born when Lavoisier introduced the experimental-quantitative principle of mass conservation.   By systematically using the balance, he destroyed the phlogiston theory: "that vague principle," as he himself put it, "which is not strictly defined and which consequently fits all the explanations demanded of it." [97]   In short, there is no exaggeration in Bernard's forceful statement:

In a word, the greatest scientific truths are rooted in details of experimental investigation which form, as it were, the soil in which these truths develop. [98]

---

[95] Quoted in M. Boas, *The Scientific Renaissance: 1450-1630* (New York: Harper & Row, 192), p. 111.

[96] Quoted in M. Caspar, *Kepler*, trans. C. D. Hellman (London: Abelard-Schuman, 1959), p. 128.

[97] Quoted in D. McKie, *Antoine Lavoisier: Scientist, Economist, Social Reformer* (New York: Collier, 1962), p. 110.   See the whole chapter "Protean Phlogiston," pp. 104-112.

[98] Bernard (note 1), p. 15.   For more details on this basic topic, see S. L. Jaki, "The

To sum up, scientific precision is a form of docility of man toward reality. As such it is far from unbecoming to the dignity of man. It is rather a better realization of man's dignity, which is founded on respect for reality rather than on wishful thinking. In this connection, a word must be added about another manifestation of the same mentality. I refer to *scientific specialization.* Many people are critical of specialization as though it would be a dehumanizing attitude. They consider the specialist a kind of escapist—a person who is unable or unwilling to face the world and seeks refuge in the ivory tower of research. Although some scientists may be made less than human by their specialization, this result should not be imputed to specialization as such. For the opposite is true. Specialization is of itself humanizing. It is so, first of all, because it implies humility. No one can know everything he would like to know. Scientific specialization is but the acknowledgement of such a situation. "The scientific attitude always means a resignation and a self-denial—even for the most fertile scientist. This self-denial is the source of specialization." [99] But scientific specialization is also humanizing in a more specific sense. It gives to man a sense of vocation or mission in the service of an ideal. Man in general can attain the fullness of his humanity only by dedicating himself wholly to an ideal. But if this applies in other fields of the human endeavor, it must apply also in the case of scientific specialization. This point has been stressed particularly by Max Weber in a famous essay discussing science as a "vocation."

In the field of science only he who is devoted solely to the work at hand has "personality." And this holds not only for the field of science: we know of no great artist who has ever done anything but serve his work and only his work. [100]

(c) Interpersonal Control

A further aspect characterizes the dependency typical of science. It is the universal practice of submitting scientific results to the critique of experts in the field. That is to say, the scientific researcher is not satisfied with just putting his own results to the test of nature. Over and above that, he welcomes the efforts of other people to check the validity of his views.

This point is important because it manifests one essential distinguishing feature of science as a form of knowledge. Science, of its nature,

Edge of Precision," in his *The Relevance of Physics* (Chicago: University of Chicago Press, 1966), pp. 236-280.

[99] C. F. von Weizsäcker, *The History of Nature,* trans. F. D. Wieck (Chicago: University of Chicago Press, 1949), p. 2.

[100] M. Weber, "Science as a Vocation," in *Essays in Sociology,* trans. and ed. H. H. Gerth and C. W. Mills (New York: Oxford, 1958), p. 137.

is not personal or private knowing. On the contrary, it is *public* or, to put it with Ziman, *consensible* knowledge. [101] This neologism is apt because it points out what, in the general estimation, makes science truly such. This factor is the general agreement or consensus of experts on the validity of a given proposition, hence the eagerness usually shown by researchers to have their work tested by the scientific community.

It is easy to see how the acceptance of interpersonal control is particularly clear evidence for the spirit of creative dependency of science. The scientist does literally everything possible to make his science faithful to reality. The pursuit of objectivity does not allow him to shrink back from any test, however painful. The most grueling test is, of course, the critiques of one's peers. In fact it is well known that, in general, there are no harsher critics than colleagues and competitors. As for the widespread opinion that, contrary to other bodies of specialists, the scientific community is unusually friendly toward new ideas, history proves it to be an illusion. Scientists also remain human in the sense that, as a group, they do their best to defend established views and suppress novelty. [102] As an example, Planck, the great innovator of physics, used to say that startling discoveries like his could gain universal acceptance only because of the death of the older generations of scientists. To sum up, if science eventually succeeds, this is truly a triumph of the creative dependency of man.

3. SCIENCE AS AN EXPERIENCE OF KNOWLEDGE

We have examined scientific knowledge from the two complementary viewpoints of creativity and dependency. We can now try to understand the essence of the scientific experience of knowledge by discussing synthetically some of its characteristic features.

(a) A Novel Way of Knowing

The first result we gather from our investigation is that science is a cognitive endeavor truly novel and unprecedented when contrasted with the prescientific mentality. This result emerges from our entire discussion, but it can be seen best by examining some common misinterpretations of science itself. Indeed, why would public opinion so widely misinterpret science if not because science is an original and different way of knowing?

---

[101] J. M. Ziman, *Public Knowledge: An Essay Concerning the Social Dimension of Science* (London: Cambridge University Press, 1968), p. 11 and *passim*.

[102] See B. Barber, "Resistance by Scientists to Scientific Discovery," in B. Barber and W. Hirsch, eds., *The Sociology of Science* (New York: Free Press, 1962), pp. 539-556.

The view of science as a superhuman feat is a widespread cliché. Science is seen as consisting basically of intuition—penetrating, precise, rigorous. The development of scientific research is accordingly conceived as a procedure dominated by logical concerns. The scientist is thought to be interested only in rigorous proofs and irresistible demonstrations. Thus he appears not only as superhuman, but also inhuman: "... a truth-finding machine steered by intuitive sensibility." [103]   Unfortunately, this distorted view is still reinforced by a certain public image cultivated by many scientific workers themselves. They emphasize, in various ways, impersonality and aloofness in their behavior. In particular, scientific papers—with their studied effort to eliminate completely the personality of the writer—can be quite misleading. Medawar's mischievous remark that they "not merely conceal but actively misrepresent the reasoning that goes into the work they describe" is certainly not without foundation. [104] For the layman tends naturally to take such papers as genuine expressions of the scientific attitude as a whole.

The rejection of logic and rationality as preeminent manifestations of the scientific mentality could not be more explicit on the part of those who most clearly deserve the name of scientist, namely the great creators. Heisenberg, quoting with approval the mathematician Hardy, puts it bluntly: "Those things in science which you can do by rational arguments are really not worthwhile doing." [105]   The basic motivation of this conviction is implicitly present in our preceding investigation and lies in the fact that logic or rational deduction is a totally inadequate approach when one intends to know how nature really is. In Einstein's words:

Pure logical thinking cannot yield us any knowledge of the empirical world; all knowledge of reality starts from experience and ends in it. Propositions arrived at by purely logical means are completely empty as regards reality. [106]

We find the identification of logic and science repeatedly rejected by creative scientists thoughout history. Galileo was already unequivocally clear on this point:

It seems to me that logic teaches to know whether discourses and demonstrations already done and found actually proceed in a conclusive way. But that it may teach how to find conclusive discourses and demonstrations, this verily I do not believe. [107]

[103] M. Polanyi (note 78), p. 15.
[104] Medawar (note 37), p. 151.
[105] In Kuhn *et al.* (note 17); interview February 7, 1963, p. 16.
[106] A. Einstein (note 68), p. 271.
[107] Translated from Galileo Galilei, *Opere*, F. Flora, ed. (Milan: Ricciardi, 1953), p. 1074, note.

The same conviction has been uttered by others—for instance, Poincaré and Pauli. [108] In this connection, a comparison of Brunschvicg's, cited with approval by Piaget, clarifies the difference between the intelligence, which creates, and the logicomathematical deduction, which systematizes what has been created.

... intelligence wins battles or indulges, like poetry, in a continuous work of creation, while logico-mathematical deduction is comparable only to treatises on strategy and to manuals of "poetic art," which codify past victories of action or mind but do not ensure their future conquests. [109]

The view of science as fundamentally a technical ability comes to the fore in the widely used, stereotyped expression "scientific method," which is normally defended by appealing to the authority of Francis Bacon. Science is taken here essentially as a form of doing. The researcher is supposed to be nothing but a faithful recorder of sense impressions or a diligent collector of facts. Discovery is then supposed to be the psychologically ineluctable outcome of such a procedure. The standard, foolproof tool of discovery is precisely seen as consisting of the so-called method.

In the light of the preceding, it is clear that this second view of science is no less a caricature than is the foregoing one. The basic reason is obvious. There is no such thing as method when creative science—that is, discovery—is concerned. To be sure, one can speak of some sort of methodological approach in the scientific investigation of nature. Its various steps are observation, theorization, experimentation, and criticism by the community of experts. But this procedure is no more than a generic guideline—psychologically necessary, but totally unable to ensure discovery. In other words, method can be critically important, but it is by itself sterile insofar as scientific creation is involved. Bernard, the great discoverer, put it aptly:

The experimental method, then, cannot give new and fruitful ideas to men who have none; it can serve only to guide the ideas of men who have them.... The idea is a seed; the method is the earth furnishing the conditions in which it may develop, flourish and give the best of fruit according to its nature.... The method itself gives birth to nothing. [110]

Working scientists are, as a rule, quite critical of Bacon's views. The following caustic remark by Bernard is typical: "... those who make the most discoveries in science know Bacon least, while those who read and

---

[108] H. Poincaré (note 34), p. 126. Concerning Pauli's views, see W. Heisenberg, "Wolfgang Paulis philosophische Auffassungen," *Die Naturwissenschaften* **46** (1959), 661-663.

[109] J. Piaget, *The Psychology of Intelligence,* trans. M. Piercy and D. E. Berlyne (Totowa, N.J.: Littlefield Adams, 1966), p. 31.

[110] Bernard (note 1), p. 34.

ponder him, like Bacon himself, have poor success." [111]   The reason for this blunt condemnation is, as Bernard states it elsewhere: "Bacon was not a man of science, and he did not understand the mechanism of the experimental method." [112]   Thus, masters who intend to train aspiring researchers caution them against being misled by the Baconian mentality. For instance, Freedman speaks of "three fallacious suppositions" implied in the Baconian views. [113]   This, however, should not be construed as a personal denigration of Bacon himself.   For, if Bacon was no scientist, he was the first to admit it.   "I am only a trumpeter" (*Ego enim buccinator tantum*), he himself used to say. [114]   His great merit was a ceaseless propaganda in favor of science.   This, among other results, seems to have been determinant in bringing about the foundation of the Royal Society, one of the greatest scientific bodies of all time. [115]

If we wish to characterize positively the novelty of scientific knowing, we can describe it as a complex development involving the whole personality of the researcher.   The *complexity of science* is obvious after our preceding discussion.   Nonetheless, it is important to emphasize it, because the humanistic misunderstanding of science arises mostly when one forgets to take this complexity into due account.   Since science is so complex, the temptation is always present to reduce it to some of its component elements, especially theory or experiment.   But then, by losing sight of the whole, science itself is reduced to nothing. Indeed: "Isolated, theory is empty and experience blind; and both are useless and of no interest alone." [116]   Thus it is necessary to resist positively the widespread tendency toward the superficialization of science, as Einstein stresses by pointing to the example of Galileo.

It has often been maintained that Galileo became the father of modern science by replacing the speculative, deductive method with the empirical, experimental method.   I believe, however, that this interpretation would not stand close scrutiny.   There is no empirical method without speculative concepts and systems; and there is no speculative thinking whose concepts do not reveal, on closer investigation, the empirical material from which they stem. [117]

The *developmental character of scientific knowledge* is another property that appears clearly from our investigation.   Indeed, if discovery

---

[111] *Ibid.*, p. 225.

[112] *Ibid.*, p. 51.

[113] P. Freedman, *The Principles of Scientific Research* (New York: Pergamon, 1960), pp. 22f.

[114] Dugas (note 28), p. 323.

[115] For details see M. Purver, *The Royal Society: Concept and Creation* (Cambridge, Mass.: MIT Press, 1967).

[116] H. Poincaré (note 34), p. 275.

[117] A. Einstein, Introduction to Galileo's *Dialogue* (note 25), p. xvii.

issues only from both observation and theorization, it is unavoidable that science will require a long time of development before achieving maturity. But, here again, it is not without reason that one should insist on this aspect of science, because the public continues to be impressed by the worn-out cliché. The figure of Archimedes running naked from his bath and shouting "Eureka" still exerts a very powerful attraction. Thus, if one wants to understand science, one should make an effort to see it as a gradual growth, with all its slowness and tentativeness. Also, one should see science as a genuine conquest to be attained only after much effort. For there is generally a great delay in time between the first bright idea and science as a mature reality. History could not be more explicit on this point. We know, for instance, that Galileo was already in possession of the exact law concerning the constant acceleration of falling bodies in 1604, but he was not able to prove it satisfactorily until much later. He first published it in his *Two New Sciences,* which appeared in 1638. [118] Something similar happened to Newton. Twenty years had to go by between his first insight on universal gravitation—symbolized in the tale of the falling apple—and the time when he finally published his *Principia.* [119] Nor are these famous cases exceptions.

Counter to a widespread view, it is not true that discovery is mostly a matter of youth and brightness of mind. In fact, even researchers who were most productive in their early years could not be creative without a long period of investigation. An apposite example is that of Einstein. He published his fundamental paper on relativity when he was just 23 years old, but work on this basic discovery had been going on for seven unbroken years before that date. [120] In general, scientific thinking demands a good deal of time to develop. [121] Nor is it always a smooth ride. For, even when a researcher has succeeded in formulating his guiding hypothesis or theory, much remains to be done in the line of clarification and correction of details. Thus he must make himself adaptable and supple all the time. A telling testimony is that of Darwin.

I have steadily endeavoured to keep my mind free, so as to give up any hypothesis, however much beloved (and I cannot resist forming one on every subject), as soon as facts are shown to be opposed to it.... with the exception of the Coral Reefs, I cannot remember a single first-formed hypothesis which had not after a time to be given up or greatly modified. [122]

---

[118] See historical details in Carugo and Geymonat (note 69), pp. 768f.

[119] See details in Manuel (note 12), p. 151.

[120] See details in Wertheimer (above, n. 23), pp. 213-233, especially pp. 214f.

[121] See P. Duhem, "Hypotheses Are Not the Product of Sudden Creation, but the Result of Progressive Evolution. An Example Drawn from Universal Attraction," in Duhem (note 50), pp. 220-252.

[122] C. Darwin, in Barlow (note 8), p. 141.

A final feature characterizes the novel way of knowing typical of science. It is the *total personal involvement* of the researcher. This characteristic is evident at this point because it summarizes what we have been discussing. But its humanistic relevance is great, and thus it should be mentioned explicitly. If we want to understand the humanistic significance of science, we must realize that science demands the total engagement of the scientist. For science is, above all, a way of conceiving reality, an idea. But an idea comes to light and bears fruit only if a person consecrates all his resources to it.

The idea is not a substitute for work and work, in turn, cannot substitute for or compel an idea, just as little as enthusiasm can. Both, enthusiasm and work, and above all both of them *jointly*, can entice the idea. [123]

(b) The Dignity of the Active Mind

A second fundamental aspect marks science as an experience of knowledge. It is a new awareness of the dignity of man's own mind. An example can serve to introduce the discussion of this subject. We find it is Galileo's *Dialogue*. Sagredo, the enthusiastic admirer of the new science, is so moved by the arguments in favor of the heliocentric theory that he feels surprised that so few people have come to accept such a convincing doctrine. To this Salviati, the scientific spokesman for Galileo, retorts:

No, Sagredo, my surprise is very different from yours. You wonder that there are so few followers of the Pythagorean opinion, whereas I am astonished that there have been any up to this day who have embraced and followed it. [124]

After this unexpected assertion, Salviati goes on to explain the reason for his position. This reason is the recognition of the great mental effort that man must make to be scientific.

Nor can I ever sufficiently admire the outstanding acumen of those who have taken hold of this opinion and accepted it as true; they have through sheer force of intellect done such violence to their own senses as to prefer what reason told them over that which sensible experience plainly showed them to the contrary.

Scientific man finds an understandable reason for pride in the power of his mind. Science requires that a person be able to check his senses through his intelligence, instead of allowing himself to be guided blindly by the impressions of the senses. As Einstein and Heisenberg put it, it is theory that decides what is observed (see Chapter 1, Section 1bb).

---

[123] Weber (note 100), p. 136.
[124] Galileo (note 25), pp. 327f.

Galileo is untiring in stressing the need for dominating the senses by means of the mind in order to achieve genuine knowledge of nature. Writing against the self-complacent phenomenism of a philosophical adversary, he expresses his experiential view:

...circumstances do occur, in which the sense—in its first apprehension—can err and be in need of correction, and this can only be obtained with the help of straight rational discourse. [125]

In the same context, Galileo goes on to conclude: "Let then cease in this case the trust which the intellect ought to have in the sense." [126]

The heliocentric hypothesis is a most convincing illustration of the greatness of the scientific mind because its adoption demands that man break loose from what is most instinctive and pleasant—to live in a world that can be seen and imagined. Against this tendency, science insists on the daring and the austerity of purely mental vision. It is such a vision that constitutes the essence of science. The philosopher Hanson summarized the situation well when discussing the work of Galileo himself:

To comprehend the structural plan of the physical world required not busy elbows, dextrous fingers, and sharp eyes. It required hard thinking about the nature of Nature—about the *essential* form and format of physical processes and phenomena. [127]

The impressive power of the scientific mind can be realized at a glance if one surveys the process leading to discovery. When the scientist starts his research, he is confronted with so many distracting details that his mind must set busily to work and clarify the situation. This he does by way of abstraction and idealization. The researcher idealizes the object of his study by extracting from it the features that are relevant and constructing a schematic representation of the object itself. As Galileo remarked when speaking of the study of motion, this is an absolutely necessary preliminary step.

About those accidents of gravity, velocity, and also figure [of bodies in motion]—since they vary in infinite ways, no making of real science is possible. Therefore, in order to deal scientifically with such a matter [the study of motion], one must abstract from them....[128]

But if activity is needed to abstract and schematize, it is required even more in order to attain theoretical understanding; for there is no automatic way of passing from observed data to theoretical principle. As Einstein stressed, discussing this matter:

---

[125] Translated from Galileo (note 107), p. 1091.
[126] *Ibid.*, p. 1092.
[127] In Kaplon (note 76), p. 44.
[128] Galileo (note 69), p. 303.

Here there is no method capable of being learned and systematically applied so that it leads to the goal. The scientist has to worm these general principles out of nature by perceiving in comprehensive complexes of empirical facts certain general features which permit of precise formulation. [129]

The central experience of the scientist, then, is exactly that of *creativity*. Sometimes this experience is expressed in paradoxical form, as in the often-quoted words by Einstein:

...the concepts which arise in our thought and in our linguistic expressions are all—when viewed logically—the free creations of thought which cannot inductively be gained from sense experiences. [130]

The creative character of science should not be misinterpreted. It certainly does not imply that science is merely a construction of the mind. In the passage cited, for instance, Einstein clearly wants to deny the possibility of a "logical" passage from sense data to theoretical concepts. Elsewhere he insists explicitly on the necessity of sensible experience. "Concepts can only acquire content when they are connected, however indirectly, with sensible experience." [131]   But the point is that science deserves to be called a "creative act," as Einstein himself says when discussing atomism. [132]

Another way of expressing the consciousness of mental creativity typical of science is contained in the frequent metaphor of the scientist as a builder. Kepler spurred himself to work on Tycho's vast amount of observational data by thinking of being an architect called to erect a magnificent structure. In his own words: "Tycho possesses ... the material for the erection of a new structure ... he lacks only the architect who uses all this according to plan." [133]   This Keplerian remark has a universal validity. The aim and pride of the genuine scientist is that of rebuilding, through his mental effort, the intelligible structure of nature. As Bentley Glass puts it:

...the scientist seeks for facts—or better, he starts with observations.... But I would say that the real scientist ... is no quarryman, but is precisely and exactly a builder—a builder of facts and observations into conceptual schems and intellectual models that attempt to present the realities of nature. [134]

---

[129] Einstein (note 68), p. 221.
[130] *Ibid.*, p. 22.
[131] *Ibid.*, p. 277.
[132] *Ibid.*, p. 343.
[133] J. Kepler, in Caspar (note 96), p. 102.
[134] B. Glass, *Science and Ethical Values* (Chapel Hill: University of North Carolina Press, 1965), p. 94.

The result of the scientific experience of creativity is a heightened awareness of the dignity of man's mind. This point is delicate, for scientists are frequently accused of being arrogant and impatient of other people's opinions as a result of their doing science. But arrogance is just the manifestation of human weakness, not the fault of science as such. For genuine science trains man to modesty, not arrogance. As Bernard pointed out, science effectively reveals to man his own ignorance; that is, "We really know very little, and we are all fallible when facing the immense difficulties presented by investigation of natural phenomena." [135] Genuine modesty, however, is not only not incompatible with a sense of dignity, but rather demands it. The scientific mind is characterized by such a dual attitude of modesty and dignity. The whole of science is really based on a sense of dignity in that science is nothing but an active search for truth, away from any external imposition of accepted opinions.

Speaking of the awareness of creativity engendered by science, the most prominent figure remains that of Galileo, who first gave a brilliant and forceful expression to such a manifestation of the scientific experience. Galileo was utterly convinced that the scientist is a creator—just as the artist is. Some polemical notes of his, written against one of his self-righteous critics, stress this illuminating comparison. "The difference between philosophizing and studying philosophy is that which exists between drawing from nature and copying pictures." [136] Thus Galileo insists that man, in his search for truth (which he calls "philosophy") should not be satisfied with studying what other people have said, but should rather try to find out the truth for himself. [137] Of course, he goes on in explaining his comparison, a novice painter has to learn his trade by copying from other painters. However, he could never become a genuine painter were he to continue to copy pictures forever.

In the same way, a man will never become a philosopher by worrying forever about the writings of other men, without ever raising his own eyes to nature's works in the attempt to recognize there the truths already known and to investigate some of the infinite number that remain to be discovered. [138]

The awareness of intellectual dignity fostered by scientific research can hardly be overrated. For it is one of the fundamental contributions of science to the self-understanding of man. By doing science, man comes

---

[135] Bernard (note 1), p. 39.
[136] In Drake (note 7), p. 224.
[137] It is important to notice the usage of the term "philosophy" by Galileo. It stands neither for philosophy in the specialized sense of today, nor for science (as many interpreters assume). Rather, it stands for total search for truth, thus for both science *and* philosophy.
[138] In Drake (note 7), p. 225.

to perceive the power of his mind. As a result, he realizes that he has not only a right, but also a duty, to use it. Thus we can understand why, in particular, Galileo waged an unrelenting war against all sorts of pedants and dogmatists. To him this was ultimately a moral obligation: the duty of being faithful to truth and God himself. Thus, for example, he writes in his celebrated attack on Sarsi (Orazio Grassi):

Sarsi says he does not wish to be numbered among those who affront the sages by disbelieving and contradicting them. I say I do not wish to be counted as an ignoramus and an ingrate toward Nature and toward God; for if they have given me my senses and my reason, why should I defer such great gifts to the errors of some man? Why should I believe blindly and stupidly what I wish to believe, and subject the freedom of my intellect to someone else who is just as liable to error as I am? [139]

To sum up, doing science molds the mind of man and inspires him with a new sense of personal worth. This point must be borne in mind in order to understand the genuine meaning of science. Many a polemical reaction of scientists, in fact, stems from the righteous indignation they experience when their passionate search for truth is haughtily dismissed or questioned. If they can exaggerate in their polemics, we should at least recognize the seriousness of their motivation. [140]

(c) A New Perception of Nature

A third aspect stands out in the scientific experience of knowledge. It consists in a literally new way of perceiving observable reality or nature. Typically the real world presents itself differently to the scientific person than to the nonscientist.

Essentially, science gives to man a new awareness of the objective intelligibility of nature. This point is illuminated by some considerations of Heisenberg's. In discussing the nature of science with a colleague, he stresses that one can have two kinds of cognitive experience while doing theoretical physics. The theoretician must always begin his work by giving a compact formulation to the observational data hitherto available. But, by so doing, the scientist realizes that this is just the work of man.

...one has the feeling that one has invented these formulas himself, one has invented them with more or less satisfactory success. [141]

[139] "*Assayer*," in Drake, *op. cit.*, p. 272.
[140] Among the various reasons for polemic between Newton and Robert Hooke there was also a very superficial interpretation of science by the latter, who reduced success in research to "but a luckey bitt of chance." Newton could not help but react with immense scorn to such a view. For details see A. Koyré, *Newtonian Studies* (Chicago: University of Chicago Press, 1968) pp. 220-260, especially pp. 235 and 257.
[141] Translated from *Der Teil* (note 21), p. 139.

But then, he continues, when one comes upon "the quite simple great interconnections of nature (*ganz einfachen grossen Zusammenhänge*) which can be fixed in axiomatic form," at this moment "the situation presents itself as completely different." Heisenberg describes the characteristic experience of scientific objectivity as follows:

Then appears suddenly before our inner eye an interconnection, which also without us has always been there and which quite obviously has not been made by man. These interconnections are surely the proper content of our science. Only when one has been fully permeated by such interconnections can one really understand our science.

The novelty with which the scientist perceives nature is twofold. The first sense refers to observation. As a result of his industrious familiarization with nature, the scientist succeeds in seeing with his own eyes entirely unsuspected manifestations of nature itself (see Chapter 1, Section 1ab). This is the spectacular aspect of science that appeals so strongly to public imagination. As an example, the first scientific best seller was Galileo's little book entitled *Starry Messenger*. Five hundred copies of the book had been printed. They were sold out at once. Within three months, orders for just as many more had poured in from all over Europe. [142] The reason for such popular commotion was exactly the novelty of nature that science—in this case, telescopic astronomy—had suddenly disclosed to the human eye. This novelty consisted in the enormous number of stars, the geography of the moon, the four Jovian planets—all things "never seen from the creation of the world up to our time." [143]

The main sense in which science deserves to be called a new perception of reality lies not in seeing more things, but in seeing things with new eyes. The new vision is strictly of an intellectual type. As an example, we can realize why Galileo was so proud of his science of mechanics. His study of motion, as he himself pointed out, was "a most new science about a most ancient subject." [144] The novelty of science, in general, consists in giving man an unsuspected perception of the observable world as a whole—while simultaneously revealing to him the unexpected power of his mind. Galileo never tires of stressing this aspect of science. Sagredo, for example, remarks admiringly of Salviati, the personification of the new science: "Most of the time he solves questions which appear not only obscure, but repugnant to nature and truth. And this he does with reasons or observations or experiments which are most trite and familiar to everybody." [145] Sagredo's words are a charming expression

---

[142] Details in Drake (note 7), p. 59.
[143] "*Starry Messenger,*" in Drake, *op. cit.*, p. 50.
[144] Translated from *Discorsi* (note 69), p. 168.
[145] *Ibid.*, p. 100.

of the intellectual enthusiasm aroused by the ability of science to present the ordinary world in a completely new light.

(d) Unshakable Certainty

Another aspect characterizes the experiential nature of scientific knowledge. It is the steadfast conviction and unshakable certainty frequently shown by researchers long before their discovery can be proved with empirically apodictic arguments. This aspect of the scientific mind tends to irritate nonscientists, as though it were due to conceit and arrogance. It is rather a revealing manifestation of the scientific mind. The working scientist is far from being a coldly intellectual type who is totally detached from his views and evaluates them only on the strength of unassailable critical evidence. On the contrary, he is a person passionately committed to his own insights. This is a fact, well attested by historical evidence.

As an illustration, we may consider the attitudes of two scientists who were psychologically as different from one another as possible: Kepler and Galileo. Galileo was the forceful, outgoing, and indomitable type. Kepler was the opposite. His chief biographer assures us that he "did not want to be an innovator, nor to separate himself from others and found it painful to have to go his own way." [146] Kepler was not able to hide his timidity, even in his scientific writings. In the preface to his *Epitome Astronomiae Copernicanae* he self-deprecatingly confides to the reader that he inclines to be a conformist: "I like to be on the side of the majority." And yet, we see that both Kepler and Galileo showed remarkable assurance in defending the Copernican view at a time when there was not yet any compelling evidence available. In point of fact, Kepler was writing the self-denigrating words cited in the very work in which he was most forcefully vindicating the Copernican theory. This theory was at the time anathema in the Protestant learned circles in which the pious and timid Kepler had been educated and to which he intimately yearned to belong. This notwithstanding, Kepler felt that he had no choice. As a scientist, he was convinced he had to overcome his own feelings and the embarrassment of displeasing other people in order to uphold the truth of the hitherto unprovable Copernican hypothesis.

Concerning Galileo, his determined support of the Copernican view needs no documentation. However, to penetrate somewhat deeper into his scientific mind, we may cite a famous passage in his writings that is frequently quoted as an indication of mental arrogance and even of sovereign disdain toward the touchstone of science—the experimental test of theory. I refer to the inertial discussion, in the *Dialogue*, of the trajectory described by a stone falling from the top of a ship's mast.

---

[146] Caspar (note 96), p. 371.

Salviati asserts categorically: "... experiment ... will show that the stone always falls in the same place on the ship, whether the ship is standing still or moving with any speed you please." [147] Simplicio boggles at this statement which contradicts flatly the Aristotelian conception of motion. Nevertheless, he declares himself willing to accept this startling result if experiment proves that this is really the case. To this Salviati replies that he neither has made any experiment nor is there need of experiment to be certain about the proposition enunciated. "Without experiment, I am sure that the effect will happen as I tell you, because it must happen that way. ..." This example, and another famous passage about the uniform velocity of falling bodies, may well seem to indicate that Galileo liked to flabbergast his critics through a mischievous and highly equivocal flourish of wits. [148] But as often happens when dealing with a superior mind, the apparently obvious interpretation is wrong. Galileo was just putting in striking rhetorical form a common attitude of the scientific mind. The scientist can be certain even without apodictic evidence.

Historical data are sufficient to justify a general statement. The unshakable certainty of the creative scientist is not really based on the undisputableness of experimental confirmation. Conduitt, for instance, reports in his biographical notes about Newton:

He said that he first proved his inventions by geometry and only made use of experiments to make them intelligible, and to convince the vulgar. [149]

Other testimony is that of Einstein, as related by Moszkowski. This writer once inquired how he would have felt had his relativistic calculations regarding the orbit of Mercury been proved wrong by experimental evidence. Einstein's reply was entirely in the Galilean spirit:

Such questions did not lie in my path. That result could not be otherwise than right. I was only concerned in putting the result into a lucid form. I did not for one second doubt that it would agree with observation. There was no sense in getting excited about what was self-evident. [150]

Many similar examples could be cited. They refer to researchers in both the theoretical and observational fields. The celebrated pathologist Charles Nicolle expressed this experience well. Describing his own reaction when

---

[147] Galileo, *Dialogue* (note 25), pp. 144f.

[148] For the discussion of uniform velocity of falling bodies see Galileo, *Discorsi* (note 69), pp. 74-76.

[149] Quoted in L. T. More, *Isaac Newton: A Biography* (New York: Dover, 1962), p. 610.

[150] A. Moszkowski, *Einstein the Searcher: His Work Explained from Dialogues with Einstein,* trans. H. L. Brose (New York: Dutton), p. 5.

he became aware, all of a sudden, that typhoidal contagion he had been investigating was propagated by lice, he wrote:

In the course of this very brief period I experienced what many other discoverers must undoubtedly have experienced also, viz. strange sentiments of the pointlessness of any demonstration, of complete detachment of the mind and of wearisome boredom. The evidence was so strong that it was impossible for me to take any interest in the experiments. Had it been of no concern to anybody but myself, I well believe that I should not have pursued this course.[151]

The testimony cited is instructive because it illumines the experiential nature of scientific certainty. Experiment, of course, is needed to test publicly the validity of a theory. But the certainty that the creative scientist perceives is not a matter of rigorous and systematic experimental demonstration. It is rather a personal, intimate experience of the objective intelligibility of nature as concretely perceived in personal contact with nature. Heisenberg emphasizes this point by relating a conversation of his with Einstein. Einstein had always felt uncomfortable about the new quantum theory. Thus he had bluntly asked his youthful visitor why he adhered so strongly to his theory "while so many and central questions are still completely unclarified."[152] Heisenberg could but reply in experiential terms—by pointing to his own experience and that of all scientific discoverers. To him it was clear that a certain basic simplicity and beauty, which could be expressed in theoretical formulation, "represent a genuine trait of nature." Of course, he admitted, the mathematical formulation of the theory also contains a subjective aspect because it expresses the relation of man to nature.

But since one would never have arrived at these forms by himself, since they are disclosed to us by nature in the first place—this makes clear that they too belong to reality, not just to our thoughts about reality.

As for the obvious objection that his view tended to reduce the certainty of science to an aesthetic criterion, Heisenberg could but reply by stressing the experiential character of his conviction and by appealing, for understanding, to the experience of other creative people, including Einstein himself.

I must confess that for me a very great convincing power originates from the simplicity and beauty of the mathematical scheme that here is presented to us by nature. You must have experienced it, too—one is almost frightened by the simplicity and compactness of the intercon-

---

[151] Quoted in R. Taton, *Reason and Chance in Scientific Discovery* (New York: Science, 1962), pp. 77f.
[152] Translated from W. Heisenberg, *Der Teil* (note 21), pp. 98f.

5

nections (*Geschlossenheit der Zusammenhänge*) that nature all of a sudden spreads before him and for which he was not in the least prepared. The feeling which comes over a person at that sight is just entirely different from the joy that one feels when one believes to have completed particularly well a piece of workmanship—of physical or nonphysical type.

To sum up, the unshakable certainty of scientific discoverers grants us a valuable insight into the nature of scientific knowing. Science is essentially a matter of personal experience, of intimate awareness. Experiment and public criticism may be indispensable, but they do not constitute science as such. This resides rather in the intimate consciousness of perceived truth.

(e) The Communal Spirit

Up to this point we have been discussing scientific knowledge as experienced by the individual scientist. One final aspect remains to be examined: the experience of science as a corporate undertaking. Science is typically a communal enterprise. The physicist Ziman put it well:

The scientific enterprise is corporate. It is not merely, in Newton's incomparable phrase, that one stands on the shoulders of giants, and hence can see a little farther. Every scientist sees through his own eyes—and also through the eyes of his predecessors and colleagues. It is never one individual that goes through all the steps in the logico-inductive chain; it is a group of individuals, dividing their labor but continuously and jealously checking each other's contributions. [153]

In our effort to understand scientific knowledge, we must try to realize this additional dimension in its true light.

To begin with, it is necessary to eliminate a misunderstanding. The view is occasionally heard that science is but the product of an anonymous group. This view is false because it is incompatible with the creative character of science itself. Indeed, no scientific discovery can be made unless there is some individual ready to pursue it actively. The contribution of the individual is decisive in the creation of science. As has been wittily said, "All truth begins in the minority of one." [154] This is a fact. History confirms it—especially in regard to the so-called simultaneous discoveries and inventions.[155] It is indeed true that numerous discoveries and inventions of similar kind have occurred practically at the same time in different places. Thus the impression can be engendered that scientific

---

[153] Ziman (note 101), p. 9.

[154] D. L. Watson, *Scientists Are Human* (London: Watts, 1938), p. 24.

[155] For a list of such discoveries and inventions see, for instance, B. Barber, *Science and the Social Order* (New York: Free Press, 1952), pp. 199f.

achievement is but a sort of ineluctable social happening. But the impression is misleading, as shown by historical investigation. History, in fact, proves that the so-called simultaneous achievements were either not identical or that they took place as a result of mutual influence among contemporary researchers.[156]

Although the originality of the individual is indispensable to science, it is true that science is a communal undertaking. This is due to the complexity of scientific research and the very nature of the human mind. No individual scientist can see all the issues involved in a particular piece of investigation. He needs the help of other people and, in turn, contributes to the work of other people. One example, the discovery of oxygen, illumines this state of affairs. This discovery gave occasion to endless controversies, compounded by nationalistic feelings. History shows that the full-fledged attainment was not the achievement of any single person, but rather the outcome of various, interlocking contributions. Indeed, who was the real discoverer of oxygen? Was it Priestley, who first produced oxygen in 1774 but thought that it was just "dephlogisticated air"—that is, ordinary air from which philogiston had been removed? Or was it Lavoisier, who, having heard of Priestley's experiment, produced oxygen again in 1776 and correctly pronounced it a self-standing element? Priestley cannot be considered the true discoverer because he failed to understand the genuine nature of oxygen. But neither can Lavoisier because his success clearly depended on that of Priestley. In conclusion:

The story of oxygen shows better than anything else how a scientific discovery is made in stages.... Hence it is idle to call either Priestley or Lavoisier the "real" discoverer of oxygen.[157]

In general, the communal nature of science becomes more and more prominent, the greater the complexity of the subject matter involved. One outstanding example, documented in great detail, is offered by the development of quantum physics.[158]

To realize how the communal spirit of science is experienced by the scientist, we must consider its two complementary factors. The first is the personal influence exerted by the great scientific creator. History shows that such a creator is essentially an *intellectual leader*. He is the pioneer and trailblazer who first brings to the public attention a new and exciting world. His example and enthusiasm attract other people to fol-

---

[156] For a summary of the question, with indication of relevant literature, see, for instance, R. K. Merton, *On Theoretical Sociology: Five Essays, Old and New* (New York: Free Press, 1967), pp. 9f.

[157] M. Daumas, "The Birth of Modern Chemistry," in R. Taton, ed., *A General History of Science*, trans. A. J. Pomerans, vol. II (London: Thames and Hudson, 1964), pp. 497f.

[158] See the many interviews collected and edited by T. S. Kuhn *et al.* (note 17).

low in his footsteps, by continuing and developing his own explorations. This situation explains the seminal efficacy of the creators themselves. Examples are numerous: it may suffice to say a word about the influence exerted by Galileo and Newton. Galileo is great as a scientific discoverer in his own right, but he is certainly greater as the father of science itself. His general influence is unsurveyable, but his specific leadership is well documented. In a famous letter, written in 1610, Galileo listed a large number of research projects that he, with the optimism of genius, hoped to carry out in his own lifetime. In actual fact, he was able to realize only part of his program. And yet his plans did not prove idle. On the contrary, they led to decisive breakthroughs, during his life and shortly afterward, at the hands of such gifted disciples as Bonaventura Cavalieri and Gian Alfonso Borelli.[159] Concerning the inspiration and leadership exerted by Newton on scientific research, many books have been written, with detailed documentation. The subjects that came to be explored following his speculations and suggestions range from optics and mechanics to chemistry, physiology, medicine, electricity, and atomism. The number of researchers who appealed to him as their leader is legion.[160]

The aspect that complements the intellectual leadership exerted by the great creators is the sense of *living tradition* that has come to establish itself in the scientific world. It is a well-attested psychological datum: the great masters of the past are far from dead and gone. They are rather continually influential on scientific researchers, generation after generation. This is the evidence collected by the contemporary psychologist of science.

Analysis of self-images shows that the scientists draw their main identity from their affiliation with the great discoverers, the great contributors to scientific knowledge, the men who have given us the picture of the world we have today.[161]

However, if this is the case, one should not interpret this attitude as a kind of servile veneration for the past masters. The contrary is true. For science is characterized by the urge to move forward, to progress continuously toward the new and unexplored. Tradition has to be understood here in its literal meaning as the handing down of a skill and connoisseurship, while pointing the way toward continual progress. It is a transmission of living attitudes and convictions, and an exhortation toward

---

[159] See Galileo's letter to Belisario Vinta and historical commentary in Galileo Galilei, *Opere* (note 107), pp. 890f.

[160] For an excellent analysis of the work of the principal experimentalists and theoreticians who followed Newton's lead in the 18th century, up to B. Franklin, see I. B. Cohen, *Franklin and Newton: An Inquiry into the Speculative Newtonian Experimental Science and Franklin's Work in Electricity as an Example Thereof* (Philadelphia: American Philosophical Society, 1956), especially pp. 203-284.

[161] Eiduson (note 80), pp. 152f.

unending advance. The leaders and inspirers are the great creators. But science results from the contributions of all those who share in their spirit. Claude Bernard has given a classical formulation to this characteristic phenomenon, by speaking in terms of giants and pygmies:

Great men may be compared to torches shining at long intervals, to guide the advance of science ... each great man makes the science he vitalizes take a long step forward, he never presumes to fix its final boundaries.... Great men have been compared to giants upon whose shoulders pygmies have climbed.... This simply means that science makes progress subsequently to appearance of great men, and precisely because of their influence. The result is that their successors know many more scientific facts than the great men themselves had in their day. But a great man is, none the less, still a great man, that is to say— a giant. [162]

The combination of leadership and tradition gives rise to that typical social phenomenon which is the *scientific community*, with its distinctive mentality. The community arises because practicing scientists realize that they belong together. They may live apart from each other in space and time. They may be marked by strong individual traits. Nonetheless, they know that they constitute the "invisible college." The communal or collegial sense is originated by the actual sharing in the same spirit, characterized by well-defined attitudes and goals. In summary, then, this is the first comprehensive manifestation of the humanistic significance of science. Man, by doing science, comes to experience knowledge—and himself as a knower—in a typical and unprecedented way.

---

[162] Bernard (note 1), pp. 41f.

CHAPTER 2

DYNAMICAL ORDERLINESS:
THE SCIENTIFIC EXPERIENCE OF NATURE

Science is knowledge. But knowledge encompasses two basic poles: on the one side the knowing subject, on the other side the object known. Thus, to understand science as an experience of the whole man, we must now undertake a second step in our investigation. We have studied science as an activity of the knowing subject. We must now consider science as a source of information about the object. This is the thread we are going to follow. Nature is significant as the object of science because of its orderliness. Hence, first of all, we shall discuss the discovery of natural orderliness as a characteristic contribution of science. But order of nature, as discovered by science, presents the distinctive feature of being a dynamical rather than a static reality. Hence, second, we shall examine the dynamical properties of natural orderliness. Third, we shall discuss some synthetical traits of the scientific experience of nature.

1. THE ORDERLINESS OF NATURE

If we consider science as a lived experience, there is no doubt that science itself is centrally aimed at finding order in nature. For order is the feature according to which natural things and events becomes understandable. There is order, in fact, where many form one—that is, where there is some sort of unitary pattern graspable by the human mind. But science is the attempt to understand nature. However, if the search for orderliness is central to science, it is necessary to explore what sort of orderliness science itself has brought to the fore. For the term "order" is capable of numerous acceptations, hence it remains vague and unenlightening unless studied in its various manifestations. As a consequence, the aim of this first section will be an exploration of the three main forms of order which are brought to light by scientific research.

(a) Order as Factual Occurrence

The first type of order significant to science is the factual occurrence of regularities in nature. Man starts to be a scientist the moment he perceives that some observable events, which hitherto had appeared disconnected, actually manifest a unitary pattern of behavior. Such patterns, once their existence has been publicly tested by experiment, constitute the so-called scientific *laws of nature.* The term is used to signify simply the recognition of a fact: nature is effectively orderly.

Discovery of factual orderliness is only a limited and modest goal for man in his effort to understand nature. In fact, science does not content itself with this elementary stage. Nonetheless, one would seriously fail to understand the scientific experience were he to overlook the importance of this basic achievement. For the realization of factual orderliness is not merely an indispensable phase on the way toward deeper insights. Rather, it is a great attainment in itself, whose importance can be assessed only by taking into account the difficulties one has to master in order to conquer it. The difficulties are great because, when man first confronts nature, he is faced with a bewildering variety of events whose behavior appears incomprehensible to the mind. Thus the scientific discovery of factual orderliness in nature constitutes in itself a milestone in man's effort to understand the world he lives in.

As an illustration of the significance of the first step of science, we may consider why Galileo came to deserve the lofty title of "father of science." The obvious reason was his ability to prove that nature was effectively understandable. Up to his time, the general tendency had been to speculate grandly about the intelligibility of the world, but without coming to grips with its effective structure as observable. Galileo brought all his enormous energies to bear on this one point: he tried systematically to find out how things actually were. His work was criticized as too narrow by contemporary philosophers, notably Descartes. In actual fact, Galileo's contribution was invaluable because it gave science a genuine start. Galileo himself was well aware of the discipline of mind required by scientific research as opposed to the enthusiastic expectations of the public. His nuanced but determined position is expressed by Salviati, the spokesman of science, in replying to a suggestion of Sagredo. Sagredo is typical of the cultivated and open-minded layman. Thus he finds it natural to remark—in the course of the discussion of constantly accelerated motion—that one could well use the opportunity to explore a much-disputed question. The question regarded the cause of the acceleration itself. To this, Salviati replies by rejecting the proposal. And he adds the methodologically important reason:

For the present it suffices to our author [Galileo] that we understand that he wants us to investigate and prove some properties of accelerated motion (whatever the cause of its acceleration may be).[1]

No doubt, Galileo would never have opposed explanation as a proper goal of science. However he, as a pioneer, had to insist that one cannot hope to understand nature unless one starts by exploring the observable regularities of nature itself.

The basic stage of science we are considering is the discovery of precision and regularity in the observable world where to all appearances there is only approximate recurrence or meaningless variety. Let us survey briefly the contribution of this approach to the enlightenment of man. The original field of exploration was motion. This was, obviously enough, the first area in which man had to engage himself in his effort to understand nature. Aristotle had already insisted, and the Middle Ages wholeheartedly accepted, that "one cannot understand nature, if one does not understand motion" (*ignorato motu, ignoratur natura*).[2] The reason for the importance of the study is the simple fact that motion is the most prominent manifestation of what happens in nature, from the majestic movements of stars and planets to the most banal everyday happening. In actual practice, the study of astronomical motion goes back to time immemorial. However, science could arise only when man started to be interested enough in regularities of motion that he was dissatisfied with mere approximations or purely phenomenistic descriptions. The breakthrough in scientific astronomy was the combined feat of Tycho the observer and Kepler the theoretician. Their work proved convincingly that the skies were really the home of intelligible order. Back on earth, the study of motion was subject to far greater difficulties. For, if everything moved, it seemed that nothing moved according to a precise regularity. It was the glory of Galileo to succeed in disentangling intelligible order from such apparently hopeless phenomena as the fall of ordinary bodies and the trajectory of missiles. The other branches of science developed by following the same procedure. Thus, for instance, optics was born when Newton proved that light was transmitted in a precisely defined way. Likewise, chemistry became scientific when Lavoisier demonstrated the conservation of mass in chemical reactions. These examples could be easily augmented.

To give a somewhat concrete view of the character and implications of the factual discovery of orderliness brought about by science, I wish to

---

[1] Translated from A. Carugo and L. Geymonat, eds. G. Galilei, *Discorsi e Dimostrazioni Matematiche intorno a Due Nuove Scienze* (Turin: Boringhieri, 1958), p. 186, cf. pp. 184-186.

[2] For the context in Aristotle see F. Solmsen, *Aristotle's System of the Physical World: A Comparison with His Predecessors* (Ithaca, N. Y.: Cornell University Press, 1960), especially p. 174.

sketch rapidly here an entire sequence of such discoveries. I refer to atomic research—one of the most stirring investigations ever entered into by the human mind. [3] Every step in this development is marked by the realization of some factual type of order—and each new form of order presents itself as more surprising and unexpected than the previous one. To begin with, the celebrated achievements of Dalton that gave rise to atomic physics stress the importance—and difficulty—of finding precise order in nature. The difficulty was great because Dalton's discovery of the existence of atoms was, properly speaking, not based on some new experimental data. The evidence was rather publicly available, printed in scientific journals. What was needed to transform those data into science was the creativity of a mind able to read precise regularities from a mass of empirical measurements. This achievement was made by Dalton and constitutes his immortal glory. However, once the first laws about atomic regularities had been found, the existence of atoms could merely be said to have been made plausible. Many objections remained. This situation gave rise to an increasingly large number of investigations, all striving to lay bare the intimate structure of matter. Step by step, factual orderliness was found in the behavior of gases, in the structure of crystals, in the chemicophysical properties of elements. Atomic order, as a macroscopic fact, was embodied in the Periodic Table. Atoms were universally accepted by scientists as actually existing. However, if there were atoms, it was necessary to find what properties they had and how they interacted. Once again, the answer had to be sought through patient and ingenious investigation. Particularly informative proved to be the radiation arising from the interior of atoms and molecules. The information was contained in the regular patterns that are called spectra. The complexity of spectra, however, posed new puzzles. The final discovery was the most surprising and revealing of all. It proved that there was indeed orderliness at the atomic level, but of a totally unanticipated type. This orderliness is expressed in those baffling properties of microparticles that are called corpuscular and wavelike. At this point, the search for atomic order was virtually completed. The investigation had demanded about a century and a quarter of hard work on the part of some of the most penetrating minds and a host of collaborators. The sequence outlined illustrates convincingly the momentousness of the basic, relatively limited stage of science we have been discussing.

---

[3] For details see E. Cantore, *Atomic Order: An Introduction to the Philosophy of Microphysics* (Cambridge, Mass.: MIT Press), pp. 17-132.

(b) Order as Intrinsic Principle

The discovery of factual order, for all its importance, is just the beginning of the scientific endeavor. Once the mind has realized that a certain regularity occurs, it desires to know why that regularity takes place. Hence, scientific man tries to penetrate the intimate structure of observable reality and grasp the source of the regularity he has discovered. This is the theoretical phase of science. It consists—as Born put it, speaking of the interpretation of atomic spectra—of reading and understanding a message that is obviously there, but which at first appears unintelligible.

Here one had a direct message from the interior of the atom ... and this message did not sound at all like gibberish, but rather like an orderly language—except that it was unintelligible.... It was the same situation as with the extinct Maya peoples, of whose script numerous specimens have been found in ruined cities of Yucatan; unfortunately, nobody can read them. [4]

The first idea of nature as possessing in itself an intrinsic source of order capable of accounting for observable regularities is to be traced back to Kepler. Up to his time, the general view had been that there were a number of transparent, concentric spheres rotating around the earth. The spheres were postulated in order to account for the motion of the planets. Each planet was supposed to be rigidly attached to one of the spheres. Kepler rejected the hypothesis of the spheres and drew the obvious inference. He speculated that the planets had in themselves a moving force (*vis motrix*). [5] But Kepler had no clear idea about the subject, nor were the scientific data available to him sufficient to permit discovery of the nature of the moving force. As is well known, success in discovering the intrinsic source of mechanical regularity in nature was to constitute the immortal glory of Newton.

Newton, naturally, depended on his predecessors. In particular, he studied the work of Kepler and Galileo on heavenly and earthly motions. But then—in a soaring flight of mind—he perceived that all of these regularities, apparently unrelated, were but the particular expressions of one fundamental orderliness. They resulted from the universal gravitational attraction that is typical of ponderable matter as such. The tale of the falling apple—no matter whether historical or apocryphal—illustrates this essential insight beautifully. For Newton had precisely the insight that only one agency was responsible for the many types of observable motion, be it the fall of an apple or the perambulation of the

---

[4] M. Born, *Physics in My Generation: A Selection of Papers* (New York: Pergamon, 1956), pp. 27f.

[5] For details, see E. J. Dijksterhuis, *The Mechanization of the World Picture*, trans. C. Dikshoorn (Oxford: Clarendon Press, 1961), p. 310.

moon. This one agency was universal gravitation. The greatness of Newton's achievement can hardly be realized by nonscientists, who have been accustomed to take it for granted from their childhood. But a brief historical reflection permits one to perceive the reasons for the enormous enthusiasm that swept educated Europe when Newton's synthesis became known. [6] Of course, one should discount the fulsome adulation of which Newton became the object. Yet, there was ample justification for people to be aroused to enthusiasm by the brilliance of Newton's attainment. The reason was the sudden awareness that nature was intrinsically intelligible. What hitherto had appeared to consist of disparate, hopelessly complicated phenomena proved now to be but one comprehensive phenomenon whose intrinsic source of unity was penetrable by the mind of man. It was the first instance in which man became aware of his ability to fathom observable reality to its very bottom. Einstein emphasized that Newton was great in his own right, but he is even greater because of the historical role he played. His success has become paradigmatic for science as such. In Einstein's words:

The figure of Newton has however an even greater importance than his genius warrants because destiny placed him at a turning point in the history of the human intellect.... The logical completeness of Newton's conceptual system lay in this, that the only causes of the acceleration of the masses of a system are *these masses themselves*.... The discovery that the cause of the motions of the heavenly bodies is identical with the gravity with which we are so familiar from everyday life must have been particularly impressive. [7]

Science has constantly followed the Newtonian lead in seeking to explain the factual regularities observed in nature. It would be pointless to mention further examples, which are endless. To illustrate this aspect of the scientific endeavor let us just add a word about atomic physics. Research in the atomic realm was, from its very beginning, an attempt at explanation. Man wanted to know why things presented characteristic regularities—why water was wet and diamond hard. As we have seen, however, the regularities traceable to atoms both on the macroscopic and the microscopic levels were so numerous and complex that the preliminary task of observing and describing them required much time and effort for its successful completion. Yet scientists went through all that labor

---

[6] For a synthesis of the enormous impression caused by Newton's achievement on public opinion see, for instance, "The Principate of Newton," in A. R. Hall, *The Scientific Revolution (1500-1800): The Formation of the Modern Scientific Attitude* (Boston: Beacon, 1956), pp. 244-274.

[7] A. Einstein, *Ideas and Opinions,* trans. S. Bergmann (New York: Crown, 1954), pp. 254-256.

because they were ultimately interested in the explanation of the regularities observed. This finally took place, at least in principle, in the 1920s and 1930s. [8]   At present we possess all the elements necessary to the understanding of the whole range of properties that characterize matter and radiation—form of bodies, their size, color, hardness, thermal features, electric conductivity, chemical interactivity, and the like.   In a true sense, we know now not only how things are, but also why they are what they are.   This achievement of atomic physics marks another great step forward in mankind's march toward understanding—of the same kind as Newton's attainment.   Ordinary matter, which from time immemorial had appeared as self-evident or natural or even meaningless to man, reveals now a profundity and significance beyond compare.

In the light of the preceding, it is clear why the humanist who lives in the scientific age should strive to realize the experience of the intrinsic intelligibility of nature perceived by the working scientist.   The reason is that we touch here upon the core of science as such.   What is deeply attractive and rewarding for man doing science is not the ability to predict events nor the power to fashion nature according to his wishes.   For these attainments—no matter how important in practice—are but by-products or marginal consequences of science itself.   Genuine science consists in understanding.   It is an unprecedented awareness that observable reality is essentially intelligible or penetrable to the mind of man (see Chapter 1, Section 1bc).   As we shall later see in detail, this affinity between mind and nature is the key to the humanistic significance of science as a whole. Heisenberg put it admirably while discussing science and international understanding.

The core of science is formed, to my mind, by the pure sciences, which are not concerned with practical applications.   They are the branches in which pure thought attempts to discover the hidden harmonies of nature. . . .   Science can contribute to the understanding between peoples. It can do so not because it can render succor to the sick, nor because of the terror which some political power may wield with its aid, but only by turning our attention to that "center" which can establish order in the world at large, perhaps simply to the fact that the world is beautiful. [9]

Before entering a direct discussion of the humanistic significance of science as such, however, we shall continue our analysis of the experience of natural orderliness contributed by science itself.

---

    [8] For details, see Cantore (note 3), pp. 133-192 and 254-293.
    [9] W. Heisenberg, *Philosophic Problems of Nuclear Science*, trans F. C. Hayes (London: Faber, 1952), p. 119.

(c) The Universality of Autonomous Orderliness

So far we have examined science as a search for order in nature according to the two main forms of order itself, the factual and the intrinsic. But speaking of order amounts to speaking of unity. For order is essentially unity. What then is the overall view of nature to which scientific research leads the human mind? "The progress of science is the discovery at each new step of a new order which gives unity to what had long seemed unlike."[10] These words by Bronowski aptly synthesize the overall result of scientific research—provided, of course, we interpret them in the light of concrete evidence as offered by science itself. For in this area, more than ever, the philosophical humanist must be cautious not to jump to unproven conclusions. With this proviso, it is true that the overall development of science leads toward a unified perception of nature.

Science began early to tear down the apparent barriers that used to separate the universe of nature. The first unification of the world took place in an observational-spatial sense. To prescientific man it had seemed an almost unavoidable necessity to distinguish sharply between the earthly or sublunar world on the one hand and the celestial world on the other. The former, as obvious, was characterized by continuous change—ceaseless production and destruction of things. The celestial world, on the contrary, appeared to be completely exempt from internal change. Hence it came to be taken for granted that the universe was twofold in essence. The sublunar world was considered to be made up by four elements or essences, while the celestial world was thought to consist of a new enigmatic element, the so-called quintessence. Science rapidly exposed the emptiness of this phenomenistic systematization. When Galileo trained his telescope on the moon, he noticed at once that the moon itself was by no means made up of some esoteric material, but that it was something like the earth—with mountains and valleys, and the vicissitudes that mark our planet. Shortly afterward, he and Scheiner discovered that not even the sun was perfect and unchangeable because it manifested continually changing spots. Thus, rapidly, the idea of the unchangeability of the heavens came to fall into disrepute and was replaced by the conviction that the universe was truly such, namely one all-encompassing structure. However, the decisive success in unifying the universe was attained only by Newton. For he was the first to prove that the same type of interaction—gravitation—was really universal, in that it accounted for the motion of both heavenly and earthly bodies. This was the stirring aspect of the Newtonian synthesis: the manifold world of observation had become one universe through the penetrating power of the human mind.

---

[10] J. Bronowski, *Science and Human Values* (New York: Harper Torchbooks, 1956), p. 15.

Following the Newtonian synthesis, many were inclined to think that man had already succeeded—precisely through Newton's discovery—in unifying perfectly his world of experience. It appeared that gravitation was capable of accounting in principle for all regularities of nature. But this was never borne out by scientific evidence. We are now much more circumspect in our expectations of unifying the world through science. In fact, up to the present—even if we consider the physical realm alone— no complete unification is possible. Several basically different and hitherto mutually irreducible forms of interaction are needed for our physical understanding of the world. So it would be unwarranted to assert that science has unified nature in the strict sense of the term. Nevertheless, it is undeniable that science has already contributed immensely to giving us a perception of nature as a unified whole.

To have an idea of this achievement, let us start from the inanimate world, or the realm of physics. We notice immediately that vastly different phenomena are brought to unity in the sense that they are all reduced to some, very few forms of interaction (gravitational, electromagnetic, and so forth). When we move to the field of life we find that here, too, what appeared formerly to be a disparate and irreducible multiplicity is now fundamentally reduced to the unity of precisely defined and constantly occurring interactions. But science, in its search for order and unity, does not stop at the investigation of animal life. Rather, it boldly tackles the problem of man himself—insofar as he is accessible to observation. Hence we have the fairly recent sciences of sociology, psychology, and anthropology. They are sciences in the genuine sense of the term because they investigate universally occurring regularities and try to explain them by pointing to various types of interaction between man and man, and between man and environment.

In conclusion, we can say that science has contributed uniquely to bringing about a new conception of the order of nature. We can now speak of the universality of autonomous orderliness because the orderliness discovered by science really affects the whole of observable reality. To be sure, as mentioned, the unification due to science is far from perfect, nor should we allow ourselves to take it as a foregone conclusion. For nature constantly defies any form of rationalistic schematization. However, it is a well-proven fact that, through science, the universe has become unified as never before, and seems to be on the way toward an increasing unification. In other words, the world shows that it truly constitutes an interconnected whole—and that the mind of man can penetrate the structure of this whole more and more. Even if we do not abandon ourselves to wishful thinking but insist on the hard data of science, it seems illuminating to compare the development of science to the growth of a continent. The simile is Taton's.

In the past, science was comparable to a group of islands standing in an ocean of ignorance; in our day, the islands have been transformed into one vast continental network, and the ocean into a series of inland lakes. [11]

The comparison is apt in that it points not only to a fact, but also to an overall trend. The trend of science is toward an increasing discovery of universal autonomous orderliness in nature.

2. THE DYNAMISM OF NATURE

The discussion of order we have conducted up to the present has only covered one basic aspect of the orderliness of nature as brought to light by scientific investigation. We have considered order as a factual manifestation and an intrinsic property of nature. Now we must take into account another aspect on which science sheds further and quite significant light. This new aspect is the dynamism of natural orderliness—a very important factor in the understanding of the humanistic experience of the scientist. We shall examine the issue briefly but systematically, beginning with the static conception of the order of nature.

(a) Order as Repetition

The basic form of order accessible to the human mind is the static or repetitive one. Speaking of order, in fact, we tend at first to think of a fixed or unchangeable pattern. But this is precisely the fundamental discovery of science—that there are fixed and unchangeable patterns in nature. Thus, before entering a discussion of the dynamism of natural orderliness, we must attempt to assess the significance of this basic contribution of science to the understanding of nature. In particular, we must try to see why the discovery of static order is a valuable scientific achievement in its own right—and not merely a preliminary, though indispensable, step in the scientific exploration of nature.

The valuableness of the scientific discovery of order as something fixed and unchangeable in nature has often been pointed out by biologists, with good reason. In fact, as history shows, the conception of fixity is fundamental to an understanding of life. Rostand, for instance, when discussing the origin of scientific biology, stresses the essential role played by the idea of fixity of species. In his own words:

---

[11] R. Taton, in R. Taton, ed., *A General History of Science*, trans. A. J. Pomerans (London: Thames and Hudson, 1966), vol. IV, p. xxiii.

...far from being an obstacle to the progress of science, this idea fulfilled a pressing need and introduced some order where utter confusion
had reigned before. [12]

To illustrate his view, Rostand cites several examples that show that until
the idea of fixed species was introduced there was no possibility of achieving any biological understanding. What is common to these examples is
that, in the realm of life, everything was considered capable of happening.
Thus, for instance, a renowned surgeon of the eighteenth century—
Nathanael de Saint-André—accepted as plausible the story that a woman
had given birth to a rabbit. Another quite able scientist—the microscopist
J. T. Needham—maintained that a mold could turn into a little animal.
From this kind of situation, it is clear that no understanding of life could
be obtained before man was able to discover some permanent order in
nature. Eiseley has summarized this state of affairs aptly:

Before life and its changes and transmutations can be pursued in the
past, the orders of complexity in the living world must be thoroughly
grasped. [13]

In this connection, a word must be said about that countless army
of scientific workers whose central activity is to observe, measure, and
compare objects of nature. Sometimes they are ridiculed as dull and pedes-
trian in their approach. Such an accusation is most unfair, because their
work is essential to the establishment of the very foundations of scientific
understanding. We can see this point in a particularly clear manner when
considering the contributions of preevolutionary taxonomists in the realm
of biology. Those whose names have been recorded by history were as
a rule gifted and broad-minded persons. But what proved their greatness was their courage to observe, classify, and organize what appeared to
be a hopelessly complicated multitude of living forms. Through their
study of affinities and relationships among species and genera, these early
biologists contributed to make biology truly scientific. As Linnaeus wrote
in connection with his own lifelong contribution to biological nomenclature:

The first step of science is to know one thing from another. This
knowledge consists in their specific distinctions; but in order that it
may be fixed and permanent, distinct names must be given to different
things, and those names must be recorded and remembered. [14]

Besides being objectively important, the discovery of fixed or re-

---

[12] J. Rostand, "The Formation of Species," in R. Taton, ed., *A General History of Science,* trans. A. J. Pomerans (London: Thames and Hudson, 1964), vol. II, p. 514.

[13] L. Eiseley, *Darwin's Century: Evolution and the Men Who Discovered It* (New York: Doubleday Anchor Books, 1961), p. 15.

[14] Quoted in Eiseley, *loc. cit.*

petitive orderliness in nature is also subjectively rewarding. It is a conquest through which the mind succeeds in attaining understanding. Obviously enough, the outcome is a sense of joyous exultation. We have already mentioned the enthusiasm aroused by the achievements of such pioneers as Galileo and Newton. Many more examples could be given. For most discoveries are of the kind we are discussing. All of them are felt as rewarding by man because they make him see nature in a wholly new light.

To illustrate the novel way of experiencing nature, I wish to insist briefly on the results of atomic physics. When speaking of orderliness and precision, people tend normally to think of macroscopic bodies—the regularity of astronomic phenomena or the accuracy of precise standards of measurement, such as the meter, whose prototype is preserved with great precautions at Sèvres, France. But the regularities of atomic phenomena outstrip by far—both in reliability and reproducibility—any macroscopic standard of precision. When we observe, for instance, a certain portion of the cadmium spectrum or a specific line emitted by a well-defined mercury isotope, we find frequencies that are constantly accurate, well beyond any possibility of measurement on the macroscopic scale. Hence, a truly accurate clock is considered to be, at present, not one that relies on astronomical observations but one based on atomic oscillations. Atomic phenomena also present spatial regularities whose accuracy is beyond compare. As a standard of length, for instance, there is nothing as reliably precise as the distance between lattice planes of crystals. Furthermore, what is remarkable in the accuracy of atomic events is that they are reproducible at will—and this despite the obvious fact that individual atoms have different individual histories and are continually exposed to all sorts of external disturbances. When the thoughtful person, therefore, reflects on such manifestations of natural orderliness, he cannot fail to perceive nature in a new way. He feels rewarded and heartened by such a discovery. He rejoices that science has extended so enormously the realm of precision, all the way down to the smallest components of matter. For he knows now that matter is not something merely factual and even brutish in its factuality. Rather, matter bears clearly the typical traits that relate it to mind, namely orderliness and intelligibility. As a charming instance of the new way a scientist comes to perceive nature it would be enough to think of a snowflake. The ordinary man sees a white bit of matter; the atomic scientist sees an orderly crystal and a whole world of harmony and beauty. As Weisskopf put it aptly:

The simple beauty of a crystal reflects on a larger scale the fundamental shapes of the atomic patterns. Ultimately all the regularities of form and structure that we see in nature, ranging from the hexagonal shape of a snow flake to the intricate symmetries of living forms in flowers and animals, are based upon the symmetries of these atomic patterns. [15]

(b) Order as Dynamism

The discovery of fixed orderliness in nature, for all its importance, is still only the first step in the scientific understanding of the world. The second step is the discovery of orderliness as a dynamical, instead of a static, feature of nature. We must now examine the import of such a discovery and the experience that flows from it to the scientific mind. To study this subject, we shall concentrate on two main points, namely an investigation of the way science came to this insight, and how dynamism can be thought to be a universal feature of nature.

The path followed by science in the discovery of natural dynamism passes through biology. Taxonomists labored long to bring order among the countless forms of living structures. The first attempts of systematic classification were based on the properties of single organs—the sexual parts, for instance, in the work of Linnaeus. This approach was only partially successful because the structure of a given organ can change through adaptive variation, and thus a purely morphological criterion can easily lead the taxonomist to wrong conclusions. Further attempts of systematic classification tried to take into account the entire structure of the living forms being investigated. These new attempts aimed at what was called a natural classification, in opposition to the former type of classification, which was called artificial. Such classification, as has been said, had the advantage of taking into account the living organism as a whole. But, of course, to achieve such a classification, it was necessary to compare among themselves many organisms studied as holistic structures. Furthermore, it was necessary to detect the intrinsic reasons capable of accounting for the different forms assumed by corresponding organs in different species, still extant or already extinct. This painstaking investigation had a twofold result. It showed conclusively that life is so manifold and rich that it cannot be imprisoned in any number of fixed categories or species. In addition, by studying similarities and differences of living forms scattered through space and time, the scientific mind was led to discover the internal mechanism accounting for the changes observed in living structures.

As has been repeatedly said, biology reached maturity through the work of Darwin just as physics did at the hands of Newton. The comparison is accurate in that both men succeeded in transforming their science from description to intrinsic explanation. Before Newton, physics had

---

[15] V. F. Weisskopf, *Knowledge and Wonder: The Natural World as Man Knows It* (New York: Doubleday, 1962), p. 98.

already been able to describe accurately a large number of mechanical motions.   The great contribution of Newton was to unify the study of motion by pointing out the observable agency, inherent in nature, that gives rise to the regularities of mechanical phenomena.   The contribution of Darwin was of the same kind.   Before him, biology had already accurately studied a large number of regularities manifested by living structures.   Darwin discovered that the agency that accounts for the similarities and differences of living forms is a factor that is inherent in nature itself and which operates according to well-determined rules.   This agency he called natural selection.   To be sure, in a way Darwin's achievement cannot be compared completely with that of Newton, because the Darwinian conception of biological evolution was not entirely adequate to account for all the complexity of the phenomenon. [16]   Nonetheless, Darwin's merit is supreme, because he was the first researcher who succeeded in unifying biology from within.   Following his lead, man could begin to understand life as a whole.

The discovery of biological evolution proved to have a wide philosophical significance.   The reason is that, through evolution, the human mind for the first time realized the occurrence of a kind of order that combines two features, which until then had appeared incompatible, namely regularity and changeability.   Normally it would seem that regularity is something that requires perpetuity and repetitiveness, whereas changeability appears to be pure unexpectedness and chaos.   Darwin proved that the natural orderliness of life presents the two features simultaneously.   One can speak of regularity and of changeability or novelty at the same time because biological order is dynamical instead of static, but its dynamism is itself ordered.   This is the discovery: the realization that nature possesses the inborn ability to produce unexpected forms and structures— not in a capricious way but according to precisely defined rules.   Darwin showed that nature is able to produce truly new forms in the sense that their properties are not latent either in the forms themselves alone or in the environmental conditions alone.   The novelty comes about through an unpredictable interplay between the living forms and the environment. However, the novelty is not simply such—namely, pure change—but is an orderly novelty.   For Darwin indicated the main factor accounting for the change.   This factor is the so-called natural selection.   Whenever a living structure happens to undergo a change or mutation that suits the environment better, the structure reproduces itself more abundantly and thus comes ultimately to be dominant in its environment.   Consequently,

---

[16] Besides natural selection, the main factors of evolution are now acknowledged to be mutation, sampling errors, and genetic migration.   For details see, for instance, G. G. Simpson, *This View of Life: The World of an Evolutionist* (New York: Harcourt, 1964), pp. 73-77.

through the interplay of structures and environment, one can understand how a given structure can persist unchanged or be replaced by another structure which is itself well defined and orderly.

Further light has been shed on the processes that constitute nature's dynamical orderliness by the discovery of the quantum properties of matter. Before such properties came to the fore, science could explain change only by invoking the factor of chance. But chance, by itself, cannot really explain how and why order comes about. For chance is essentially the antithesis of order. Quantum theory makes dynamical orderliness intelligible because it shows how order and change are truly compatible; in fact, natural change is itself orderly. Quantum theory actually proves that nature has two main characteristics. [17] In the first place, identity, wholeness, and specificity are by no means exclusive properties of living structures, but belong to atomic and molecular structures as well. In the second place, interaction between structures and environment takes place according to precise and known rules. Since structures constitute autonomous totalities, they remain internally unchanged as long as the energy acting on them does not reach a certain threshold typical of the structure in question. When the energy exceeds the threshold, it is the entire structure that reacts as a whole. Thus it can only undergo well-defined mutations that are typical of the structure involved. As a consequence, we can understand why things can change in such a way that their change remains an intelligible event, not simply to be accounted for by ascribing it to nonintelligible chance. The element of chance continues to be present, no doubt. But it is reduced to an extrinsic factor. Actually, what is unexpected and totally unforeseeable is merely the strength of individual energetic situations in which a given structure happens to find itself. But change as such is now shown to take place according to well-defined rules.

(c) The Universality of Orderly Dynamism

A further consideration brings out more fully the significance of dynamical order disclosed by science. Orderly dynamism is revealed as a feature that affects the entire observable reality. We can discuss this theme under two headings: first, the growing expansiveness of nature; second, the historical or developmental character of nature itself.

By *expansiveness of nature* is meant the observable property according to which nature tends to grow, branch out, fill up the universe—while at the same time enriching itself with new forms and unprecedented structures. This phenomenon is most visible in the evolution of life.

---

[17] For details see, for instance, Cantore (note 3), especially "The Ontological Implications of Atomic Physics: The Meaning of Atomic Order," pp. 254-280.

T. H. Huxley uses an illuminating comparison. [18]   He likens the filling of
the earth with life to the filling of a barrel.   To get an idea of how life
is plentiful and resourceful in taking advantage of any possibility available,
Huxley suggests that we imagine the earth as a barrel being filled with
apples until they heap over the brim.   At this point plenty of space is
still available.   Thus he suggests we think of adding pebbles until they
overflow.   Then, to fill the interstices among apples and pebbles, we
should add as much sand as possible.   Finally we should pour in as much
water as possible.   This analogy is illuminating because it represents
graphically the urge of living forms to fill the space and use the resources
of the earth to their very limits.   It rightfully stresses the teeming abundance that characterizes life as studied by science.   A nonscientific person
could hardly imagine how a handful of soil or a breath of air are literally
crowded with multiform living structures.

Expansiveness, however, is not just a property of life.   On the
contrary, it is a feature of nature as such.   Nature as a whole is a huge
dynamical process, orderly and expansive.   Quantum physics is particularly
illuminating here.   It shows that inanimate matter, too, tends to grow
in complexity and richness by taking advantage of all possibilities allowed
by the environment.   Weisskopf speaks in this connection of the "quantum ladder." [19]   With this expression he intends to convey the insight
that the entire realm of nature can be thought of as a series of steps or
rungs.   Each rung is occupied by well-defined structures.   But the matter
of such structures is not static.   It tends to pass from one rung to the next,
and thus to increase in complexity.   The factor that determines on which
rung matter finds itself is the energetic environment.   The image of the
quantum ladder, therefore, gives an overall view of nature as a single
ongoing development in which order prevails alongside variety and complexity.   We can now understand how our world, with its manifold
features, comes about.   From the most dispersed form of aggregation
—called the plasma state—nature proceeds to constitute increasingly
complex structures: nuclei, atoms, molecules, crystals, and living forms.
Accordingly, the world becomes increasingly meaningful in its unity as
well as multiplicity.

The other aspect of universal dynamism brought to light by science
is the *historical or developmental character of nature* as such. [20]   Of course,
it is not possible to speak here of history in the strict sense of the term.

---

[18] The analogy is discussed by G. G. Simpson, *The Meaning of Evolution: A Study
of the History of Life and of Its Significance for Man* (New Haven: Yale University Press,
1967), pp. 113-115.

[19] Weisskopf (note 15), pp. 142-151.

[20] See, for instance, C. F. von Weizsäcker, *The History of Nature,* trans. F. D. Wieck
(Chicago: University of Chicago Press, 1949).

For history, strictly, refers to the temporal unfolding of human events. Nevertheless, it is a fact that nature shares with history the property of temporal development. Nature, too, is time-oriented, irreversible, and indelible. In other words, nature is far from perennial sameness and endless repetition. It is rather a time-directed process.

The historicity of nature can be most easily understood by starting from biological considerations. The changes brought about by biological evolution are unique and irreversible happenings. They are unique in that it takes a singular convergence of circumstances—environmental conditions and suitable mutations of the living forms involved—to produce viable evolutionary changes. They are irreversible in that they are transmitted from parents to offspring, generation after generation. But the historical or developmental character applies also to nonliving matter. For every physicochemical interaction requires a suitable amount of energy to occur. But, as the second law of thermodynamics shows, free energy is consumed in the process. For, once it has become dissipated in the form of heat, energy cannot be recaptured and recycled again. Thus, from the energetic point of view, it is not possible to think of physicochemical events as endless repetitions in which the time factor has no role to play. They are, rather, basically irreversible happenings.

To sum up, science reveals a new fascinating dimension of natural orderliness. The universe constitutes a comprehensive whole, which is growing in time. Not only do individuals develop and unfold, but so does the universe itself, through the continual interactions of its components.

3. Science as an Experience of Nature

We have outlined the main aspects that constitute the structure of nature as of interest to science. We complete our investigation by summarizing the typical new ways in which the scientific mind comes to experience nature in its entirety. A few words will be said about the three principal categories of humanistic relevance, namely intelligibility, goodness, and beauty.

(a) Accessibility to the Mind

The fundamental aspect of nature that strikes the scientific mind is the intelligibility, or accessibility to the mind, of nature itself. The entire enterprise of science, in fact, is a striving after understanding. To realize concretely this characteristic aspect of the scientific experience, let us survey rapidly the two main stages of the scientific endeavor as a search for knowledge.

The experience of nature's intelligibility, as perceived by the scientific mind, is twofold. The first is the realization that, contrary to appearances, nature is not unattainable to man, but rather accessible to him. We can relive this experience by recalling briefly the history of astronomy. Pre-scientific man was convinced that the heavenly bodies were definitely beyond his reach as an earthbound creature. The sky was the symbol of inaccessibility. But science has even enabled man to reach the sky, beginning with the exploration of the moon. One of the first achievements of Galileo was the observation of the moon. The telescope put the moon within reach not only of man's mind but also of man's senses. The surprise and enthusiasm of Galileo at this achievement shows through at the beginning of his *Starry Messenger*, when he says that man can now know the moon "with all the certainty of sense evidence." [21]

Scientific astronomy has continually increased the range of the universe accessible to the mind of man. One great breakthrough was due to William Herschel. His epitaph literally states that he "broke through the barriers of the heavens" (*coelorum perrupit claustra*). These words are justified in that, before Herschel, the knowledge man had of the skies was still limited to the solar system. As a consequence, the outer stars were virtually inaccessible and, as a result, meaningless to man. Man could see only that they were there. He gazed at them, but largely for such practical purposes as navigation and astrology. For the stars seemed to be just fixed points of reference, mere dots embedded in the heavenly vault. The scientific work of Herschel enabled the mind of man to penetrate the universe of the outer stars and discover their manifold intelligibility. To summarize with Lodge, as cited by Whitrow:

Herschel changed all this. Instead of sameness, he found variety; instead of uniformity of distance, limitless and utterly limitless fields and boundless distances; instead of rest and quiescence, motion and activity; instead of stagnation, life. [22]

Another astronomic breakthrough, again aptly expressed by an epitaph, is the one achieved by Joseph Fraunhofer. His epitaph states that he "brought the stars near" (*approximavit sidera*). [23] This statement is factual in that, through Fraunhofer's work, the mind of man could

---

[21] In S. Drake, ed. and trans., *Discoveries and Opinions of Galileo* (New York, N.Y.: Doubleday Anchor Books, 1957), p. 28.

[22] Herschel's epitaph is quoted in G. J. Whitrow, *The Structure and Evolution of the Universe: An Introduction to Cosmology* (London: Hutchinson, 1959), p. 22. The comment about the significance of Herschel's work is due to O. Lodge and is also cited by Witrow.

[23] Fraunhofer's epitaph is quoted in J. T. Merz, *A History of European Scientific Thought in the Nineteenth Century* (New York: Dover, 1965), vol. II, p. 47, note 2.

penetrate the very chemical structure of the stars. Up to his time, science had revealed the accessibility of the stars, but only outwardly. Man was able to study only their motions. Fraunhofer, through the spectral study of radiation, supplied the key to understand what materials the stars were made of, and also how they evolved in time. The power of science to make nature accessible to man can perhaps be exemplified in no more striking manner. As a historical footnote we may add that science, by so doing, surprised even some of its most enthusiastic philosophical supporters. Fraunhofer made his results known shortly after Auguste Comte, the self-styled champion of science, had contended publicly that man would never be able to know the internal structure of the stars.

The second sense in which the scientist experiences the accessibility of nature to the mind is even more profound and rewarding. Science is not content merely to extend the area of knowledge to farther and more recondite objects. Over and above that, it strives after theoretical discovery or intrinsic explanation. This is, then, the more profound way in which the scientist perceives the accessibility of nature. He realizes that nature is intrinsically understandable, intimately penetrable by the mind of man. Science, thus, fulfils more and more the ancient wish of man, which had long appeared to be a mere dream. Aristotle had already expressed this intimate yearning for knowledge when he said that "ignorance of motion was ignorance of nature." (See Chapter 2, Section 1a.) Science satifies this desire beyond all expectations. Everything moves in nature, and science makes these many kinds of motion understandable to man. From the early achievements of Galileo and Kepler, to the theoretical insights of Newton and Darwin, on and on to our own times—the march of science has been a continual increase in the explanation of motion. Nature, through science, is thus made accessible to the mind of man in the genuine sense of intrinsic intelligibility.

(b) Superabundant Richness

A second characteristic aspect marks the scientific experience of nature. It is the awareness of an unsuspected and ever-increasing plentifulness. We can speak of a sense of superabundant richness with which the scientist perceives nature. Such a richness has two principal manifestations, one extensive and one intensive.

The extension of the wealth of nature brought to light by science becomes immediately apparent if one reflects on the historical development of science itself. Science has taken man farther and farther, and deeper and deeper, into the universe. The horizon, which seemed close at hand at the beginning, has kept expanding, endlessly. The history of science is, without exaggeration, a history of receding backgrounds. Examples abound, but none is so convincing as the continual expansion of astronomy. For it shows concretely how the world we live in is enormously rich and varied and immense. This is clear even if we restrict our consideration to the solar system, the limited sky region that forms our little corner of the universe. To the nonspecialist it may appear that the solar system is an uncomplicated and easily understandable piece of reality. After all, only a handful of planets are present in it, most of which are visible to the naked eye. Yet even here we notice a surprising richness, which comes to light only gradually through the patient investigation of science. In the first place, consider the number of heavenly bodies present. Up to 1900, observational astronomers had counted 449 minor planets and asteroids. In the following 50 years, the number of such known bodies increased by about 2000 more. In the second place, consider the extension of the solar system itself. Up to 1930, man thought he knew the size of the system. But then the very remote planet Pluto was discovered through a combination of theoretical calculation and observational sagacity. At present we know that the solar system extends to at least 40 times the distance that intervenes between the earth and the sun. [24]

When we go beyond the solar system, quantities and dimensions truly stagger the imagination. Up to the present we can photograph some 100,000,000 galaxies or stellar aggregates. But each of these is a world in itself—just as the Milky Way, of which our solar system constitutes but a negligible part. As for the spatial dimensions involved, only numbers can be used to express them. The farthest reaches of the universe detectable to man up to now are no less than 5 billion light-years away from us. [25] To sum up these few hints about the extensive richness of nature brought to light by science, we can see the experiential reason for expressing the results of science in quantitative terms. The reason is the richness of nature which totally exceeds the power of our imagination. Thus numbers alone can somehow convey the largeness and plentifulness of the universe. Indeed, by doing science man becomes increasingly aware that nature is literally immense or boundless. It cannot be measured and cannot be assigned boundaries simply because it keeps growing apace with the growth of science.

A simple numerical summary can give an idea of the extension of nature to which science has penetrated so far. The summary is Taton's in his comprehensive description of contemporary science.

This [the expansion of science] is perhaps best appreciated from the

---

[24] Data given by J. Rösch, in Taton, ed. (above, note 11), vol. IV, p. 333.

[25] Data given by J. L. Greenstein, in *The Scientific Endeavor: Centennial Celebration of the National Academy of Sciences* (New York: Rockefeller University Press, 1965), p. 58.

fact that in 1900-1960 the greatest measurable length increased from 10^{26} to 10^{40} units, time from 10^{10} to 10^{16} units, temperature from 10^5 to 10^{11} units, and pressure from 10^{10} to 10^{16} units. [26]

The extensive consideration of science is not adequate to convey the experience of nature's richness as disclosed by science itself. In fact, science is not satisfied with discovering more and more objects or with making increasingly precise measurements of objects already known. More significant, science constantly increases the number of aspects or features it detects in already known objects. In the exploration of the stars, for instance, an astronomer is now able to obtain no less than 30 different pieces of information from the study of a single star image. [27] But, of course, we do not need to contemplate the heavens to realize the surprising variety of properties detected by science in nature. What amazes the thoughtful person is still what caused such a great surprise to Galileo and his contemporaries. It is the fact that such a richly complex orderliness is present in the most obvious and everyday phenomena. As Sagredo put it while addressing himself to Salviati:

Sir, you certainly give me frequent occasion of admiring the richness and with it the supreme liberality of nature. You do this because, starting from things that are so common, and I would even say somewhat trivial, you extract pieces of information which are very surprising and novel, and very frequently remote from every imagination. [28]

The only difference, for the present-day scientific mind, is the obvious one that derives from the enormous increase of information supplied by science since Galileo's own times. At present we can literally say that there is nothing trifling, nothing meaningless or unimportant in any of nature's manifestations. For all details of nature present distinctive features of orderliness and intelligibility. Thus the superabundant richness of nature is one of the basic experiences that strike the reflective scientific person more and more deeply.

In this connection a word should be said about a much disputed issue, which finds here an illuminating interpretation. I refer to the frequently mentioned distinction between *quality* and *quantity*. The view is widespread among humanists that science is concerned only with quantity, whereas the humanities are interested in quality. In the light of the preceding, not much remains to be said to clarify this point. It is certainly true that science is concerned with quantity. The reasons, as mentioned, are respect for observable reality (Chapter 1, Section 2b) and also the

---

[26] Data given by Taton, in *General History* (note 11), vol. IV, p. xxii.

[27] Details in H. Shapley, *Beyond the Observatory* (New York: Scribner, 1967), pp. 130-132.

[28] Translated from Galileo, *Discorsi* (note 1), p. 110.

public nature of science (Chapter 1, Section 2c). But this clearly does not mean that science is merely interested in quantity. Because of its concern for objectivity, science is interested in all the diversity and manifoldness of natural order. That is, science is interested in quality as well as quantity. Nor can science be said to reduce everything to quantity simply because it tries to understand the manifold phenomena of nature by tracing them back to the measurable or quantitatively observable aspects of their components. For what science aims at by so doing is not to explain away the multiplicity of qualities by making them the pure expression of quantity. Rather, it tries to discover how the varied qualities that one can perceive in everyday experience result themselves from some elementary or basic qualities of nature itself. These basic, qualitatively distinct features of matter are, for instance, gravitational interaction, electromagnetic characteristics, and the quantum features—corpuscular and wavelike manifestations—of matter at the atomic level. As a consequence, it cannot be said that science does not care for qualities. Actually, it cares so much for them that it tries to arrive at an intrinsic understanding of them. It wants to know why things present certain color, hardness, electric and thermal features, and so on. [29] To conclude, the superficial distinction between science and the humanities as the study of, respectively, quantity and quality should be avoided because it is misleading and ultimately false. The richness of nature, in fact, is *both* quality and quantity. Thus a genuine knowledge of nature can only be concerned with the two of them at the same time.

(c) Harmonious Beauty

A third basic aspect characterizes the scientific experience of nature. It is the realization that harmony and beauty pervade nature as such.

Scientists speak frequently of beauty. Sometimes they ascribe beauty to scientific theories. Beauty is then the property according to which a theory can be perceived by the mind as forming a harmonious whole embodied in many details. The theory—in this conception—is implicitly compared to an artistic masterpiece. In an artistic creation—for instance, a medieval cathedral—the mind perceives a unitary insight or inspiration present as a unifying factor in all its parts. In the same way, the mind realizes a harmonious unity present in all the parts of a successful scientific theory. [30]

However, scientists as a rule are not primarily concerned with the

---

[29] See V. F. Weisskopf, "Quality and Quantity in Quantum Physics," *Daedalus* **88** (1959), pp. 592-605.

[30] See a brief discussion in L. de Broglie, *Continu et Discontinu en Physique Moderne* (Paris: A. Michel, 1941), p. 88.

beauty of theories. For science is, by definition, objective study of nature. Hence scientists are centrally concerned in their researches with the objective meaning of nature itself. In particular, they are interested in beauty as an essential manifestation of such a meaning. After all, the objective beauty of nature is what ultimately explains the beauty of the scientific theories themselves. The attraction exerted by the beauty of nature is one of the main manifestations of the scientific experience. Poincaré put it aptly when he wrote:

The scientist does not study nature because it is useful to do so. He studies it because he takes pleasure in it, and he takes pleasure in it because it is beautiful. If nature were not beautiful, it would not be worth knowing, and life would not be worth living. [31]

What is the essence of the beauty the scientist feels moved by and tries to capture in the theories he formulates? It is not, properly speaking, the beauty of various shapes and colors that man can perceive through observation. It is rather, to continue with Poincaré in the context cited: "...the more intimate beauty which comes from the harmonious order of its [nature's] parts, and which a pure intelligence can grasp." Beauty is therefore something typically intellectual: it is a harmony that the mind alone can perceive. However, one should not infer therefrom that beauty, as experienced by science, is something abstract or merely intellectual. Scientific beauty is not abstract but very concrete because it is detected in effectively existing, observable objects. Furthermore, scientific beauty is not merely intellectual because it consists in patterns that can be perceived by the trained senses, at least indirectly. The originality of science consists in its ability to bring to the fore a type of beauty that is different from that ordinarily perceived by man. It is a beauty that can be adequately realized only by means of intellectual search—and yet a beauty that underlies and makes possible the ordinary, sensible beauty itself. Poincaré stressed this contribution of science to beauty in the passage quoted above:

It is this that gives a body, a structure so to speak, to the shimmering visions that flatter our senses, and without this support the beauty of these fleeting dreams would be imperfect, because it would be indefinite and ever elusive.

Once the character of scientific beauty is understood, there should be no quarrel with the contention often made by scientists that science

---

[31] H. Poincaré, *Science and Method*, trans. F. Maitland (New York: Dover), p. 22. On the same subject, see the entire essay "Die Bedeutung des Schönen in der exakten Naturwissenschaft" by W. Heisenberg in his *Schritte über Grenzen: Gesammelte Reden und Aufsätze* (Munich: Piper, 1971), pp. 288-305. The same essay has been published in book form, in bilingual edition (English translation by Cantore) by Belser Verlag, Stuttgart, 1971.

has brought a new sense of beauty to man. This is a fact, in a twofold regard. In the first place, science has immensely enlarged the realm of beauty accessible to man even when the term is taken in the ordinary acceptation of a pleasant sensible appearance. Nonscientific man, in fact, could never succeed in imagining the breathtaking grace and symmetries, for instance, that are revealed by the microscope. In this first sense, one must certainly agree with the zoologist Simpson when he stresses that "No poet or seer has ever contemplated wonders as deep as those revealed to the scientist." [32] But, as has been intimated, there is also a second and stronger reason for speaking of science as a revealer of the beauty of nature. In fact, science has revealed that nature's beauty is not merely something factual, to be perceived by the aesthetic sense, but also something intellectual, to be understood by the searching mind.

The contribution of science to beauty can be made a little more concrete by discussing briefly the theme of the *harmony of the world*. It is an indisputable fact that man has always been enchanted by the recurring regularities of nature, especially by the motions of the celestial bodies. The ancients even came to imagine the motions of such bodies as due to very complex mechanism of crystalline spheres, perfectly synchronized and absolutely precise. Pythagoras was probably the first who gave a name to such a poetical construction. He spoke of "music of the spheres" as though these hypothetical entities formed a cosmic orchestra, with sound inaudible to the ear but all the more satisfying to the mind. This attractive metaphor inspired countless poets, and even contributed to the birth of science. Kepler, in fact, was moved to search for his famous laws by the conviction that there was some sort of musical harmony involved in the mathematical proportions of the trajectories of the planets. However, as is well known, in the end it became clear that the celestial spheres were no more than a man-made contraption, the wishful projection of the uninformed imagination. Should one infer therefrom that the cosmic experience of man, embodied in the metaphor of the harmony of the spheres, has simply been exploded by science and reduced to nothing? The assertion is frequently made, but it does not stand serious scrutiny. For science has not just destroyed an imaginary universe. Rather, it has replaced it with another universe. To be sure, the new universe can no longer be imagined, as the former could, but it is no less beautiful or inspiring because of that. The new universe cannot be imagined simply because it is immensely rich and continually developing. But it is even more beautiful and inspiring than its prescientific counterpart because man now knows that the harmony of the world is not just an appealing poetical image, but an ascertained manifestation of nature itself. Indeed, the

[32] Simpson (note 16), p. 233.

harmony of the world is but another name for the all-encompassing inter-connectedness that pervades all of nature as disclosed by science. It is this universal orderliness that lies at the roots of the many beauties we can now enjoy through understanding and even perceive through the senses, at least indirectly. Such beauties are, for instance, the regularities of the atomic world with their macroscopic manifestations in crystals and molecular structures, the countless species of animals and plants forming but one immense stream of evolving life, the birth and expansion of the universe itself. In short, modern science still discloses harmony and inaudible music in the cosmos, but on a different and higher level than that accessible to prescientific man. It is a harmony that pervades the whole of nature, that we can now understand from within, and that we should enjoy all the more.

To synthesize our discussion of the scientific experience of nature, a backward glance may suffice. Man has always felt that the world of nature was meaningful to him only if he could speak of nature as overall orderliness, accessible to his mind. The Greeks devised the idea of the *cosmos,* the Romans spoke of the *universe.* In its own typical new way science has confirmed that the world of nature is indeed cosmic, harmonious, beautiful. Furthermore, it has stressed beyond any possibility of doubt that the world constitutes truly an all-encompassing structure penetrable by the mind of man. To be sure, many problems do arise because of these astounding revelations of science. But the recognition of problems should not prevent man from gratefully acknowledging that the scientifically discovered universe has an original humanistic significance of its own.

CHAPTER 3

WONDER AND AWE: THE SCIENTIFIC EXPERIENCE OF ULTIMATES

We have finished exploring the scientific experience under its two most prominent aspects. We have considered science as a subjective experience of knowledge and as an objective experience of nature. Should we at this point think that our experiential investigation of science is virtually completed? It might seem that there are plausible reasons for such a view. In fact, if science is knowledge, and knowledge consists essentially of two polarities, the subjective and the objective, what else remains to be studied after one has examined the subjective and objective sides of scientific knowledge? Despite its plausibility, I am convinced that this position would not be justified. In fact, if we were to terminate our experiential study of science at this point, we would actually fail to understand the most profound aspects of the scientific experience itself. For we would not know why science arises in the first place nor what its overall significance for man is. In other words, were we not to press our exploration further, we would fail to understand science as an experience of the whole man.

To attain a deep and comprehensive view of the experiential nature of science it suffices that we continue to develop the line of thought followed so far. We have seen that science is a characteristic form of knowledge whose intrinsic aim is the discovery of natural orderliness. We must now inquire, to begin with, why scientists come to do science. Then we must explore to what ultimate experiential conclusions scientists are led by their own success in doing science. These are then the themes that constitute the subject matter of the present chapter. Our sources here, more than ever, will be the creative scientists themselves. Our effort will be to understand the actual meaning of their testimony by situating it in the overall dynamism of the scientific endeavor.

1. The Wonder of Science

When we inquire of the thoughtful scientist about his first motivation for doing science, we practically always hear him speak of a sense of wonder.

Thus we can begin our study of the all-encompassing experience of science by analyzing the role played by wonder in the scientific process. But wonder is formally defined in a threefold way: as something that surprises and attracts, something that causes admiration and delight, and something that leaves one astonished and with a sense of questioning. Accordingly, this threefold division offered by the semantics of wonder may serve as a guideline in our attempt to understand why science comes about and to what conclusions it leads.

(a) The Surprise at Natural Order

To begin from the beginning, what triggers the scientific endeavor in the first place? In other words, how is the human mind prompted to seek for that knowledge which, in the end, will constitute science? If we ask those who know from experience, namely the creative scientists themselves, their answer is unanimous. What first moves them to do science is nothing but a sense of surprise.

Psychologically speaking it is obvious that man needs to be surprised in order to start to do science. For science is a search for new knowledge. But no one searches for new knowledge unless one is surprised at the way things are, instead of taking them for granted. The popular consensus puts it unequivocally: *ab assuetis non fit passio, assueta vilescunt.* What we are accustomed to does not strike us, it appears to us trivial. Science actually begins when one is struck by the fact that the world of observation is the way it is. That is to say, the first dawning of science is an awareness of novelty and unexpectedness experienced in connection with one's everyday world. This surprise is of a cognitive type, in a twofold sense. The first sense consists in the sudden realization that the observable world is actually knowable to man. The second sense consists in the realization that, despite its being knowable, the world is not yet known to man. What moves man to do science, then, is the attraction to explore an intelligibility that is obviously there, but has been overlooked or ignored so far.

For an example of the role played by wonder in giving rise to science, let us turn to the testimony of two of the most creative contributors to quantum physics, Bohr and Heisenberg. As is well known, this branch of science is characterized by a highly abstract mathematical formalism and deals with entities that cannot be directly seen by man. Thus one would expect that the researchers who dedicated their lives to it were motivated by some recondite reason, especially by mathematical ingenuity and concern with logic. And yet, when we ask Bohr and Heisenberg why they became quantum physicists, their candid answer is: a sense of wonder. They were surprised that the things of ordinary experience were the way they were, they were surprised that people did not find anything striking in this situation, and thus they decided to dedicate their lives to the task of finding out. Heisenberg gives a particularly charming analysis of this state of affairs.

When he first met Bohr, he was somewhat suspicious of the latter's motivation in doing science. Bohr was continually speaking in terms of modelistic and planetary representations of atoms. Was he perhaps not moved by the subconscious desire to impose a rationalistic-mechanistic scheme on nature, rather than just trying to find out how things were? Bohr was quick in dispelling this suspicion from the mind of his youthful interlocutor. Recalling the genesis of his famous theory, he rejected emphatically the idea that his starting point might have been the desire to mechanize the structure of the atom. In his own words: "But for me the starting point was the stability of matter which from the standpoint of traditional physics is a pure wonder." [1] Going on to explain his thought, Bohr stressed that the stability of matter is something quite obvious but at the same time very surprising. The obviousness consists in the fact that matter always presents the same features to man. The surprise was caused for him by the fact that no one had as yet been able to understand why matter was stable. For, on the basis of the hitherto accepted Newtonian interpretation of nature, the stability of matter was simply not understandable.

By stability I mean that the same substances always occur with the same properties, that the same crystals are formed, that the same chemical compounds arise, etc.... There is then in nature the tendency to produce specific forms—I employ the word forms now in the most general sense—and to always reproduce anew these forms, even when they have been disturbed or destroyed.... All this is by no means self-evident, but on the contrary it seems not to be understandable when one accepts the axioms of Newtonian physics....

As regards Heisenberg himself, he relates in detail his first motivation for doing physics. He basic question (*Grundfrage*) was to know "why there are, in the material world, forms and qualities that always recur." [2] The examples that he gives are homely. He speaks of the surprise generated by the fact that water always presents the same characteristics, despite the different processes by which it may be produced such as melting of ice, condensation of vapor, or burning of hydrogen. To him it was surprising that people had been taking all of this for granted.

---

[1] Translated from W. Heisenberg, *Der Teil und das Ganze: Gespräche im Umkreis der Atomphysik* (Munich: Piper, 1969), pp. 60f.
[2] *Ibid.*, pp. 37f.

In traditional physics this has always been presupposed, but never understood.... In this area, then, natural laws of quite different kind must be at work to account for the fact that atoms arrange themselves and move always in the same ways, so that always the same substances with the same stable properties arise.... Here, then, an enormous virgin country discloses itself to us in which man perhaps for decades will be able to discover new interconnections.

To sum up, what first moves man to become a scientist is not, as frequently stated, organized skepticism or a sense of rebellion against hitherto accepted views. Conceptions of the kind do not apply to the beginning of science because they are just negative and, as a consequence, incapable of bringing about any new knowledge about the world. What animates the aspiring scientist, on the contrary, is something quite positive. It is the surprise that arises from the direct experiential contact with nature. This surprise lies in the awareness that nature is there to be known. The scientist is the person who takes such an awareness seriously and, as a consequence, engages himself in the search for new knowledge. This being so, one realizes at once why scientists prize originality of mind so highly. To be original and open-minded constitutes the basic feature of the person who has a taste for science. For science reveals that nature is so rich and unexpected in its manifestations that a person should never rest self-satisfied on the laurels of past intellectual attainments. Planck expressed well this basic trait of the scientific attitude:

...compared with immeasurably rich, ever young nature, advanced as man may be in scientific knowledge and insight, he must forever remain the wondering child and must constantly be prepared for new surprises. [3]

(b) The Admiration of Discovery

A second acceptation of the term "wonder" is applicable to the scientific experience. The scientist does not experience surprise only at the beginning of his work, when he realizes that the world is knowable and yet not known. He is even more surprised when he attains the goal of his striving—that is, when he effectively discovers the intelligibility of nature he had been pursuing. The surprise is now that of being confronted with something that is greater and more exciting than anticipated. To distinguish from the first kind of wonder, we can speak in this connection of admiration and its accompanying emotions, such as delight and a sense of rewarding gratification.

---

[3] M. Planck, *Scientific Autobiography and Other Papers,* trans. F. Gaynor (New York: Philosophical Library, 1949), p. 117.

To realize directly this characteristic aspect of science, let us consider some of the concrete evidence available. In the first place, contrary to a widespread cliché, the emotions of the scientist are totally involved in the experience of discovery. Scientists are far from being the cool, unemotional people—all brain and no heart—who are frequently imagined by laymen. Scientists themselves continually speak of thrill and excitement. And they are ingenious in resorting to comparisons to convey their feelings. The psychologist Eiduson has collected instructive data in this regard. One leading physicist, for instance, compared the scientific quest for knowledge to the prospector's search for gold. [4] The prospector shovels dirt in a gold field. The work is monotonous and wearying. But then, all of a sudden, a gold nugget is there. The thrill of the finding amply compensates the wearisome quest. A similar thrill rewards the scientific researcher. The physicist cited concludes by avowing candidly:

...to me the biggest thrill is seeing a new effect for the first time. It may happen only once or twice a year, but it's worth all the drudgery that precedes it.

Other scientists make the same point. For instance, a chemist affirmed significantly:

One thing about science is that the search for understanding is far more exciting than I had ever expected it to be when I was young. [5]

The emotional aspect of admiration clearly plays a major role in the scientific enterprise.

But if the emotions of the scientist are involved in the scientific process, how should one evaluate the situation? Is the reaction of the scientist but something subjective, a release of psychological tension which has nothing to do with the objective content of science itself? Numerous philosophers seem to think so, but scientists are of quite a different opinion. One chemist, for instance, expressed to Eiduson in the following illuminating way the joy he was experiencing because of science.

Well, to me, science is terribly exciting. I'm not saying that the satisfaction comes from just having solved a problem; the satisfaction really comes—as far as I'm concerned—in the achievement of understanding. [6]

This testimony is illuminating because it shows why scientists find science rewarding. What they experience is not a mere psychological reaction

[4] Quoted by F. Bello in P. C. Obler and H. Estrin, eds., *The New Scientists: Essays on the Methods and Values of Modern Science* (New York: Doubleday Anchor Books, 1962), p. 81.

[5] B. T. Eiduson, *Scientists: Their Psychological World* (New York: Basic Books, 1962), p. 157.

[6] *Ibid.*, p. 110.

—something entirely subjective. Actually, their joy and exultation are aroused by the discovery of truth—something entirely objective. Such an experience is expressed by the scientists in various ways. For instance, the great neurologist Ramón y Cajal compares the joy of scientific discovery to that of bringing forth a new life:

Such supreme joy and satisfaction makes all other pleasures appear as pale sensations and compensates the scientist for the hard, constant, analytical work, like childbirth labor involved in achieving a new truth. [7]

Born adds another consideration in the same vein by asserting that the pleasure of discovery is not merely subjective, like that enjoyed, for instance, in solving a crossword puzzle. For the subjective feeling is accompanied here by the consciousness of objective creativity, just as it happens in the creativity of art and philosophy.

This pleasure is a little like that known to anyone who solves crossword puzzles. Yet it is much more than that, perhaps even more than the joy of doing creative work in other professions except art. It consists in the feeling of penetrating the mystery of nature, discovering a secret of creation, and bringing some sense and order into a part of the chaotic world. It is a philosophical satisfaction. [8]

In the light of the preceding, we realize why creative scientists insist much that science is essentially a theoretical vision—as opposed to any practical results that man can expect from science itself (Chapter 1, Section 1bc; Chapter 2, Section 1b). Louis de Broglie summarizes the situation well by declaring that the joy of objective discovery is what motivates genuine scientists in their research.

The great epoch-making discoveries in the history of science (think, for example, of that of universal gravitation) have been like sudden lightning flashes, making us perceive in one single glance a harmony up till then unsuspected, and it is to have, from time to time, the divine joy of discovering such harmonies that pure science works without sparing its toil or seeking for profit. [9]

(c) The Astonishment at Natural Intelligibility

We have seen that wonder continually attends science in the making. It triggers the interest of the researcher and rewards his effort. However,

---

[7] S. Ramón y Cajal, in E. H. Craigie and W. C. Gibson, *The World of Ramón y Cajal with Selections from His Nonscientific Writings* (Springfield, Ill.: Charles C. Thomas, 1968), p. 191.

[8] M. Born, *My Life and Views* (New York: Scribner, 1968), pp. 47f.

[9] L. de Broglie, *Physics and Microphysics*, trans. M. Davidson (New York: Grosset & Dunlap, 1966), p. 208.

once a scientific view of reality has successfully established itself, is there any room left for wonder? In other terms, should one think that science and wonder are compatible or, rather, that the scientific attitude entails the exclusion of wonder from the mind of man? This is the further issue we have to examine in order to realize the role played by wonder in relation to science.

To clarify our issue from the start, it is obvious that there is a kind of wonder that is incompatible with science. It is the surprise that results from ignorance. Thus, for instance, children or simpletons are surprised at the tricks of a prestidigitator, while reflective grown-ups are not. But if this is the case, what is the situation as regards science? Should one say that it kills wonder by banishing ignorance from the mind of man or, rather, that the success of science is itself the source of wonder, of a kind unsuspected to nonscientific man? This is our question. To discuss it I shall refer to the statements of two main antagonists, both of whom claim to be speaking in the name of science: Mach and Einstein.

Mach was himself a scientist, a distinguished experimental physicist. But, as will soon be clear, his ideas on this subject were mainly of philosophical origin. Mach's stand on wonder in relation to science is uncompromising. To him there is no doubt that wonder is just a subjective feeling, a manifestation of personal ignorance. Thus science must be the enemy of wonder. In his own words:

Novelty excites wonder in persons whose habits of thought are shaken and disarranged by what they see. But the element of wonder never lies in the phenomenon or event observed; its place is in the person observing. People of more vigorous mental type aim at once at an *adaptation of thought* that will conform to what they have observed. Thus does science eventually become the natural foe of the wonderful. [10]

Why did Mach take this position? The philosophical origin of his conviction becomes clear when we study the arguments he presents to defend his thesis. His mentality is dominated by the empiricist conception of knowledge and the consequent logicomathematical interpretation of science. Not for nothing, in fact, does he tend to reduce science as such to classical mechanics, but he also explains mechanics itself merely in terms of sense impressions. As an example, we can refer to his own lengthy discussion of the mechanical principle of virtual displacements, which he presents as typical of the scientific understanding of nature. He concludes his discussion with a statement that is quite illuminating:

---

[10] E. Mach, "On Transformation and Adaptation of Scientific Thought," in his *Popular Scientific Lectures,* trans. T. J. McCormack (La Salle, Ill.: Open Court, 1943), pp. 214-235, especially p. 224.

PHILLIPS MEMORIAL
LIBRARY
PROVIDENCE COLLEGE

Collecting all that has been presented, we see that there is contained in the principle of virtual displacements simply the recognition of a fact that was instinctively familiar to us long previously, only that we had not apprehended it so precisely and clearly. This fact consists in the circumstance that heavy bodies, of themselves, move only downwards. [11]

This statement is important because it exemplifies Mach's general conviction about the nature of science as such. As he puts it elsewhere in his philosophical writings:

The greatest advances of science have always consisted in some successful formulation, in clear, abstract, and communicable terms, of what was instinctively known long before and of thus making it the permanent possession of humanity. [12]

In brief, science is for Mach simply a reception and organization of sense impressions or, as he put it, "adaptation of thought." But, as such, science must be incompatible with wonder. For, clearly, there can be no room for surprise left once the mind of man has become adapted to some sense impressions. Rather, under this assumption, the opposite must be true. That is, science must not only destroy wonder, but also leave man with a sense of disillusionment. For adaptation amounts ultimately to showing that things are self-evident, practically trivial. This is indeed Mach's doctrine. In his own words:

As a fact, every enlightening progress made in science is accompanied with a certain feeling of disillusionment. We discover that that which appeared wonderful to us is no more wonderful than other things which we know instinctively and regard as self-evident; nay, that the contrary would be much more wonderful; that everywhere the same fact expresses itself. Our puzzle turns out then to be a puzzle no more; it vanishes into nothingness, and takes its place among the shadows of history. [13]

To sum up, Mach's position is clear and forthright, but is certainly not derived directly from his own experience as a scientist. Rather it is a statement stemming from his philosophical interpretation of the science of mechanics.

When we turn to Einstein's position, we are in an entirely different climate. Einstein knew Mach's historical and philosophical work quite well. He even ascribed to it a certain positive influence concerning the origin of his own theory of relativity. [14] And yet, not only did Einstein

---

[11] E. Mach, *The Science of Mechanics: A Critical and Historical Account of Its Development*, trans. T. J. McCormack (La Salle, Ill.: Open Court, 1960), p. 87.

[12] E. Mach, "The Economical Nature of Physical Inquiry," in *Lectures* (note 10), pp. 186-213, especially p. 191.

[13] Mach (note 11), p. 41.

[14] See Einstein's Autobiographical Notes, in P. A. Schilpp, ed., *Albert Einstein, Philosopher-Scientist* (New York: Harper Torchbooks, 1959), p. 21.

not share Mach's view on wonder, but he explicitly maintained the exactly opposite thesis. To Einstein, there can be no doubt that science, far from destroying wonder, actually contributes to it—the more the wonder, the greater the progress of science itself. What was the reason for his conviction? We can safely say that such a reason was his own experience as a scientific creator. Particularly one text, contained in a private letter to a friend, manifests the experiential source of Einstein's conviction clearly.

You consider it strange that I sense the comprehensibility of the world (in so far as we are justified to speak of it) as a wonder (*Wunder*) or an eternal mystery (*ewiges Geheimnis*). Now, a priori one should expect a chaotic world which can in no way be grasped by thought. One could (better, *should*) expect that the world should prove subject to law only in so far as we intervene by putting order ourselves. It would be a type of orderliness like the alphabetic order of the words of a language. On the contrary, the kind of orderliness which results, for instance, from Newton's theory of gravitation is of an entirely different character. Even though the axioms of the theory are set down by man, the success of such an enterprise presupposes a high level of order in the objective world which we have no a priori right to expect. There lies the "Wonder" which increases steadily with the development of our knowledge. Here lies the weak point of the positivists and professional atheists who feel happy in the consciousness that they have not only successfully dedivinized (*entgöttert*) but even dewonderized (*entwundert*) the world. [15]

The lines of the controversy could not be more sharply drawn. Which of the two great antagonists should be seen as the genuine representative of the scientific mind? If we remain consistent with our experiential approach in the study of science and its significance, the answer cannot be doubtful. Einstein is the authentic spokesman of science here. In fact, several reasons support his thesis.

The fundamental reason why successful science increases wonder rather than suppressing it is touched upon in the foregoing quotation. It is the success of science itself. Indeed, the achievement of science is the detection of objective intelligibility in nature. But such detection cannot help being surprising if one reflects on it. In fact, on the one hand, science is an idealized and logical structure due to the effort of man. Its universality and rigor are but a man-made formulation. On the other hand, nature exists in complete independence of man. It is far from ideal and abstractly universal. Rather, it consists of countless individual beings which are subject to continual, unforeseeable change. Thus it would appear that chaos, not orderliness, should be the prevailing feature of nature.

---

[15] Translated from A. Einstein, *Lettres à Maurice Solovine* (Paris: Gauthier-Villars, 1956), p. 114; letter dated March 30, 1952.

This is then the main reason why man should be surprised at the success of science. It is the convergence of man's science and nature's behavior. This convergence is a fact, but is also a question, because its explanation is far from self-evident. Consequently, the more science advances, the more the wonder aroused by it increases rather than disappears. Einstein insisted repeatedly that there is motive for astonishment in this connection. As he put it in a writing discussing the work of Planck:

He [the physicist] is astonished to notice how sublime order emerges from what appeared to be chaos. And this cannot be traced back to the workings of his own mind but is due to a quality that is inherent in the world of perception. [16]

Variations on the reason given can be found in the literature. Planck, for instance, finds a source of surprise in the fact that men doing science are quite different in many ways, while science is just "one." Moreover, the discoveries of science prove to have a scope that far outstrips the expectations of those who first made them. In his own words:

Rightly viewed, the real marvel is that we encounter natural laws at all which are the same for men of all races and nations. This is a fact which is by no means a matter of course. And the subsequent marvel is that for the most part these laws have a scope which could not have been anticipated in advance. Thus the element of the wondrous in the structure of the world picture increases with the discovery of every new law. [17]

Still another reason for surprise at the success of science is indicated by Planck. It has to do with the peculiar way science itself changes over the centuries. In principle, no one should be surprised at the changeability of science. For things do change in time, and men change, too. Hence it is only to be expected that man's knowledge of nature is subject to change. And yet, the changeability of science is surprising in that all the novelties that science discovers do not wipe out the previous scientific worldview, but simply perfect and complement it. This is surprising because it appears to be a unique phenomenon in the history of mankind. To cite Planck again:

But the circumstance which calls for ever greater wonderment, because it is not self-evidently a matter of course by any means, is that the new world picture does not wipe out the old one, but permits it to stand in its entirety, and merely adds a special condition for it.... As the multitude of the natural phenomena observed in all fields unfolds in an ever

[16] A. Einstein, Prologue to M. Planck, *Where Is Science Going?*, trans. J. Murphy (New York: Norton, 1932), p. 11.
[17] Planck (note 3), p. 93.

richer and more variegated profusion, the scientific world picture, which is derived from them, assumes an always clearer and more definite form. The continuing changes in the world picture do not therefore signify an erratic oscillation in a zigzag line, but a progress, an improvement, a completion. [18]

To be sure, the objection against the wonder arising from the achievements of science can be pressed by insisting on the Machian conception of knowledge. If knowledge is pure adaptation of man to his environment as perceived through the senses, the success of science is the most natural thing in the world. For science is merely the product of man's conditioning as exerted upon him by the environment. Hence, as Louis de Broglie sums up the objection,

...it [humanity] must not be astonished to recover in the material world the logic and the rules of reasoning that it has extracted from it. [19]

However, if this objection can be worth examining by the philosopher, it only meets with scorn on the part of the creative scientist as such. For, if there is anything that is not an adaptation of man to what he instinctively perceives through the senses, it is science. Galileo, for one, was already extolling the dignity of the creative mind. Science at the beginning consisted in the ability to realize as true with the mind what appeared absurd to the senses (Chapter 1, Section 3b). But the advances of science into new areas of research have always been of the same kind. This is particularly clear when we consider the attainments of contemporary physics. For the results of relativity concerning the interconnection of space and time, and those of quantum physics concerning the corpuscular and wavelike properties of microparticles, are something that is diametrically opposed to what man naturally perceives through the senses. Thus the wonder of science stands. As de Broglie put it, in the context cited:

We are not sufficiently astonished by the fact that any science may be possible, that is, that our reason should provide us with the means of understanding at least certain aspects of what happens around us in nature.

To close, it is obvious that wonder and science are deeply interconnected. This is so not only in the psychological sense that wonder is the necessary starting point for science. Far more significant is the fact that science, on the strength of his very success, leads the reflective person to a renewed sense of wonder. As we have seen, it is especially the great creator who is affected by the experience of wonder. The reason is clearly the penetrating power of his mind; he cannot help inquiring about the

---

[18] *Ibid.*, p. 98; cf. pp. 98-100.
[19] de Broglie (note 9), p. 209.

ultimate meaning of the effective intelligibility of nature, which he un-covers through his science. This adds an unexpected new dimension to science as an experience of the whole man. We cannot here pursue the issue in its further fascinating developments. Nevertheless, the outcome of our inquiry is heartening for the humanist who respects science. For he realizes that the world, through science, has not been robbed of its wonder. On the contrary, nature and the world are even more wonderful than before—as a consequence of the success of science.

2. The Awesomeness of Nature

The realization of wonder as the crowning of the scientific experience is necessary but not sufficient if one wants to perceive the whole experiential situation of the successful scientific person. For wonder, though related to an object, is mainly a subjective reaction, a state of questioning. Thus me must now continue our discussion of the scientific experience by con-centrating more on the objective side of the issue. That is to say, we must now investigate how the scientist comes to experience nature and reality in general as a result of his success in doing science. To designate this type of experience I employ the term *awe*. Awe is, according to ordinary parlance, the respect and reverence that accompany wonder when man is faced with the sublime. But this seems to be an apt description of the state of mind experienced by the scientist when confronted with the intelligibility of nature. We are going to survey the evidence bearing on this statement.

(a) The Shock of Unexpectedness

We can begin our investigation by examining the objective counterpart of admiration. Admiration, as the term has been employed previously (Chapter 3, Section 1b), is a subjective reaction—the delighted rejoicing that stems from discovery. But discovery has obviously an objective side, too. This we intend to analyze here. Discovery, being the realization of something new, consists in the sudden awareness of an unexpected situa-tion. The experience is such that the researcher all at once feels himself confronted with a manifestation of nature that is, for him, unanticipated and unforeseeable. His normal reaction is one of commotion and even perturbation. To realize this aspect of the scientific experience, let us consider a couple of examples taken from the history of science.

The first example is of an experimental type: it refers to the discovery of the atomic nucleus. Rutherford, with his team of assistants, was study-ing the atom experimentally. To train two of these assistants, Geiger and Marsden, he suggested that they probe the interior of the atom with alpha particles. On the basis of all the evidence available, he did not anticipate any important results from their experiments. The alpha particles were known to be so powerful that they could only be expected to traverse the atom, more or less undeflected. In fact, for a while nothing special was reported and Rutherford felt confirmed in his expectations. But then, all of a sudden, the experimenters told him the great news. They had observed that a sizable proportion of the particles, far from moving straight through the atom, were deflected at large angles: for many the angle of deflection was more than 90° while some were even scattered backward. Rutherford was shaken at the announcement. The startled excitement with which he greeted the news was still vibrating in the words he used to recall the episode 20 years afterward.

It was quite the most incredible event that has ever happened to me in my life. It was almost as incredible as if you fired a 15-inch shell at a piece of tissue paper and it came back and hit you. [20]

It seems justified that we use the term "shock" to characterize the situation described. For Rutherford's own comparison emphasizes the jolting impact of scientific discovery. Shock here is the reaction of the person who is stunned by the discovery he has made. Suddenly, what he had hitherto considered obvious loses its compact solidity. The world, at least for a moment, appears to him as shaken at its foundations. But, of course, if one can speak of shock as produced by the astonishing unexpectedness of nature, one can also speak of awe with relation to nature itself. For the nature that is shocking reveals itself as something that deserves the greatest respect.

The startled respect for nature inspired by science can be seen even more impressively in the case of theoretical discovery, for the experience is now one of a sudden, penetrating insight into the most intimate structure of reality. Man has the impression of glimpsing, with the eyes of his mind, the basic principles that make nature understandable. The obvious consequence is that man feels shaken in his whole person, being intimately overwhelmed by the staggering richness of nature's own intelligibility. To realize this type of reaction, we can consider one autobiographical example given by Heisenberg. He speaks of his preoccupied searching for a theoretical understanding of atomic spectra. After the long struggle follows the great illumination. He found the clue he was looking for in energetic considerations. He felt able to express the intelligibility of atoms in precise mathematical language. At that moment,

---

[20] E. Rutherford, in J. Needham and W. Pagel, eds., *Background of Modern Science* (New York: Macmillan, 1938), p. 68.

he felt simply overwhelmed by the overpowering wealth of intelligibility that nature was displaying under his very eyes. Many years later, he still recalled this experience vividly.

In the first moment I was deeply frightened (*erschrocken*). I had the feeling that, through the surface of atomic phenomena, I was looking at a deeply lying bottom of remarkable internal beauty. I felt almost giddy at the thought that I had now to probe this wealth of mathematical structures that nature down there had spread before me. [21]

The awesomeness of nature experienced by the scientist comes most strikingly to the fore in the testimony just cited. In this connection it is remarkable that Heisenberg expresses himself in terms of fright. The word is so strong that one would almost feel tempted to consider it an emotional exaggeration. Yet Heisenberg is well known for the accurate care with which he chooses his words. In addition, there is no reason to suspect emotionalism in the description of an event when the description was made several decades after the event itself. A lapse of time weakens rather than strengthens the intensity of an emotional reaction. Thus, apparently, the reason Heisenberg speaks of fright here is that, in his own estimation, no other term can adequately convey the upsetting character of his experience. Our interpretation is confirmed by the fact that Heisenberg found this experience so convincing that he used it to persuade even his most redoubtable critic. Replying to the objections of Einstein against his quantum theory (Chapter 1, Section 3d), he appealed to the irresistible power of the experience cited.

You must have experienced it, too—one is almost frightened (*man fast erschrickt*) in front of the simplicity and compactness of the interconnections that nature all of a sudden spreads before him and for which he was not in the least prepared. [22]

(b) The Awareness of Inexhaustibility

The analysis of individual discoveries gives an idea of the attitude of awesome respect instilled in man by nature. However, to understand the general reaction of the scientist toward nature, additional considerations are needed. For discoveries are scattered and passing events in the life of a scientist, whereas nature is a permanent whole. To designate the all-embracing experience with which the scientist comes to perceive the whole of nature one can speak in particular of an awareness of inexhaustibility.

A celebrated example from the life of Newton concretizes this aspect

---

[21] Translated from W. Heisenberg, *Der Teil* (note 1), pp. 89f.
[22] *Ibid.*, p. 99.

of the scientific experience. Newton was not a person who took science lightly or who suffered from exaggerated modesty. His bitter polemics, especially against Hooke and Leibniz, are convincing evidence of his great esteem for science, beginning with his own discoveries. Finally Newton, particularly late in his life, rather inclined to vainglory: he obviously enjoyed the lavish praise heaped on him. Thus when he, in his old age, summed up the significance of his life endeavor, one would have expected some utterance of high praise for science. On the contrary, as is well known, Newton uttered the startling words:

I do not know what I may appear to the world; but to myself I seem to have been only like a boy, playing on the sea shore, and diverting myself, in now and then finding a smoother pebble or a prettier shell than ordinary, whilst the great ocean of truth lay all undiscovered before me. [23]

This famous statement has been interpreted in various widely different ways. To understand it properly, we must situate it in the entire framework of Newton's thought. In particular, we must take into account the mentality of the scientific creator as such. To this end, we may begin by discounting falsity or pretension on Newton's part. The person who recorded the words cited, John Conduitt, was a would-be biographer. It is hardly likely that Newton decided to risk appearing a failure to posterity just for the sake of fake modesty. For, clearly, his words can easily be construed as an expression of dejected despondency. What was then the reason for Newton's statement? I think that it was essentially an experiential awareness, though perhaps somewhat tinged by polemical undertones. The experiential motivation was paramount. Newton did not mean to say that his lifelong scientific endeavor had been a failure. Just the opposite. By means of his comparison he wanted to express the rewarding experience he had had by doing science. Indeed, the little boy who plays on the shore of the ocean of truth is having a quite pleasant time. For it is obviously gratifying for him to realize the immensity of the truth that lies there, in front of him, even though much of this truth remains as yet undiscovered. However, if such an experience is rewarding, it is so not in a narrow, man-centered sense. For the gratification stems from the consciousness of being in touch with something that is unbounded and awesomely majestic. Thus, the feeling that Newton wanted to convey was the profound and quiet joy that prevailed in his mind as a result of having perceived the boundless cognoscibility of nature.

Concerning a possible polemical motivation of Newton's words, it suffices to recall his lifelong combat against the Cartesian mentality, which

---

[23] Quoted in F. E. Manuel, *A Portrait of Isaac Newton* (Cambridge, Mass.: Harvard University Press, 1968), p. 388.

was then dominant in philosophical as well as scientific circles. According to Descartes, knowledge could be considered satisfactory only if man grasped with his mind the essence of the object he was studying. But Newton, precisely because of his experience as a knower, could not help rejecting vehemently such a view. For Descartes' idea was reducing knowledge to an intellectual-rationalistic process, of the aprioristic-deductive type. The mind of man was supposed to possess, through a single intuition, all the intelligibility of things. As a consequence, the scientific investigation of nature had to be nothing but a logical deduction from a priori self-evident notions. Newton had learned from his own work that knowledge was of a totally different kind. It was not an intuition, but rather an awareness gradually attained through the engagement of man's entire personality. Above all, knowledge was never the mental possession of the total intelligibility of nature. For man, through the increase of his own knowledge, was becoming more and more aware of the inexhaustible intelligibility of reality. Thus, in sum, we can realize why, in Newton's thought, the simile of the little boy was significant. It was so, centrally, because it embodied accurately his own experience as a knower. Additionally, the simile was important because it exposed the arrogance of the rationalistic mind, which Newton found quite distasteful.

Newton's attitude, as described, is far from unique among scientists. Great creators, when questioned about the importance of their researches, tend to reply that they know very little about the area of nature they are investigating. They often even go so far as to state that their ignorance grows apace with the success of their researches. Can this reaction be dismissed as mere pretense or, if genuine, as being due to pessimistic cynicism? No, because many defend it as an authentic achievement of the human mind. In this connection, one convincing example is offered by Galileo in his most enthusiastic book about the importance of science, his celebrated *Dialogue*. There he fights all along the presumption of contemporary philosophers who claimed they knew enough about nature. To him, there is no better proof of ignorance than a self-complacent assertion of knowing. In a climactic passage Sagredo, the scientifically enlightened layman, scoffs at the arrogance of those who

... want to make human abilities the measure of what nature can do. ... This vain presumption of understanding everything can have no other basis than never having understood anything. For anyone who has experienced just once the perfect understanding of one single thing, and has truly tasted what knowledge consists of, would recognize that of the infinity of other truths he understands nothing. [24]

[24] G. Galilei, *Dialogue Concerning the Two Chief World Systems, Ptolemaic and Copernican*, trans. S. Drake (Berkeley: University of California Press, 1962), p. 101. I have corrected the translation somewhat to make it closer to the original Italian.

To this outburst Salviati, the personification of the creative scientist, replies by stressing the teaching of experience and of history itself. He points to the shining example of Socrates as the best evidence that confession of ignorance is the profoundest manifestation of knowledge.

Your argument is quite conclusive; in confirmation of it we have the evidence of those who do understand or have understood some thing; the more such men have known, the more they have recognized and freely confessed their little knowledge. And the wisest of the Greeks, so adjudged by the oracle, said openly that he recognized that he knew nothing.

On the basis of the foregoing analysis, we can understand the scientific awareness of inexhaustibility. It is not by itself a negative reaction. It is, rather, a deeper experience of reality. The working scientist is more aware than anyone else of the limitations of human conceptual knowledge when measured against the richness of the reality he is exploring. For conceptual knowledge can only contain a tiny aspect of reality—that is, what can be expressed by means of clear and distinct ideas. But reality infinitely exceeds what man can say about it. Thus human knowledge can never be absolute or ultimate. In particular, scientific knowledge can never exhaust the intelligibility of even the tiniest components of nature. A striking justification of this attitude is, for instance, the discovery of the so-called wavelike properties of atomic particles. Up to the mid-1920s it appeared self-evident that atomic particles possessed only corpuscular features. Experiment shattered that illusion. [25]

Awareness of inexhaustibility seems to be a typical manifestation of the creative scientist. What is humanistically remarkable is the occurrence of a similar experience in other fields of human creativity. Great creators are normally conscious of the excellence of their gifts as well as of their own achievements. However, they usually emphasize also the limitations of their creations. It is as though they perceive an impassable gulf between even the most perfect of their works and the reality they are trying to express. To realize this phenomenon, we may consider two examples. Michelangelo is undisputedly great as a sculptor. He himself had great esteem for his work. In one of his poems he claims to have been born "to sculpt divine things" (*a scolpir cose divine*). And yet he seemed to despise his own productions. Approximately at the same time he was writing the poem quoted, he used a scornful word to designate the countless blocks of marble into which he had breathed immortal life. He called

---

[25] For details about the significance of such surprising properties see, for instance, E. Cantore, *Atomic Order: An Introduction to the Philosophy of Microphysics* (Cambridge, Mass.: MIT Press, 1969), pp. 116-132.

them "big plump dolls" (*bambocci*). [26]   Another striking example of the same attitude of genius is offered by the life of Thomas Aquinas.   This master theologian, at the height of his creative powers, was engaged in writing the third and final part of his masterpiece, the *Summa Theologica*. All of a sudden, he laid down his pen determinedly, never to take it up again.   His assistant, Reginald of Priverno, inquired whether he had been taken ill.   No, Thomas was feeling well.   But then, the other scolded him, how did he dare to disappoint the expectation of so many people who counted on his intellectual leadership?   Thomas was adamant.   "The end of my writing has definitely come" (*venit finis scripturae meae*). Pressed for a reason, Thomas could only say that whatever he had written so far appeared to him "like a bunch of straw" (*tamquam paleas*). [27]

To sum up, the experience of inexhaustibility seems to be a distinctive mark of the great human creator as such.   It takes greatness to be creative. But it takes even more greatness to acknowledge the limitations of one's own achievements while rejoicing at the surpassing greatness of reality. Science is humanistically significant in that it enables man to attain such greatness.

(c) The Perception of Mystery

Awareness of inexhaustibility is one of the highest points of the scientific experience.   However, to capture this experience in its fullness, we must press our investigation further.   For the very term of inexhaustibility is a negative one.   But, of course, what makes reality inexhaustible to man is something positive rather than negative in itself.   To designate the ultimate manifestation of the scientific experience as documented in the scientists' writings we can speak of perception of mystery.

The term mystery may appear out of place and even shocking when employed in connection with science.   For science to many seems to be the direct antithesis of mystery.   The motto of Marcelin Berthelot, the celebrated experimental chemist, still rings convincing to many ears: "The world is today without mysteries." [28]   Nevertheless, mention of mystery in connection with science is by no means an intrusion due to philosophical bias.   It is rather a consistent characterization of the scientific experience realized in depth.   Einstein, for one, made this point strongly.   In a radio

---

[26] For details see G. Papini, *Vita di Michelangelo nella Vita del Suo Tempo* (Milan: Mondadori, 1964), pp. 605f.

[27] We owe these particulars to Reginald of Priverno himself as contained in his Memoirs. They are incorporated in any complete biography of Aquinas.

[28] The dictum was uttered in 1884.   See quotation and discussion in G. Sarton, Introduction to reprint of J. Needham, ed. *Science, Religion and Reality* (New York: Braziller, 1955), p. 8.

broadcast significantly entitled "My Confession of Faith" (*mein Glaubens-bekenntnis*), he emphasized the experiential importance of mystery.

The most beautiful and most profound experience that man can have is the sense of the mysterious. This constitutes the foundation of religion and of all other profound striving in art and science. He who has not experienced it seems to me—if not dead—at least blind. [29]

If the term "mystery" is deemed acceptable in connection with science, how should we define the term itself? For mystery can be interpreted in various ways. Etymologically, that is mysterious which is hidden from the sight of the multitude. But, then, the word mystery is frequently used to designate what is simply unknown—especially if it is difficult for man to attain an adequate knowledge of the thing in question. Thus, for instance, ordinary people speak of the mysteries of electricity, meaning thereby that electricity is something difficult to know, but certainly not unknowable to man. In line with tradition, we should take the term strictly. *Mystery* is something that is not only unknown, but unknowable. That is, mysterious applies to something whose existence can be realized but whose nature cannot be fully understood, adequately penetrated, and exhausted by the human mind. Mysteriousness, therefore, is predicated on a reality—whatever it may be—that is truly there and positively exists, but is such that it exceeds the capability of man's conceptual knowledge. Hence the common practice of talking of mysteriousness by using such synonyms as ineffability, unutterableness, incomprehensibility, and the like.

As shown by Einstein in the text cited, the fundamental reason for speaking of mystery in connection with science is experiential. We can begin therefore by analyzing the experiential structure of mystery in general. Then we shall see whether the idea of mystery truly applies to the experience of the scientist. A good description of mystery as experientially perceived is contained in a passage by Rudolf Otto, the well-known philosopher and religious anthropologist:

The truly "mysterious" object is beyond our apprehension and comprehension, not only because our knowledge has certain irremovable limits, but because in it we come upon something inherently "wholly other," whose kind and character are incommensurable with our own.... [30]

Otto's description is illuminating because it removes the apparent contradiction that seems inherent in the definition of mystery. We have defined mystery as unknowable. How can one become aware of something

---

[29] Translated from text in F. Herneck, "Albert Einsteins gesprochenes Glaubensbekenntnis," *Die Naturwissenschaften* 53 (1966), 198.

[30] R. Otto, *The Idea of the Holy*, trans. J. W. Harvey (Baltimore: Penguin, 1959), p. 42.

that is unknowable?   Because of the experiential nature of knowledge.
Otto rightly describes man's awareness of the mysterious as something
"we come upon."   Awareness of mystery, therefore, is genuine knowledge,
not pure nonknowledge or ignorance.   For, clearly, if we are conscious of
coming upon something, we possess some knowledge about that something.
We know at least that that something is really there, really exists.
However, the reality that is mysterious gives rise to a special kind of
knowledge.   For, no sooner than we realize its presence, we become aware
of our inability to grasp it fully with our minds.   Otto rightly says that
the "kind and character [of the mysterious] are incommensurable with
our own."   As a consequence, the cognitive peculiarity that characterizes
the realization of mystery stands out clearly.   It is a form of awareness
that is as removed as possible from common ignorance.   And yet it is
also a consciousness of the inherent limitations of man's conceptual knowledge.   This situation can be explained by means of an analogy, in terms
of light and darkness.   Common ignorance is like the ordinary awareness
of darkness.   Nothing is seen because there is no light to see anything.
Perception of mystery, on the contrary, is a totally different awareness of
darkness.   It is like the darkness that is felt when one is overwhelmed
by dazzling light.   Perception of mystery, therefore, is the highest form
of knowledge attainable to man.   For, through it, man becomes experientially aware of something that so much exceeds the power of the human
mind that this cannot comprehend or encompass it.   And yet, man has
the joy of feeling inundated by the dazzling light of this exceeding intelligibility.

     Does mysteriousness, as analyzed, apply to the scientific experience
of reality?   As against the widespread assumption that mystery and
science are incompatible, numerous creative scientists declare that the
highest point of science is precisely the awareness of mystery.   We have
already read a passage from Einstein's " confession of faith."   In the
same text he goes on to state that the perception of mystery is the highest
point to be attained by the human mind.   He maintains that this is the
essence of religion, and the very essence of science.

To perceive that, behind what can be experienced, something is hidden
which is unattainable for our spirit—something whose beauty and sublimity reach us only indirectly and by way of a pale reflection—*this*
is religiousness. In *this* sense I am religious. It is enough for me to
sense these mysteries with astonishment and to attempt, in humility, to
formulate with my mind a scanty representation of the sublime structure
of reality. [31]

     Einstein is not alone in upholding mysteriousness in relation to science.

[31] A. Einstein (note 29).

Another outstanding physicist, Max Planck, emphasizes that the rewarding experience of science is to know the explorable and venerate the inexplorable. In his own words:

And he, whom good fortune has permitted to cooperate in the erection of the edifice of exact science, will find his satisfaction and inner happiness, with our great poet Goethe, in the knowledge that he has explored the explorable and quietly venerates the inexplorable. [32]

Similar assertions are made by contemporary leading biologists. Simpson, for instance, stresses the beauty that science reveals to man and the sense of awe which results from this experience:

No poet or seer has ever contemplated wonders as deep as those revealed to the scientist. Few can be so dull as not to react to our *material* knowledge of this world with a sense of awe that merits designation as religious. [33]

Then, quoting and commenting on an expression of Joseph Needham, the famous Cambridge biologist, he insists that the marvelous discoveries of contemporary biology lead to a recognition of mysteriousness that is forever beyond the possibility of scientific investigation.

...all of us want... to ask why living beings should exist and should act as they do. Clearly the scientific method can tell us nothing about that. They are what they are because the properties of force and matter are what they are, and at that point scientific thought has to hand the problem over to philosophical and religious thought. There lie the ultimate mysteries, the ones that science will never solve. [34]

We can conclude our investigation of the perception of mystery as experienced by the creative scientific mind. It is an experience of the ultimate. Nature reveals itself to man as something that man himself will never be able to fully penetrate with his mind nor to express in a conceptual formulation. This mysteriousness is something that the scientist feels bound to acknowledge as truly real, and as a rewarding attainment for the mind that succeeds in realizing its existence. In other words, mysteriousness as experienced by the scientist is not something negative —a dead end, an impenetrable darkness. It is rather something eminently positive, as overwhelming as dazzling light. For the scientist knows that, when he speaks of mystery in connection with nature, he speaks of something that can be personally experienced as being really there. The experience, however, is a direct contact—one that can hardly be described

---

[32] Planck, *Autobiography* (note 3), p. 120.
[33] G. G. Simpson, *This View of Life: The World of an Evolutionist* (New York: Harcourt, 1964), p. 233.
[34] The essay of Needham referred to is contained in J. Needham, ed. (note 28).

in words. We have, therefore, in the perception of mystery, the highest form of awesome respect with which nature inspires the mind of the person who sets out to explore it.

3. Science as an Experience of the Absolute

We have already gone far in our analysis of the scientific experience. Should we develop further this sensitive and delicate investigation? In a way it would seem that we cannot discover anything worthwhile beyond what we have already established. For mystery is a term that designates something unexceedable. Thus one could simply conclude with Einstein, when speaking of the wonder and eternal mystery generated by science: "The point is that we are forced to be satisfied with the recognition of the 'Wonder' without finding a legitimate way out of it." [35] And yet we cannot simply abandon our investigation at this point. The reason is that, as we have already had the occasion to notice, scientists themselves occasionally inject a further theme in their discussion of scientific experience. They often connect their experience of science to that of religion. But religion is a highly complex subject and a controversial one. Thus, to do justice to the profound significance of science for the whole man, it is obviously necessary that we explore in some detail the scientific experience as bearing on religion or the perception of the absolute.

There is no doubt, of course, that a discussion of the relationships between science and religion is a very difficult matter. For religion and science have been so constantly at odds that they appear to be best left alone in order to avoid painful misunderstandings. [36] Nevertheless, as mentioned, this refusal to face the issue would not be consistent on our part, since the scientists themselves are the first to bring up the subject of religion when discussing their experience in depth. Thus, out of respect for science and concern for man, we cannot honestly avoid tackling this issue. Nor, on the other hand, should we fear that our research might lead to needlessly painful controversies. For the great scientists themselves, as usual, will be our guides in our experiential investigation. Thus there can be good reason for expecting that our effort may be useful in the end to both science and religion instead of contributing to a further split between the two.

---

[35] Conclusion to the text quoted above (pertaining to note 15). Einstein's letter to Solovine, dated March 30, 1952.

[36] For an extensive, though superficial and much dated survey of this topic see A. D. White, *A History of the Warfare of Science with Theology in Christendom* (1896) (New York: Dover, 1960), 2 vols.

To attain the aim outlined, I shall adhere closely to the experiential attitude which has guided our investigation so far. Thus I shall avoid discussing any theoretical question, and concentrate my attention on what scientists say, and their reasons for saying it. To develop this theme, I shall begin by analyzing the testimony of some contemporary scientists which present an unmistakable connection with what is ordinarily called religion or experience of the absolute. Then I shall discuss synthetically the characteristic traits of the religious experience of nature as emerging from the investigations of scientific anthropology. Finally I shall examine directly the sense, if any, in which one can speak of an intrinsic connection between science and religion.

(a) A Personal Encounter

If there can be any justification for speaking of religion in connection with science as a lived experience, this can only be because science ultimately leads man to a personal encounter with the absolute or the ultimate source of reality. Is one entitled to employ such a terminology when referring to science? The controversial character of our issue comes here quite clearly to the fore. We shall try to avoid the difficulty by adhering closely to our experiential guideline. What is in actual fact the content of the scientific experience as bearing on the question of the absolute? This is the only question we want to explore here.

Some forthright and enlightening testimony concerning our subject is offered by Heisenberg when narrating a dialogue of his with Pauli. The two had been lifelong friends and had a great respect for each other's sharpness of mind and outspoken sincerity. The episode took place in early 1952. The two friends were Bohr's guests in Copenhagen. As usual with Bohr, they had spent the best part of the day discussing philosophical problems raised by quantum physics. The burden of the conversation had been a sympathetic, but unmistakably critical, analysis of the positivist position concerning the search for truth and meaning in science. Finally, in the evening, Heisenberg and Pauli decided to take a walk toward the harbor while continuing their discussion along the way. The general topic of their conversation was still the same of the day: the intrinsic intelligibility of nature and its implications. After a long walk and much talk, they paused in silence at the end of the harbor jetty. Then, quite unexpectedly, Pauli posed the direct question: "Do you believe in a personal God?" Heisenberg, taken by surprise, decided to answer the question carefully. Thus, instead of replying directly, he reformulated the question by picking his words with care. The following are his words, and Pauli's comments, in alternation.

Can you or can man move so close, come into such an immediate contact with the central orderliness of things or of events—about which
[orderliness] no doubt is possible—as one can do with the soul of another
person? I employ here expressly the difficult to explain term "soul" in
order not to be misunderstood. If you ask this way, then I would
answer "Yes..."

You mean that the central orderliness can be present with the same
intensity as the soul of another person?

Possibly.

Why have you employed here the term "soul" and not spoken simply
of another person?

Because the word "soul"—precisely here the central orderliness—designates the core of a being which in its outward manifestations can be
quite manifold and unsurveyable.

I do not know whether I can go along with you. One should not overrate one's own experiences.

Certainly not. But, then, in science too one appeals to one's own experiences or to those of other people which are reliably reported. [37]

This episode is particularly significant for our effort to understand
the scientific experience of the absolute. For it expresses a perception
whose personally lived sincerity, clarity and sense of measure are remarkable. As such, it may be seen as a characteristic manifestation of
the scientific mind pursuing to the end its search for knowledge and
meaning. In particular, we can notice here several illuminating features.
The first and last of these features is the experiential character of the
perception. There is nothing of reasoning here, but only an experiential
awareness. This awareness is the source of certitude, just as the source
of certitude in science is experiential. The second feature is the gradual-
ness of the perception. One becomes aware of the soul or inner personality
of a human being only little by little, through a continual effort of respectful
attention and persistent interiorizing reflection. In the same way one can
become aware of God through science only by way of a patient and persistent striving. The third feature is the personlike characteristics of the
absolute or God as perceived by the reflective scientist. What we designate
with the term person in ordinary parlance is a being defined by intelligence and creative power. Thus God, as presented here, is a personlike
being—the creative source from which the observable orderliness of things
and events emerges in all its manifoldness and unsurveyable richness. The
fourth feature is the mysteriousness of God, which is implicit in this

---

[37] Translated from Heisenberg, *Der Teil* (note 1), p. 293.

presentation. The soul or inner personality of man is unavoidably mysterious to any observer. The reason is that man is inwardly creative. Hence the manifestations of his personality can never be anticipated, much less be expressed in conceptual language. But if the soul of man is mysterious because of its creativity, the ultimate source of reality is infinitely more mysterious, since it is infinitely more creative.

Heisenberg's testimony stands out in its profundity and precision. Can it be taken as genuinely representative of the scientific mind reaching its most profound insight into the meaning of reality? Evidence from scientists' writings seems to indicate that this is the case. A famous text in the vein cited is the passage that concludes Darwin's *Origin of Species*. From Darwin's autobiographical writings we can easily relive the experience to which he refers. It is the experience that imprinted itself indelibly on his mind when first confronted with the luxuriant flora and fauna of Brazilian forests. He was struck by the richness and interconnectedness of the scene observed, which manifested a harmony of intelligible law surpassing anything man could express in conceptual thought. Darwin could not help perceiving in this overwhelming outpouring of nature a manifestation of the Creator. Thus, although he was later plagued by religious doubts, he could never bring himself to suppress his original experience and its corresponding expression in writing. We still read it, in the last of the many, painstakingly reworded editions of his immortal work.

There is a grandeur in this view of life, with its several powers, having been originally breathed by the Creator into a few forms or into one; and that, whilst this planet has gone cycling on according to the fixed law of gravity, from so simple a beginning endless forms most beautiful and most wonderful have been, and are being evolved. [38]

Other important testimony on science, as leading to a personal encounter with the absolute, is offered by Einstein. He insists particularly on the overpowering impression produced on the mind of the researcher by the boundlessness of the objective intelligibility of nature. In his conviction, this gives rise to a sense of religious reverence:

In every true searcher of nature there is a kind of religious reverence; for he finds it impossible to imagine that he is the first to have thought out the exceedingly delicate threads that connect his perceptions. The aspect of knowledge which has not yet been laid bare gives the inves-

---

[38] C. Darwin, *The Origin of Species* (New York: Collier, 1962), pp. 484f. G. G. Simpson notes in the Foreword, p. 7: "Darwin was a painstaking, even a plodding writer, and in revision he constantly reworded and rewrote."

tigator a feeling akin to that experienced by a child who seeks to grasp the masterly way in which elders manipulate things. [39]

Thus Einstein speaks of a superior mind manifesting itself in the world discovered by the scientist.

This firm belief, a belief bound up with deep feeling, in a superior mind that reveals itself in the world of experience, represents my conception of God. [40]

To conclude, there appears to be no doubt that science can lead its most profound and consistent practitioners to an experience that deserves to be termed a personal encounter with the absolute. Of course, it is also possible for the scientist not to reach this experience. For the perception of God in scientifically known nature is the final flowering of a long search, the result of much patience and consistent engagement in response to the intelligibility of reality. What is in question here, in fact, is something that goes well beyond the ordinary scientific understanding of nature. It is a perception that the scientific mind can attain only through much reflective interiorization. It should not be surprising, therefore, that relatively few scientists attain it, at least explicitly. Nevertheless, there is no reason for denying the representativeness of the testimony of these few. For, after all, what makes the final difference in science is the intimate experience of truth—even apart from experimental confirmation (Chapter 1, Section 3d). But this applies here. As Schrödinger put it with striking succinctness: "We know, when God is experienced, this is an event as real as an immediate sense perception or as one's own personality." [41]

(b) The Religious Experience of Nature

To develop our study of science as an experience of the whole man, we must now digress. From the preceding evidence it is obvious that the scientific search for knowledge, if consistently followed through, bears some sort of connection with religion. Hence to investigate the humanistic significance of science in full depth we must now examine religion. To avoid the dangers of subjectivism and emotionalism we shall refer only to the data of scientific anthropology. This is then our topic here. We start from the assumption that religion is a phenomenon af-

---

[39] Quoted in A. Moszokowski, *Einstein the Searcher: His Work Explained from Dialogues with Einstein,* trans. H. L. Brose (New York: Dutton), p. 46.

[40] A. Einstein, *Ideas and Opinions,* trans. S. Bargmann (New York: Crown, 1954), p. 262.

[41] E. Schrödinger, *What is Life? The Physical Aspect of the Living Cell* and *Mind and Matter* (London: Cambridge University Press, 1967), p. 150.

fecting the whole being of the person involved. "For religious life . . . affects the whole of man's life, and it would be quite unreal to try to divide the mind into separate compartments." [42] On this holistic basis we examine how a religious person perceives nature. As a consequence we will be able to realize more precisely the connection between science and religion.

The most obvious manifestation of religion in relation to nature stems from the *objective surprisingness* of nature itself. Man tends to ignore religion or a personal relationship with the absolute as long as the universe in which he lives appears to him such that it can be taken for granted. But when he is jolted out of this naturalness of things, man begins to speak of religion. The anthropological evidence is clear and abundant. When, for instance, the storm lashes the waters of a lake or a sea, people speak of God walking on the waves. Similarly, the roar of thunder is called the voice of God, the lightning is God coming down to earth in anger, the earthquake is caused by his mighty footsteps. Everything that is exceptionally beautiful or powerful or useful or frightening is considered as a manifestation of the divinity. To sum up with a leading anthropologist of religion:

Everything unusual, unique, new, perfect or monstrous at once becomes imbued with magico-religious powers and an object of veneration or fear according to the circumstances (for the sacred usually produces this double reaction). [43]

The surprising character of reality, however, constitutes but the threshold or psychological gate of the religious experience of nature. Other more profound and all-inclusive features make up the typical content of such an experience. One of these basic features is the *comprehensive meaningfulness* of nature. To religious man the ultimate reason why the world is surprising and marvelous is the experiential realization that the world itself is not a collection of unrelated events, but rather that it presents an overall structure. The world makes sense; it carries and discloses a message through the very fact of its existence. That is, the world is a fact, but not merely a fact. For, in addition to its being there, the world also manifests a universal meaning. The religious person perceives nature in the light of this meaning. As Eliade synthesizes the situation:

. . . the world exists, it is there, and it has a structure; it is not a chaos but a cosmos, hence it presents itself as creation, as work of the gods. This divine work always preserves its quality of transparency, that is,

---

[42] M. Eliade, *Patterns in Comparative Religion*, trans. R. Sheed (New York: Meridian, 1963), p. 126.
[43] *Ibid.*, p. 13; cf. pp. 12-14 and 26-30.

it spontaneously reveals the many aspects of the sacred.... The cosmic rhythms manifest order, harmony, permanence, fecundity. The cosmos as a whole is an organism at once *real, living,* and *sacred*; it simultaneously reveals the modalities of being and of sacrality. Ontophany and hierophany meet. [44]

Another feature of the religious experience of nature complements the foregoing one. It is a sense of *powerful, all-encompassing dynamism.* The world that manifests itself as meaningful to religious man is not something static, like a fixed structure. It is rather a dynamism, which involves the whole of reality in its mighty process. Man can accept or reject this power, but he cannot deny its real presence. This power is then the source of fright for the rebel, while it confers a sense of security to the person who decides to cooperate voluntarily with it. For one essential aspect of the religious experience is precisely the awareness that man is a dependent, though autonomous, being. He may choose to behave as though he would be absolutely independent, but he knows in his heart that this defiant posture would in the end drain his vitality out of him. Thus religious man—particularly the primitive who has not yet been exposed to the allurements of sophisticated rationalization—tends to live as much as possible in contact with the divine as embodied in sacred objects or symbolic rituals. In summary:

The tendency is perfectly understandable because, for the primitives as for the man of all pre-modern societies, the *sacred* is equivalent to a *power,* and, in the last analysis, to *reality.* The sacred is saturated with *being.* Sacred power means reality and at the same time enduringness and efficacity. The polarity sacred-profane is often expressed as an opposition between *real* and *unreal* or pseudoreal. [45]

If religious man perceives the world as manifestation of meaning and power, it should not be inferred from this that he thinks of the world itself as the embodiment of the divinity. For the religious mind experiences the divinity as *transcending surpassingness.* That is to say, religious man perceives the meaning and power he can attain in the world as just hints or derived expressions of an ultimate source whose intrinsic meaning and power exceed his own ability to comprehend. This aspect of the religious experience is brought to the fore by the doctrine of the sky god. Practically all cultures in which the religious phenomenon has not yet been overlaid by philosophical rationalizations honor the sky god as supreme. Superficial interpretations in the past have tended to dismiss the phenomenon as nothing but a manifestation of naturism. That is, the

---

[44] M. Eliade, *The Sacred and the Profane: The Nature of Religion,* trans. W. R. Trask (New York: Harper Torchbooks, 1961), pp. 116f.

[45] *Ibid.,* pp. 12f.

sky god was considered as being just the material sky, which had been deified by the primitive mind.   This interpretation does not stand up to the facts.   For anthropological data show that religious people speak of the sky god not because they think that the sky is divine but because they perceive the sky as a specially striking manifestation of the transcendence of divinity.   To conclude with Eliade:

There is no question of naturism here.   The celestial god is not identified with the sky, for he is the same god who, creating the entire cosmos, created the sky too.   This is why he is called Creator, All-Powerful, Lord, Chief, Father, and the like.   The celestial god is a person, not a uranian epiphany.   But he lives in the sky and is manifested in meteorological phenomena—thunder, lightning, storm, meteors, and so on. [46]

As a consequence of the preceding, nature presents itself to religious man as a means toward a *personal communion* with the divinity.   To be sure, one should not misinterpret the situation by thinking that the religious person, as such, despises or at least ignores nature.   Nothing of the kind is true.   Religious man does not despise nature, because nature is for him a manifestation of the divinity.   Nor does he ignore nature.   For he is aware of the world around him, of its energies and problems, no less than is any other person.   Thus he energetically bestirs himself in study and work to attain practical aims. [47]   Nevertheless, it is true that nature is meaningful to religious man mainly because of what it stands for.   That is, the ultimate significance of nature for the religious person is its ability to establish a direct, personal relationship with God himself.   This person-to-person meeting, this direct contact, is what exceeds everything and anything in the experience of the religious person.   It is what constitutes the core of the religious experience as such.   For God is now perceived as present in all his transcendent greatness as well as supreme desirability. The traditional term to designate the divinity as experienced directly is that of holiness.   Rudolf Otto gives a beautiful synthesis of this experience:

The "holy" will then be recognized as that which commands our respect, as that whose real value is to be acknowledged inwardly.   It is not that the awe of holiness is itself simply "fear" in face of what is absolutely overpowering, before which there is no alternative to blind, awe-struck obedience.   *"Tu solus sanctus"* is rather a paean of *praise*, which, so far from being merely a faltering confession of the divine supremacy, recognizes and extols a value, precious beyond all conceiving. [48]

---

[46] *Ibid.*, p. 121; see also *Patterns* (note 42), pp. 38-41.

[47] This point is made particularly strongly by B. Malinowski against the philosophical speculations of L. Lévy-Bruhl.   Malinowski's views result from his fieldwork in the Trobriand Archipelago.   See his *Magic, Science and Religion and Other Essays* (New York: Doubleday Anchor Books, 1954), especially pp. 25-36.

[48] Otto (note 30), pp. 66f.

The final feature of the religious experience of nature is an aware-ness of *individual responsibility,* in the ethical sense of the term. To be sure, religion is not ethics, as stressed for instance by Otto in the context cited. "A profoundly humble and heartfelt recognition of the 'holy' may occur in particular experiences without being always or definitely charged or infused with the sense of moral demands." Nevertheless, normally, a sense of ethical obligation must follow from the religious experience itself. The reason is the very communion with the divinity that constitutes the essence of religion. Through his communion with God, man becomes aware of the divine plan concerning the world. He realizes that the dynamism of nature has a definite purpose. He also realizes that he him-self is inescapably involved in such a dynamism. Since he depends on nature for his life and development, he cannot avoid influencing the course of nature, for good or evil. Hence man becomes aware of his own responsi-bility concerning nature. Responsibility is to be taken here in the strict acceptation of the word. Religious man feels obliged to give an account of himself, to stand scrutiny because of his actions.

To sum up, a word may suffice to characterize the typical way religious man experiences nature. Nature, as observable through the senses and understandable through the mind, is the same for the religious and the nonreligious person. The distinguishing factor is but a matter of per-spective or additional dimension. To the nonreligious person, nature is itself the ultimate or self-evident reality. To the religious person, nature is the carrier of a message, the manifestation of a presence that exceeds nature itself. That is, the religious person perceives nature in a perspective, notices a dimension in its meaning that remains hidden to the nonreligious man.

(c) *The Humanistic Connection Between Science and Religion*

Is it possible to speak of an intrinsic connection between science and religion? Our foregoing discussion seems to indicate that this is the case. We shall complete here our experiential analysis of the scientific endeavor by taking into account some basic data supplied by the history and psy-chology of science.

(ca) *The Historical Contribution of Religion to Science*

Despite the persistent uneasiness produced by the Galilean tragedy, the historical contribution of religion—specifically, Christianity—to science has long been acknowledged by historians. One of the most penetrating discussions of the situation that made possible the rise of science is by Whitehead. He inquires why science arose only in the Christian West, which, after all, was not outstanding for either mathematics or technology or philosophy when compared with other great cultures such as the ancient Greek and the Chinese. The answer, he contends, is to be found in the influence of Christianity in molding the Western mind. The Judeo-Christian tradition did not act alone, to be sure, for it became itself compounded with the philosophical and mathematical heritage of the Greeks. However, the contribution of religion proved decisive in one central point. Whitehead stresses this point as consisting in a preliminary conviction without which science could not have arisen at all.

I mean the inexpugnable belief that every detailed occurrence can be correlated with its antecedents in a perfectly definite manner, exemplifying general principles. Without this belief the incredible labors of scientists would be without hope. It is this instinctive conviction, vividly poised before the imagination, which is the motive power of research—that there is a secret, a secret which can be unveiled. [49]

Such a preliminary conviction, although universally shared by our civilization, is truly surprising. Indeed, why should man take for granted that there is a detectable order in the world of observation, an order that penetrates every detail, when everyday experience tends to convince man of exactly the opposite? In particular, how could this attitude of mind have become generally accepted and thus gradually bring about the birth of the scientific mind? It seems clear that no other factor was at work except the intellectual influence of Christianity. In brief:

It must come from the medieval insistence on the rationality of God, conceived as with the personal energy of Jehovah and with the rationality of a Greek philosopher. Every detail was supervised and ordered: the search into nature could only result in the vindication of the faith in rationality. Remember that I am not talking of the explicit beliefs of a few individuals. What I mean is the impress on the European mind arising from the unquestioned faith of centuries. By this I mean the instinctive tone of thought and not a mere creed of words.

The historical contribution of religion to science was not just a matter of intellectual conviction. In addition, the religious motivation proper inspired the great pioneers of modern science. A particularly striking case is that of Kepler, whose whole motivation has been summed up by his greatest biographer as being essentially religious.

As the bird is created to sing, so—according to his [Kepler's] conviction—is man created for his pleasure both in contemplating the magnificence of nature and in inquiring into her secrets, not for the purpose

---

[49] A. N. Whitehead, *Science and the Modern World* (New York: Mentor Books, 1948), p. 13; cf. pp. 12-17.

of extracting practical uses but rather to arrive at a deeper knowledge of the Creator. [50]

In point of fact, Kepler's outlook was so thoroughly religious that he did not blush to consider himself and all astronomers as priests "... we astronomers are priests of the highest God in regard to the book of nature...." [51]   But Kepler's case was not unique. The early history of the Royal Society, for instance, bears evidence of such a religious motivation. Science was conceived by the members as a chief means for glorifying God and serving one's fellow man. [52]

A discussion of the way Newton achieved his immortal attainment serves to focalize the crucial contribution of religion to the success of science. Newton is rightfully celebrated for his discovery of universal gravitation. And yet it is historically clear that Newton could never have succeeded in his immense endeavor had he not been sustained by a religious conviction, which as such was totally foreign to science. Newton, in fact, manifested a disconcerting ambivalence concerning the mechanical order of the world explained by gravitation. On the one hand, he hypothesized the existence of gravitation, and thus succeeded in explaining by calculation a great number of mechanical phenomena. On the other hand, he steadfastly rejected gravitation as a sufficient cause to explain the overall mechanical orderliness of the world. Historians are usually flabbergasted by such an attitude, which shows that Newton was unable to realize fully the significance of his own achievement. If we try to understand the reason for Newton's position, we discover the decisive influence of religion on his mentality. Newton's reason for rejecting gravitation as a sufficient explanation of cosmic mechanical motion was his deep-seated aversion to action at a distance. Why then did he, despite his own unconquerable objections, work toward a synthesis in which action at a distance appears to play an indispensable role? The motivation, in Newton's thought, was theological. He was thinking of God as intervening in the course of events so as to preserve the universal orderliness of nature that science describes and explains by means of mathematically formulated theories. It goes without saying that this Newtonian position was open to very serious theological objections, as his adversaries—notably Leibniz—were only too glad to point out. However, for our purpose here the example is quite illuminating. To sum up, why did Newton succeed where other eminent thinkers had failed, and this despite his own objections? Koyré, one of the leading contemporary historians of science, has synthesized the situation admirably:

---

[50] M. Caspar, *Kepler,* trans. C. D. Hellman (London: Abelard-Schuman, 1959), p. 374.

[51] Quoted in Caspar, *op. cit.,* p. 88.

[52] For details see, for instance, M. Purver, *The Royal Society: Concept and Creation* (Cambridge, Mass.: MIT Press, 1967).

...it was his belief in an omnipresent and omniactive God that enabled him to transcend both the shallow empiricism of Boyle and Hooke and the narrow rationalism of Descartes, to renounce mechanical explanations and, in spite of his own rejection of all action at a distance, to build up his world as an interplay of forces, the mathematical laws of which natural philosophy had to establish. By induction, not by pure speculation. This because our world was created by the pure will of God; we have not, therefore, to prescribe his action for him; we have only to find out what he has done. [53]

To conclude, it is clear that, historically speaking, the influence of religion was paramount in bringing about modern science from its early inception to its definitive success in Newton's work.

(cb) The Psychological Contribution of Religion to Science

Once the scientific attitude toward nature has become accepted, is there any reason left for asserting an intrinsic connection between religion and science? People can be quite successful scientifically without being religious at all. The objection is obvious and serious. I shall try to answer it by returning once more to the testimony of creative scientists in reporting their own experiences.

Basically, the issue of a connection between science and religion poses itself even in our time on the psychological level. The preliminary conviction with which the creative investigator approaches nature remains just as puzzling now as it was at the beginning of science. The scientist starts his research with the certainty that, despite appearances to the contrary, there is an intelligible orderliness intimately present in the area of nature he sets out to explore. What is the character of this certainty? Is it simply a prejudice—namely the assumption that since nature has proved itself intelligible so far, it will continue to be so in the future? Many scientists reply that this is the case. They do research because, as the phrase goes, science works. Other scientists, usually the great creators, reject such a pragmatist interpretation. The reason is their experience of creativity. Why does man, with his mind, seek in nature something that appears impossible to find with the senses? The reason is the preliminary conviction of the absolutely reliable intrinsic intelligibility of nature. This conviction obviously should not be identified with religion as such, for it operates only at the cognitive level. And yet it seems to have a religious

---

[53] A. Koyré, *Newtonian Studies* (Chicago: University of Chicago Press, 1968), p. 114; cf. pp. 110-114. Concerning Newton's religious inspiration see also Manuel's *Portrait* (note 23): "God and the Calling of the New Philosophy," pp. 117-132, and W. C. Dampier, *A History of Science and Its Relations with Philosophy and Religion* (London: Cambridge University Press, 1966), pp. 168-176.

connotation or resonance, for it overlaps one of the main aspects of the religious experience of nature—the awareness that nature has a comprehensive meaningfulness.

The person who most persistently insisted in recent years on an intrinsic connection between science and religion, particularly in the psychological sense of the term, was Einstein. Einstein defines the term *religiousness* in a generic sense. To him it stands for "trust in the rational constitution of reality which therefore is at least in some way accessible to human reason." [54]   On the basis of this definition, Einstein declares without reservation that religiousness is an indispensable condition for anyone who wants to do science. Thus, for instance, he continues in the text just quoted: "Where this feeling is absent, there degenerates science into spiritless empiricism (*geistlose Empirie*)." To be sure, Einstein acknowledges that science and religion are totally autonomous undertakings. For, as he remarks, "scientific results are entirely independent from religious or moral considerations." [55]   Nevertheless, Einstein maintains that—psychologically speaking—one can be creative in science only by adopting an attitude that is indistinguishable from that of the religious person. Referring to his own experience, he makes this point quite strongly while discussing the obviously religious examples of Kepler and Newton:

I maintain that the cosmic religious feeling is the strongest and noblest motive for scientific research. Only those who realize the immense efforts and, above all, the devotion without which pioneer work in theoretical science cannot be achieved are able to grasp the strength of the emotion out of which alone such work, remote as it is from the immediate realities of life, can issue.... Only one who has devoted his life to similar ends can have a vivid realization of what has inspired these men and given them the strength to remain true to their purpose in spite of countless failures. It is cosmic religious feeling that gives a man such strength. [56]

As is well known, Einstein gave a pregnant, if homespun, formulation of the decisive conviction that animates the scientific mind in its research. His motto, chiseled in a foyer of the Institute for Advanced Study at Princeton, has become classical: *Raffiniert ist der Herrgott, doch boshaft ist er nicht.* In Norbert Wiener's translation: "God may be subtle, but he isn't plain mean." [57]   This expression, as Wiener hastens to comment, is not a cliché but rather "a very profound statement concerning the problems of the scientist." The reason is that man, before

---

[54] A. Einstein, in *Lettres* (note 15), p. 102; letter dated January 1, 1951.

[55] A. Einstein, *Ideas* (note 40), p. 52.

[56] *Ibid.,* pp. 39f.

[57] N. Wiener, *The Human Use of Human Beings: Cybernetics and Society* (New York: Avon Books, 1967), p. 256.

starting to do science, must be able to count on something totally reliable in the structure of reality. But the term God, as employed here, stands obviously for the ultimate source of reality.

There is also a second sense in which a psychological connection between religion and science can be detected. It regards the ultimate motivation for pursuing science itself. This, for instance, is the rather abrupt way Hoyle stated his view while delivering the keynote address to a recent meeting of the American Physical Society:

Why in fact do we do physics? ... The real motive, of course, is a religious one. ... Our aim is the same [as that of religion]: to understand the world and ourselves, not to make a profit or justify ourselves by producing an endless stream of technical gadgets. [58]

But Hoyle, of course, is not alone in maintaining this position. Einstein agrees with him when he describes the ultimate stage of science as a kind of religious amazement at the manifestation of a superior intelligence embodied in the world of observation. As he puts it:

You will hardly find one among the profounder sort of scientific minds without a religious feeling of his own. ... His religious feeling takes the form of a rapturous amazement at the harmony of natural law, which reveals an intelligence of such superiority that, compared with it, all the systematic thinking and acting of human beings is an utterly insignificant reflection. [59]

Other scientists concur. To cite just one more, Max von Laue mentions, at the beginning of his *History of Physics,* the separation between science and religion as due to the influence of Kant. However, he goes on to comment about post-Kantian scientists:

The tenet that the scientific experience of truth in any sense is "theoria," i.e., a view of God, might be said sincerely about the best of them. [60]

(cc) New Religious Awareness Due to Science: An Example

To understand a little more the connection between science and religion we can still discuss an example of a different kind. We have seen so far that religion is relevant to science. We examine now the relevance of science to religion as exemplified in Galileo's attitude.

The religious penetration of Galileo's mind is astounding. He had such a sense of the divine mysteriousness that he was the first to take

---

[58] F. Hoyle, *Physics Today* **21** (1968), 148f.

[59] A. Einstein, *Ideas* (note 40), p. 40.

[60] M. von Laue, *History of Physics,* trans. R. Oesper (New York: Academic Press, 1950), p. 4.

the apparently absurd position of upholding both the truth of the Bible
—"wherever its true meaning is understood"—and the truth of science. [61]
Accordingly, he was the lone pioneer in proposing more adequate rules
of biblical hermeneutics. And yet Galileo was so right in his position
that the theological community came gradually to assimilate his views as
perfectly obvious. The question arises: How could this man who had
no special training in theology attain a religious insight that none of the
contemporary experts was able to share? [62] The answer can only be one.
Galileo had a better awareness of God's mysterious self-revelation in the
Bible because he had a better sense of God's transcendence as manifested
by nature.

Nonscientific man tends only too easily to assume that his own way
of conceiving God is the right and adequate one. This came particularly
to the fore in the disputes between Galileo and his opponents. One of
the main objections raised against the Copernican hypothesis was theological. Many people condemned the hypothesis because it demanded the
existence of a universe that, in their view, was too vast—thus also vain
and unworthy of the divine wisdom. As usual, Simplicio is called upon
to summarize the position of Galileo's adversaries quite precisely:

But must we not admit that nothing has been created in vain, or is idle,
in the universe?... to what end would there then be interposed between
the highest of their [planetary] orbits and the stellar sphere, a vast space
without anything in it, superfluous, and vain? For the use and convenience of whom? [63]

From these words, the psychological reason many religious people opposed
science becomes obvious. It was not a matter of religion proper. It
was rather an instinctive anthropomorphism that tended to take the current
view of God as adequate. Hence it practically denied the possibility that
God could manifest himself through nature in a surprising way, completely
outstripping man's imagination and comprehension. Galileo was merciless
in exposing the arrogance of this anthropomorphic attitude, along with its
basic irreligiousness. Salviati, the spokesman of science, puts it aptly:

And finally I ask you, O foolish man: Does your imagination first comprehend some magnitude for the universe, which you then judge to be

---

[61] Galileo's exegetical views are presented by him in synthesis in his "Letter to Madame
Christina of Lorraine," which can be found in S. Drake, trans. and ed., *Discoveries and
Opinions of Galileo* (New York: Doubleday Anchor Books, 1957). Citation given in text
can be found on p. 181.

[62] A survey made recently by a Catholic exegete proves that none of the leading
theologians and exegetes of Galileo's time was able to realize fully the importance of the
issue discussed by him. See C. M. Martini, "Gli Esegeti al Tempo di Galileo," in *Nel
Quarto Centenario della Nascita di Galileo Galilei* (Milan: Vita e Pensiero, 1965), pp. 115-124.

[63] G. Galilei, *Dialogue* (note 24), p. 367.

too vast? If it does, do you like imagining that your comprehension extends beyond the divine power? Would you like to imagine to yourself things greater than God can accomplish? And if it does not comprehend this, then why do you pass judgment upon things you do not understand? [64]

As to the objection that the Copernican universe would have no use for man, Salviati retorts by reiterating his accusation of arrogance:

I say that it is brash for our feebleness to attempt to judge the reason for God's actions, and to call everything in the universe vain and superfluous which does not serve us. [65]

To these declarations Sagredo adds his caustic comment:

To me, a great ineptitude exists on the part of those who would have it that God made the universe more in proportion to the small capacity of their reason than to his immense, his infinite, power. [66]

To conclude, at this point we can understand the justification of the view that intimately connects science with religion. The reasons that support it are essentially two. In the first place, consistently interiorized science amounts to an original experience of ultimates. Science makes man perceive the wonder and awe of nature in a way that is inaccessible to the nonscientific person. In the second place, the interiorization of science is capable of making the reflective person perceive—at least vaguely—the very source of existence, which accounts ultimately for the intelligibility and wonderfulness that science continually detects in nature. Science gives to attentive and responsive man a new sensibility for the transcending mysteriousness manifested by the overwhelming richness and beauty and power of nature, which science alone is able to reveal to man.

The result we have achieved is of major importance in our effort to understand the humanistic significance of science. Precisely because of its importance, however, one should avoid exaggerating the significance of the result achieved. Thus, for instance, one should reject the thesis —though presented by a distinguished scientist—"that science is an essentially religious activity." [67] The thesis is wrong because science and religion are—and should be—autonomous undertakings. Thus a person can be scientific without caring for religion, and conversely. Nonetheless, this does not exclude that, if a person consistently seeks truth through either science or religion, he may finally realize the convergence and harmoniza-

---

[64] *Ibid.*

[65] *Ibid.*, p. 368.

[66] *Ibid.*, p. 370.

[67] C. A. Coulson, *Science and Christian Belief* (New York: Collins, 1958), p. 45: "I propose to show that science is an essentially religious activity...."

tion of the two. All the data available seem to indicate that this is the outcome to be expected. As a consequence, another exaggeration must be avoided—one frequently voiced in connection with the phrase: "Science is religion to the scientist." These words are only too often interpreted as though science represents a kind of idolatrous worship that displaces genuine religion from the mind of the scientific practitioner. Such a view is, on the basis of all the preceding evidence, biased and misleading.

In summary, a satisfactory harmonization of science and religion cannot be taken for granted. Nor can it be considered proved by our preceding considerations. Thus it is enough for us here to stress the experiential insight attained. Science and religion, far from necessarily excluding each other, summon man to a personal synthesis. The synthesis, however, can be attained only at the inmost center of the human person—the heart of man. For the synthesis itself demands much reflection, total dedication, boundless courage, and also utmost respect for the autonomy of both science and religion.

CHAPTER 4

THE HUMANISTIC EXPERIENCE OF SCIENCE: ACHIEVEMENT AND DRAMA

Up to this point we have explored successively the three main experiential aspects of science. Science is search for knowledge. But knowledge has both a subjective and an objective side. In addition it presents an internal dynamism from its first start to its full flowering. Accordingly, in the three preceding chapters we have discussed science as a subjective experience of knowing, as an objective experience of nature, and as a dynamism that arises out of wonder and ends up in the perception of ultimates. The moment has arrived to examine comprehensively our attainments in order to give a preliminary answer to our central question. Our overall concern in this book is to find the significance of science for man as such. In this first part we have investigated the question experientially. We will reach a preliminary answer about the meaning of science if we situate science itself within the overall experience of its practitioners. For even though science is an experience of the whole man, it is not necessarily the only experience of man—not even of the man who does science. In addition, science has many practical consequences of a social kind that we have not considered up to now.

Following the considerations just proposed, we shall begin by outlining the essential features of science as they emerge from our foregoing investigation supplemented by additional explicit testimony of scientists. Then we shall consider how science comes to be conceived on the broad social level once the scientific mentality becomes a mass phenomenon. Afterwards we shall investigate science as a source of many problems for the person who dedicates his life to it. Finally we shall try to assess the overall significance of science from the experiential viewpoint.

1. The Humanistic Character of Science

From our foregoing investigations we can safely infer that science, when understood as a creative activity of man, has a humanistic character of its own. Our aim here, is only to make more precise the results attained.

133

To achieve this aim we shall begin by trying to pinpoint what is the essence, if any, of genuine science. Then we shall briefly focus our attention on the initial and final experiential aspects of genuine science as an undertaking affecting the whole man.

(a) Genuine Science: The Quest for an Ideal

The issue of the essence or genuineness of science is a much controverted one. To succeed in our enterprise of bringing this essence into a focus the most promising angle is, as usual, the experiential approach. To this end, let us begin from the obvious. Science is first and foremost a matter of *research*. Why do scientists do research in the first place?

The answer to our question is expectably unanimous. All scientists agree in saying that they search for truth. To be sure, there can be different attitudes concerning this basic goal. As Rostand put it: "To each his own way of loving truth. One scientist will refer to the blisses of discovery; another to the torments of research."[1] Nevertheless, one thing is clear: research as such has an intrinsic aim. This aim has been stated by Born:

The spirit of research is the disinterested desire to disclose nature's mysteries ... this fascination of lifting the veil of mystery and of discovering harmony in apparent chaos.[2]

In other words, as the same writer put it elsewhere, research as such aims at "deciphering the secret language of nature from nature's documents, the facts of nature."[3] In fact, this is the universally shared motivation of scientists as researchers. Sometimes they even blush in confessing their intimate motivation, as one chemist interviewed by Eiduson admitted:

It sounds silly, but what I want is really an understanding of the pattern of the universe ... what I would like to do is to understand the universe for its own sake—but I am afraid that I never will.[4]

Such an embarrassment serves merely to confirm that scientific research has a goal of its own that may even overwhelm the researcher involved. The psychology of productive or creative thinking underlines this goal-directed character of research. In Wertheimer's synthetical presentation:

---

[1] J. Rostand, *The Substance of Man*, trans. I. Brandeis (New York: Doubleday, 1962), p. 273.

[2] M. Born, *The Restless Universe*, trans. W. M. Deans (New York: Dover, 1951), p. 297.

[3] M. Born, *Experiment and Theory in Physics* (New York: Dover, 1956), p. 44.

[4] B. T. Eiduson, *Scientists: Their Psychological World* (New York: Basic Books, 1962), p. 153.

In human terms there is at bottom the desire, the craving to face the true issue, the structural core, the radix of the situation; to go on from the unclear, inadequate relation to a clear, transparent, direct confrontation—straight from the heart of the thinker to the heart of his object, of his problem. [5]

To illustrate the intrinsic goal of scientific research, we can consider the example of Planck. Planck stresses that the researcher is motivated by the desire to find something permanently valid, objectively true and reliable—and this in opposition to the relative character of sense impressions, even in scientific observation itself. In his own words:

All our measurements are relative. The material that goes into our instruments varies according to its geographic source; their construction depends on the skill of the designer and tool-maker; their manipulation is contingent on the special purposes pursued by the experimenter. Our task is to find in all these factors and data, the absolute, the universally valid, the invariant, that is hidden in them. [6]

Concerning his long investigation, which ended with the discovery of the energy quantum, Planck gives as motivation this desire to find a property of nature that was universal and absolute. Reflecting on the experimental data regarding the so-called normal distribution of spectral energy, he was struck by the fact that heated bodies emitted energy in a way that was independent of the chemical composition of the bodies themselves. From this he inferred that radiated energy was an entity capable of giving information about the nature of matter as such. As he put it:

Thus, this so-called Normal Spectral Energy Distribution represents something absolute, and since I had always regarded the search for the absolute as the loftiest goal of all scientific activity, I eagerly set to work. [7]

The intrinsic goal of research is clearly the quest for truth or knowledge for its own sake. However, can we justifiably infer from it that the goal of science as such is exactly the same? In other words, can we assert that the essence of genuine science is the search for truth as an end in itself? The question is vital to our purpose, but at first sight it appears hopelessly complicated. For scientists may do science for many other purposes besides the intrinsic one of research itself. Thus it may well appear that there is no justification for speaking of *the* essence of

---

[5] M. Wertheimer, *Productive Thinking* (London: Social Science Paperbacks, 1966), p. 236.

[6] M. Planck, *Scientific Autobiography and Other Papers*, trans. F. Gaynor (New York: Philosophical Library, 1949), p. 47.

[7] *Ibid.*, pp. 34f.

science at all.   To find an answer to this crucial question we can only develop further our experiential study.

One of the most forceful declarations in favor of science as having an essence characterized exclusively by the search for knowledge is due to Einstein.   The audience to which he presented his address was impressive.   It consisted of some of the world's leading researchers: the members of the Berlin Physical Society.   The purpose was to celebrate Planck's sixtieth birthday (1918). [8]   In front of this scientific body, Einstein decided to discuss the question of why scientists do science.   To set the stage, he began by comparing science with an open temple to which many people come because of different motives.   Some of the visitors enter because they enjoy the exercise of their superior mental gifts: they look for the sensation of discovery or the satisfaction of ambition.   Some other people come in to offer the products of their brains on the altar of science for purely utilitarian purposes.   At this point, Einstein suggests, let us suppose that an angel appears and drives away from the temple all those who belong to the categories enumerated.   As a result, he continues:

...the assemblage would be seriously depleted, but there would still be some men, of both present and past times, left inside.   Our Planck is one of them, and that is why we love him.

The objection arises: Should one suppose that only the few left inside the temple deserve the name of scientist?   Not necessarily, if the term "scientist" is taken in its ordinary sense.   Einstein concedes immediately that among those who have been driven out there were "many excellent men who are largely, perhaps chiefly, responsible for the building of the temple of science."   Nonetheless, Einstein's position is firm. He wants to stress that not all scientists deserve their name in the same way.   Some deserve it partly, others fully.   The full-name scientists are only those who are left within the temple.   They alone are those who are *genuinely scientific* in the complete sense of the term.   For they alone are truly responsible for the creation and permanence of the temple of science in the first place.   As Einstein puts it poignantly:

But of one thing I feel sure: if the types we have just expelled were the only types there were, the temple would never have come to be, any more than a forest can grow which consists of nothing but creepers.

What are the reasons why genuine scientists, according to Einstein, do science?   In his mind, there is no doubt that it is a question of an ideal of the whole man.   He gives both a negative and a positive experiential description of such an ideal.   Negatively speaking:

---

[8] A. Einstein, *Ideas and Opinions*, trans. S. Bargmann (New York: Crown, 1954), pp. 224f.

...one of the strongest motives that lead men to art and science is escape from everyday life with its painful crudity and hopeless dreariness, from the fetters of one's own ever shifting desires.

But, of course, it is the positive reason that is determinant. As for it, Einstein can only speak of the desire of scientific man to create a wholly meaningful vision of reality that may satisfy the deepest aspirations of man himself.

Man tries to make for himself in the fashion that suits him best a simplified and intelligible picture of the world ... in order to find in this way the peace and security which he cannot find in the narrow whirlpool of personal experience.

To sum up, the position of Einstein is unmistakably clear, even though it may remain vague for those who are not directly acquainted with science itself. To him, science can only be genuine if its search for knowledge is not subordinated to any other aim. Of course, he admits elsewhere, there is no logical way of proving this point. Hence he can only speak out of the intimate conviction born of personal experience.

Let me then make a confession: for myself, the struggle to gain more insight and understanding is one of those independent objectives without which a thinking individual would find it impossible to have a conscious, positive attitude toward life. [9]

The position defended by Einstein is usually spoken of in terms of *pure or genuine science.* A compact definition of the mentality involved is, for instance, a statement by the famous experimental physicist Bridgman:

If there are some scientists who are sure that their most compelling motive is the acquiring of understanding, then those scientists, when acting in response to this motive, are engaged in what I call pure science. [10]

The question now is to know whether Einstein's conviction, and Bridgman's, is shared widely enough by the scientific community to be considered the characteristic motivation of the scientist as such. To find an answer, we can only resort to the data supplied by psychological and sociological investigations of the scientific community. An illuminating summary of the mentality detected is the following by the psychologist Eiduson:

Many scientists use the word "success" only *sotto voce* .... These men share some of the clichés: that a truly creative person is motivated by

---

[9] *Ibid.*, pp. 356f.

[10] P. W. Bridgman, *Reflections of a Physicist* (New York: Philosophical Library, 1955), p. 351.

pure rather than impure considerations—purity meaning that the reward should be thought of only in terms of inner satisfaction derived from arriving at the solution—and that a desire for recognition, exhibitionism, or self-aggrandizement, if it emerges at all in such a person, is only an extraneous concomitant of devotion or dedication. [11]

From the words cited it is clear that pure science presents itself to scientific researchers as an ideal to be pursued in their work. They may fail to put it into practice, but at least they recognize what they should do in principle. Purity of science—as visible especially in theoretical research—is so much seen as a normative ideal that scientists tend to establish a sort of hierarchy among themselves. They liken the great scientists to creative artists, while they ascribe a mere technical ability to the others. In Eiduson's summary:

Many describe the great scientists as "artists"—and by this they always mean the theoreticians—and the others as "guys who are painting." [12]

The objection may arise that the scientists' appeal to the purity of science is a snobbish elitism that can hardly apply to science as a mass phenomenon. But sociology denies the validity of the objection. Robert Merton, the well-known sociologist, has coined in this connection the widely accepted expression "ethos of science." He defines it as "an emotionally toned complex of rules, prescriptions, mores, beliefs, values and presuppositions which are held to be binding upon the scientist." [13] He insists that such an ethos is not perceived by the scientists as something of mere pragmatic significance, but as a matter of internal obligation. Disinterestedness or purity of science is the basic pillar of such an ethos. [14]

The tradition of the scientific community, as interpreted by its reflective members, confirms the data supplied by sociology and psychology of science. The conviction is practically unanimous that purity constitutes the essence of science. The reason adduced is that otherwise the scientific endeavor would make no sense. In particular, creation would be impossible. Thus, for instance, T. H. Huxley affirms forcefully:

The great steps in its [science's] progress have been made, are made, and will be made, by men who seek knowledge simply because they crave for it.... Nothing great in science has ever been done by men, whatever their powers, in whom the divine afflatus of the truth-seeker is wanting.... [15]

---

[11] Eiduson (note 4), pp. 178f.

[12] *Ibid.*, p. 139; cf. pp. 158 and 179.

[13] R. K. Merton, *Social Theory and Social Structure* (New York: Free Press, 1957), p. 541, note 16.

[14] For details see Merton, *op. cit.*, pp. 551-561.

[15] In C. Bibby, ed., *The Essence of T. H. Huxley: Selections from His Writings* (New York: St. Martin's, 1967), p. 43.

Polanyi, himself a distinguished scientist and philosopher, agrees entirely: "No important discovery can be made in science by anyone who does not believe that science is important—indeed supremely important—in itself." [16]   The noted biologist Jean Rostand summarizes the situation stingingly: "If a research worker did not put the truth above all else, one would have to ask why the devil he ever chose such a profession." [17]

We can conclude the first point of our summary about the humanistic character of science. Science is humanistic because it is essentially a quest for the ideal. The ideal consists in the search for truth as an end in itself. To be sure, scientific workers may do science for many other reasons as well. This is the case because the scientist may share more or less fully in the genuine spirit of science as such. But such a spirit puts science firmly into the category of humanistic undertakings. To synthesize with Born:

The scientist's urge to investigate, like the faith of the devout or the inspiration of the artist, is an expression of mankind's longing for something fixed, something at rest in the universal whirl: God, Beauty, Truth. Truth is what the scientist aims at. [18]

The ideal aspect cannot be dissociated from the conception of science without destroying science itself. This, clearly, is the prevailing conviction of the scientific community. As a further example, Heisenberg makes this point—at least by inference—when contrasting the situation of the present-day specialist with that of Kepler. Kepler enthusiastically thought of being able, some day, to master with his own mind the whole plan of creation. The modern scientist knows that things are far more complex. Nevertheless, Heisenberg concludes: "But the hope for a great interconnected whole which we can penetrate further and further remains the driving force of research for us too." [19]

(b) Personal Involvement

According to the testimony of scientists there is a second main reason why science deserves to be called a humanistic undertaking. It is the sense of total personal involvement with which science inspires its practitioners.

There is no need at this point to insist further on science as an *experience of the whole person*. However, it may be useful to recall this as a

---

[16] M. Polanyi, *Personal Knowledge: Towards a Post-Critical Philosophy* (New York: Harper Torchbooks, 1964), p. 183.

[17] Rostand (note 1), p. 262.

[18] Born (note 2), p. 278.

[19] W. Heisenberg, *Philosophic Problems of Nuclear Science,* trans. F. C. Hayes (London: Faber, 1952), p. 94.

very important datum. For the general tendency to see science as a merely intellectual endeavor is much too widespread to be easily conquered. Eiduson expresses the situation synthetically as follows:

The phrase "emotional investment" may not suggest the intense nature of intellectual experiences. These activities are described by such adjectives as "thrilling," "intimate," "completely possessing"; and the long hours, the dedication, the slavish devotion—which are part of what LaFarge has called the "emotions of science"—are only the external manifestations of the almost inexpressible affective content. [20]

In the light of the preceding it is clear why practicing scientists insist so much on *imagination and feeling*. According to Heisenberg, for instance, it is a mistake to suppose that all that matters in science is logical thinking and intellectual understanding. In point of fact, he maintains: "imagination plays a decisive role in the realm of natural science." [21] The reason, according to him, is experiential. Experience shows that one can arrive at the cognitive synthesis which marks successful science only by sensing rather than thinking one's way into phenomena. In his own words:

For, even though much experimental work is necessary, sober and accurate, in order to arrive at collecting the evidence—the synthetical organization (*Zusammenordnen*) of the evidence itself succeeds only when one is able to sense rather than to think one's way into the phenomena.

Thus, surprising as it may appear to the nonscientist, the scientific creator ascribes primacy to feeling rather than to thinking in the making of science. This is the conviction of theoreticians, such as Heisenberg, and of experimentalists, such as Bernard. In the view of the latter, feeling is the source "from which everything emanates." [22] For, as he goes on to explain:

Just as, in other human actions, feeling releases an act by putting forth the idea which gives a motive to action, so in the experimental method feeling takes the initiative through the idea. Feeling alone guides the mind and constitutes the *primum movens* of science. Genius is revealed in a delicate feeling which correctly foresees the laws of natural phenomena; but this we must never forget, that correctness of feeling and fertility of idea can be established and proved only by experiment.

One important aspect of the emotional involvement typical of science is the awareness of a *personal call*. The call is felt in the form of a power-

---

[20] Eiduson (note 4), p. 89.

[21] Translated from W. Heisenberg, *Der Teil und das Ganze: Gespräche im Umkreis der Atomphysik* (Munich: Piper, 1969), p. 254.

[22] C. Bernard, *Introduction to the Study of Experimental Medicine*, trans. H. C. Green (New York: Dover, 1957), p. 34.

ful, pervasive inclination to dedicate oneself to science. It can be defined with Louis de Broglie as a "mysterious attraction acting on certain men [that] urges them to dedicate their time and labors to works from which they themselves hardly profit." [23] What attracts scientific man can, with Kepler, be named *forma mundi.* By this term he meant the orderliness and harmony, the intelligibility and beauty that science seeks in nature and by means of which nature reveals itself to man as a meaningful totality. [24] The call to do science reaches so deep in the personality of the scientist that it can also be designated with the term "vocation." The sign of genuine vocation is that a person cannot think of any other pursuits as capable of justifying his own existence. But this seems precisely to be the conviction of the scientists who strive after the purity of science. They have the impression that nothing but a total dedication to research can give significance to their lives. A telling testimony in this regard is that of James Franck reporting about his own experience and that of his contemporaries at the turn of this century.

Everyone who went into physics at that time went into physics because he had to. He couldn't help himself. He had to. There was no attraction to go.... There was no industrial position for a physicist.... So whoever went to study physics went because he felt he could not be happy in any other way. [25]

Still another aspect characterizes the sense of total personal involvement typical of science. It is the experience of science as consisting essentially in a *wholehearted response* (Chapter 1, Section 2a). As an example, Heisenberg—in private conversation—likes to explain the situation by means of a comparison. According to him, it is not true that the creative scientist is a kind of superman who boldly takes the initiative toward unheard-of discoveries. Rather, he contends, the creative scientist is just a person "who has a more sensitive ear and listens more attentively than the average." This comparison is illuminating because it is borne out by the psychological investigation of productive or creative thinking. Wertheimer speaks of "a human attitude," which is "willingness to face issues, to deal with them frankly, honestly, and sincerely." [26] Erich Fromm summarizes this aspect by stressing that it is essentially an attitude of one who cares and responds. In his own words:

---

[23] L. de Broglie, *Physics and Microphysics,* trans. M. Davidson (New York: Grosset & Dunlap, 1966), p. 207.

[24] For details see M. Caspar, *Kepler,* trans. C. D. Hellman (London: Abelard - Schuman, 1959), pp. 376f.

[25] T. S. Kuhn, J. L. Heilbron, P. Forman, L. Allen, eds., *Sources for History of Quantum Physics* (copyright, American Philosophical Society); interview with J. Franck, July 7, 1962, p. 7.

[26] Wertheimer (note 5), p. 179.

The object is not experienced as something dead and divorced from oneself and one's life, as something about which one thinks only in self-isolated fashion; on the contrary, the subject is intensely interested in his object, and the more intimate this relation is, the more fruitful is his thinking in the first place.... In the process of productive thinking the thinker is motivated by his interest for the object; he is affected by it and reacts to it; he cares and responds. [27]

(c) A Labor of Love

The third reason why science should be called humanistic is but the consequence of the two preceding ones. We have seen that science is a quest for an ideal that involves the total personality of its practitioners. We can now add that science, if pursued consistently with its spirit, is a manifestation of love.

To avoid possible sentimentalism, let us return to Einstein's speech delivered to the Berlin Physical Society (Chapter 4, Section 2a). After having discussed the characteristics of those who deserve to inhabit permanently the temple of science, Einstein referred to Planck as one of these people. He extolled the latter's "inexhaustible patience and perseverance" in devoting himself "to the most general problems of our science, refusing to let himself be diverted to more grateful and more easily attained ends." [28] This, Einstein remarked, had often been attributed by his colleagues to extraordinary willpower and self-discipline. But, he went on to explain, such an interpretation is wrong. The reason is simply that science is a work of love. In his own words:

The state of mind which enables a man to do work of this kind is akin to that of the religious worshiper and the lover; the daily effort comes from no deliberate intention or program, but straight from the heart.

Several features manifest science as a loving enterprise. The most typical one is *passionate dedication*. As Einstein put it:

There exists a passion for comprehension, just as there exists a passion for music.... Without this passion, there would be neither mathematics nor natural science. [29]

But this passion is not something that is merely inborn. It is, rather, something that man must foster to the best of his abilities. Pavlov's words to the academic youth of his country state it aptly:

---

[27] E. Fromm, *Man for Himself: An Inquiry into the Psychology of Ethics* (Greenwich, Conn.: Fawcett, 1968), pp. 109f.

[28] Einstein (note 8), p. 227.

[29] *Ibid.*, p. 342.

Remember that science demands from a man all his life. If you had two lives that would be not enough for you. Be passionate in your work and your searchings. [30]

Two other traits are common to science and love. One is *dignified engagement*—the form of service that is the exact opposite of any slavish attitude. Wertheimer, for instance, stresses that the creative thinker is inspired by "a desire for improvement, in contrast with arbitrary, wilful, or slavish attitudes." [31] This frame of mind can be designated also as a "cold enthusiasm for truth." The expression was used by the great anthropologist Franz Boas to explain the motivation animating the famous physiologist Rudolf Virchow. [32] Another trait common to science and love is *concerned respectfulness*. Science is animated by great respect for the reality it is investigating. This is the essential reason that science itself is so insistent on objectivity. In Fromm's summary:

Objectivity does not mean detachment; it means respect; that is, the ability not to distort and to falsify things, persons, and oneself. [33]

In the light of the preceding, it is clear why we can speak of science as a labor of love without indulging in emotional exaggeration. The reason is obvious: it is the experiential similarity of love and science. Love, authentic and sincere, is not subjective partiality nor instinctive hunger for self-satisfaction. The true lover is not the person who picks and chooses what he likes in other beings to the end of gratifying himself. On the contrary, he is the person who welcomes the other in his otherness, rejoices at his uniqueness and concreteness, and seeks to enter more and more into respectful personal relationship with him. True love, in short, is essentially a matter of admiration, self-dedication, and communion. But the spirit of genuine science fits this phenomenological description of love remarkably well. For science is an expression of total openness toward observable reality. It is active responsiveness toward any manifestation of intelligibility that comes from nature. It is a responsiveness that sets no bounds to its own involvement. Above all, it is concerned respectfulness that aims at nothing but seeing reality as it is. In fine, science is love because it enables man to establish a relationship of personal communion with nature and with nature's own ultimate source of meaning.

---

[30] I. P. Pavlov, "Bequest to the Academic Youth of His Country," in A. V. Hill, *The Ethical Dilemma of Science and Other Writings* (New York: Rockefeller University Press, 1960), pp. 163f.

[31] Wertheimer (note 5), p. 243.

[32] Quoted in M. Harris, *The Rise of Anthropological Theory: A History of Theories of Culture* (New York: Crowell, 1968), pp. 257f.

[33] Fromm (note 27), p. 111.

The experience of love is what keeps the scientist going. His life is unavoidably hard. Often, it is also socially misunderstood and unrewarded. Love is the only key to understanding why the scientist feels enraptured with science. The famous neurologist Ramón y Cajal spoke for many when describing his experience of scientific investigation:

It is to me the most noble aspiration a man can follow because perhaps no other goal is so impregnated with the perfume of love and universal charity. [34]

2. THE SOCIAL SUPERFICIALIZATION OF SCIENCE

The interpretation of science we have reached is so well documented that —at least in its general lines—it can be said to respond to a situation of fact. And yet, the view of science entertained by the public and, frequently, even by scientific workers themselves is quite different. Hence the question of the actual meaning of science for man arises with new urgency. In particular, if our foregoing results are valid, we must try to find out why science is so often misunderstood by both laymen and scientific workers. To this end we are now going to consider briefly the chief manifestation of science as a social or mass phenomenon.

(a) The Attraction of Method

The first area in which the views of the creative scientist and those of the average person clash about the nature of science is theoretical. It regards the so-called *scientific method*. As we have seen (Chapter 2, Section 3a), creative scientists discount the importance of method in science. The average person, on the contrary, speaks of method as the essential feature of science. What are the reasons for these diametrically opposed positions?

The creative scientist who reflects on his own experience knows that science consists basically in an original activity of the mind. Science is a new way of seeing reality. What characterizes it is discovery. Accordingly, what is essential in science is the role of ideas. Claude Bernard, the great experimental physiologist, never tires of pointing it out:

Discovery ... is a new idea emerging in connection with a fact found by chance or otherwise. [35]

---

[34] S. Ramón y Cajal, in E. H. Craigie and W. C. Gibson, *The World of Ramón y Cajal with Selections from His Nonscientific Writings* (Springfield, Ill.: Charles C. Thomas, 1968), p. 191.

[35] Bernard (note 22), p. 35.

As a consequence, it is clear why method is not rated highly by the creative scientist. The reason is that method is sterile as far as discovery is concerned. It can only serve to check the validity of what has been discovered or extend to other areas discoveries already made. As Bernard put it, continuing the text cited:

Consequently, there can be no method for making discoveries, because philosophic theories can no more give inventive spirit and aptness of mind to men, who do not possess them, than knowledge of the laws of acoustics or optics can give correct ear or good sight to men deprived of them by nature. But good methods can teach us to develop and use to better purpose the faculties with which nature has endowed us, while poor methods may prevent us from turning them to good account.

In practice, the science known to the public and carried out at the mass level often bears little resemblance to the creative undertaking we have been discussing. The situation is unavoidable. Scientific discoveries are a comparatively rare event. Hence much of the ordinary scientific work consists in making past discoveries yield all their fruits, both theoretical and practical. This is far from being a meaningless activity. But it can easily lead to a misinterpretation of the genuine spirit of science.

The importance of drawing the theoretical consequences of a basic discovery is exemplified best by the immense labors of the mathematical physicists—especially Laplace and Lagrange—who developed and systematized Newton's original insight about gravitation. Classical mechanics is Newtonian; however, as a rigorously organized body of knowledge it is Laplacian and Lagrangian. But, of course, in the logical work of systematization, method plays a prominent role. Hence the first psychological reason emerges for the tendency to identify science with method. It is due to the impression made by the rigor of logic on the mind. It seems that by simply being sufficiently methodical, it is possible for man to conquer any field of the intelligibility of nature. This, as history proves, is what actually happened once Newtonian mechanics was brought to full development during the course of the past century. People tended more and more to think that science was just a question of a systematic application of Newton's insights to the study of nature. Even prestigious scientists went so far as to declare that physics was substantially completed: no major new discoveries need be expected. [36]

Another major form of scientific activity supplies a second psychological reason for the unwarranted identification of science and method. Science, we saw, is a new way of knowing nature. But this new know-

---

[36] For a contemporary view of the philosophical mentality in physical circles in the second half of the 19th century see J. B. Stallo, *The Concepts and Theories of Modern Physics* (1881), reprinted by Harvard University Press in 1960.

10

ledge can be directed to the solution of practical problems of current interest to man. This explains why a majority of the scientifically trained personnel is not engaged in making new discoveries, but in applying past discoveries. Their work is usually called research and development or mission-oriented science. There is no disputing the social—and, occasionally, even theoretical—importance of this kind of work. However, it is easy to see that here, again, method plays an essential part. For, to apply past discoveries, one must systematically draw the consequences that arise from them. Accordingly, the more that mission-oriented science proves successful, the stronger the inclination becomes to identify science with method. Once more, history is enlightening here. Toward the middle of the past century, society had a growing need for machines, materials, and sources of energy. Thus, it was only to be expected that it would turn to science for assistance. The first scientific laboratories arose that were geared to the solution of practical problems. Success was increasingly great. As a consequence, people began to take more and more for granted that science was able to solve all problems, provided only the right method was adopted. Some enthusiasts even thought that man could now make any discovery he pleased, because he possessed the key to discovery: method. [37] Of course, they did not pause to consider that the discoveries they were so enthusiastic about were just applications of insights that had been attained without the assistance of method.

To close, we notice here for the first time a disconcerting phenomenon. The more science becomes socially widespread and influential, the more superficial and one-sided its interpretation tends to become. Our foregoing considerations have uncovered the root of such a social superficialization of science. It consists in the psychological tendency of the masses to take science for granted because of its success. In particular, since science succeeds mostly in its practical applications because of method, people take it as self-evident that science is to be identified with method. Needless to say, such a social misinterpretation of science is not only simplistic, but false and potentially dehumanizing as well. The falsity of the opinion in question is evident in the light of our entire investigation. As for its dehumanizing potential, it becomes clear when one reflects that personal creativity is the essential feature that makes science human. Creativity gives man a sense of the ideal as well as modesty in his scientific pursuit. But if science is merely method, no room for this basic attitude is left. The ideal of scientific work as a personal contribution to truth disappears, because method is an essentially impersonal procedure. As for modesty, blind trust in method replaces it with an arrogant and dog-

---

[37] For an overall synthesis of the scientific mentality in the 19th century and the prominence of method see A. N. Whitehead, *Science and the Modern World* (New York: Mentor Books, 1948), pp. 96-114.

matic conviction about the boundless and exclusively reliable power of science itself.

(b) The Happiness of Pursuit

The second area in which the genuineness of science is frequently misunderstood concerns the emotional implications of science itself. Here again we notice a gradual shift in interpretation, beginning with the attitude of the creative scientist and ending up with the average person. The question of the emotional implications of science cannot be avoided. For science puts a premium on originality (Chapter 1, Section 3c). Also, science, as a communal enterprise, invites its practitioners to move ahead continually, without ever pausing (Chapter 1, Section 3e). As a consequence, science presents itself to man with the emotional features of the ideal (Chapter 4, Section 1a). If this is so, however, then what is the spirit that should animate the scientist in his never-ending striving? In other words, how should he conceive his quest for the ideal?

To find an answer to our questions, and to illuminate the gradual shift in the evaluation of science, we can begin with the reflections of a typically creative scientist. Heisenberg is fond of comparing the attitude of the scientist to that of Columbus. To him, Columbus is an example because he "possessed the courage to leave the known world in the almost insane hope of finding land again beyond the sea."[38] In particular, Heisenberg remarks, what was essential in Columbus' contribution to mankind was not a matter of sharpness of mind. For others before him had thought of the rotundity of the earth and of the possibility of circumnavigating the globe. Nor was it a matter of practical skill in organizing an oceanic exploration, for many others were able to do the same. The unique contribution of Columbus was rather one of character and determination.

But the most difficult aspect in this exploration voyage was certainly the decision of abandoning all land known up to then and sailing so far to the west that, given his provisions, a return could not be possible any more.[39]

But this, Heisenberg goes on to explain, is just the attitude required of anyone who intends to make science progress in the genuine sense of the term.

In similar manner, only then can unexplored territory (Neuland) be conquered in a science when one is ready—in a decisive point—to abandon the ground on which the science has rested up to then, and spring into the void, as it were.

---

[38] Heisenberg (note 19), p. 25.
[39] Translated from Heisenberg (note 21), pp. 101f.

This is a very difficult step to take, as Heisenberg continues to explain in discussing the example of relativity. The reason is that the progress of science does not only demand from its practitioners that they be ready to accept new information. What is most difficult is to change completely one's own way of thinking. In his own words:

When an unexplored territory is entered, it can happen that not only new information is to be accepted, but that the structure of thinking has to be changed, if one wants to understand what is new. Obviously, many are not ready or able to do this.

Heisenberg refers here to the case of Einstein. Einstein incurred many enmities because of the courageous novelty of his relativity theory. However, later on he was himself unable to accept the rearrangement of thought demanded by the success of quantum physics. [40]

Heisenberg's considerations bring out vividly the seriousness of the demands made on scientific man by the progress of science. But this same progress can also be considered from another point of view, namely as a source of rejoicing and self-satisfaction once the scientific enterprise has succeeded. The shift of emphasis is perhaps not always easily perceptible nor expressly meant, but it is frequently there. As an example we read the enthusiastic words of Richet, himself a distinguished physiologist, where he sings a paean to originality:

Originality is the chief virtue. To have ideas that other men have not had; to imagine unforeseen consequences; to start a new experiment; to repeat an old experiment in order to discover unexpected meanings— that is almost divine! In the history of science nobody has left his mark on the world unless he has been, in this sense, an innovator. [41]

The transition in emphasis from the self-sacrifice demanded by scientific originality to the self-enjoyment made possible by the same originality is a most natural one. Sometimes it becomes blatant or at least naïve when overstressed in front of the nonscientific public. A well-known expression of this mentality, in his naïve arrogance, is presented by Bacon. He thought of nothing less than, through science, to reform the whole structure of human knowledge. This reform project was to be his *Instauratio Magna.* A fundamental part of the *Instauratio* was the *Novum Organum,* which Bacon composed for the purpose of replacing Aristotle's basic study of cognitive categories, which was called the *Organum.* The spirit with which Bacon viewed science is graphically manifested on the title page of the *Instauratio,* whose first installment was the *Novum Organum.* The illustration portrays the ship of human knowledge moving with full

---

40 *Ibid.,* pp. 114f.
41 C. Richet, *Natural History of a Savant,* trans. O. Lodge (London: Dent, 1927), p. 38.

sails through the Pillars of Hercules—that is, going defiantly beyond what the Ancients had considered the ultimate boundaries for man. The function of Bacon's *Organum* was to make the mind explode the limits that prescientific man had thought impassable. Instead of *Ne Plus Ultra,* the motto was now *Plus Ultra.* As is well known, Bacon's enthusiasm proved infectious. As an example, one of the early historians of the Royal Society —Joseph Glanvill—entitled his work pointedly *Plus Ultra, or the Progress of Knowledge since the Days of Aristotle.* [42]

Even in our days, the conception of science as an endless quest stirs much enthusiasm. Science is now often seen as the last challenge against which man can prove himself. This view is put well by Bentley Glass:

The western frontier that once challenged adventurous and imaginative youth exists no longer; the frontier of today and tomorrow is that of science—as Vannevar Bush has called it, the "Endless Horizons." [43]

As is clear from the preceding, it is possible to strive after science as an ideal because of vastly different motives. In particular, one can be an explorer or an adverturer. The explorer risks everything to achieve a discovery. The adventurer is merely interested in the excitement that accompanies the investigation. As a matter of fact, scientific workers frequently succumb to the temptation of adventurism. This frame of mind is so widespread that the psychologist Eiduson coined a special phrase to express it.

I have transposed the expression "the happiness of pursuit" from the familiar "pursuit of happiness" to describe the pleasure scientists find in setting up stimulating situations, in meeting the challenges of the problems and the instruments, in slowly making progress on something that was very difficult. The outcome is not unimportant in the feelings of happiness and satisfaction, but it is not all-important. [44]

The happiness of pursuit can easily become, at least by implication, the main motive for doing science. A manual for the training of scientific researchers insists for instance:

The most important attribute of science is not knowledge, but its capacity for acquisition of knowledge. Knowledge which science contains is limited, frequently fragmentary and inaccurate, always liable to revision. The capacity of science to acquire knowledge is infinite. [45]

---

[42] For details see M. Purver, *The Royal Society: Concept and Creation* (Cambridge, Mass.: MIT Press, 1967), especially pp. 29 and 69.

[43] B. Glass, *Science and Liberal Education* (Baton Rouge: Louisiana State University Press, 1959), p. 61.

[44] Eiduson (note 4), p. 161.

[45] P. Freedman, *The Principles of Scientific Research* (New York: Pergamon, 1960), p. 2.

Needless to say, adventurism or the striving after the happiness of pursuit is another gross superficialization of science. For it undermines the humanistic significance of science under the guise of upholding it. Science is humanistic in that it is a striving after knowledge. Attainment of knowledge is humanizing in that, through it, man is enabled to enter an increasingly personal contact with reality, including the ultimate source of reality itself. This is the sense in which one should speak of purity of science or science as an end in itself. Adventurism claims to uphold the purity of science. But it shifts the emphasis from science as an end *in* itself to science as an end *to* itself. The result is a perversion of the spirit of science, with ominous consequences. The worst consequence is that of draining science of its cognitive significance. Indeed, adventurism implies that one does not care for research as a source of cognition but merely as an occasion for excitement. We shall revert to the demoralizing effects of this situation in discussing the widespread sense of frustration noticeable among scientific workers. Another negative consequence is that of pragmatizing the scientific enterprise as such. The researcher who gives in to adventurism tends to withdraw into his ivory tower under the pretext of preserving the purity of science. But society, which pays for his research, cannot tolerate the support of an enterprise that ends up by being an egotistic search for self-enjoyment. As a result, science is forced to justify its existence by pragmatizing itself. Instead of being respected as a source of knowledge, science tends now to be esteemed only as a source of utility.

(c) The Allure of Technicalism

The creative scientist and the average person clash still on one, comprehensive area concerning the interpretation of science. Although this disagreement is somehow the consequence of the two preceding ones, it requires specific discussion. Most people, scientists and nonscientists alike, tend to confuse science with technology. Often they go so far as to see technical products as the only relevant manifestations of science. The reflective creative scientist vehemently disputes this identification. Einstein, for instance, asserts that those "whose acquaintance with scientific research is derived chiefly from its practical results easily develop a completely false notion of the [scientific] mentality." [46] For the purpose of our investigation we must now survey the reasons and the consequences of this further misinterpretation of science.

To begin with, let us recall the obvious. As we have seen in the Preliminaries, Section 1b, the distinction between science and technology is far from arbitrary. It is founded above all on the fact that either pursuit

---

[46] Einstein (Note 8), p. 39.

has a characteristic goal of its own.   Polanyi has summarized the situation aptly:

Originality is appreciated in both, but in science originality lies in the power of seeing more deeply than others into the nature of things, while in technology it consists in the ingenuity of the artificer in turning known facts to a surprising advantage.   The heuristic passion of the technician centers therefore on his own distinctive focus.   He follows the intimations, not of a natural order, but of the possibility for making things work in a new way for an acceptable purpose, and cheaply enough to show a profit. [47]

If science and technology are distinct, what is the reason for their identification in the view of many?   The objective ground is, paradoxically, the theoretical character of science.   The point is frequently made by the creative scientists themselves.   Science, because it is theoretical, is technologically important.   Sometimes the scientists who stress this point want to boast; other times they intend to emphasize the social importance of what they are doing.   Thus, for instance, the theoretician Ludwig Boltzmann used to exclaim: "There is nothing more practical than theory!" [48]   With these words he probably wanted to refute particularly Mach's constant nagging about the uselessness of his researches in theoretical micromechanics.   The fact is, Boltzmann was right.   This situation was emphasized, for instance, by the astronomer John Herschel in a widely influential philosophical book.   He contends that science, no matter how theoretical, necessarily entails results of practical significance.

The speculations of the natural philosopher, however remote they may for a time lead him from beaten tracks and everyday uses, being grounded in the realities of nature, have all, of necessity, a practical application.... [49]

This position is widely accepted by the creative scientists.   Sometimes they even argue against their pragmatic detractors that it is to the advantage of technology that theoretical science be fostered.   The theoretician Poincaré wrote, for instance:

If we devote ourselves solely to those truths whence we expect an immediate result, the intermediary links are wanting and there will no longer be a chain.   The men most disdainful of theory get from it, without suspecting it, their daily bread; deprived of this food, progress

---

[47] Polanyi (note 16), p. 178.

[48] For details see E. Broda, *Ludwig Boltzmann: Mensch, Physiker, Philosoph* (Vienna: Deuticke, 1955), p. 96.

[49] J. F. W. Herschel, *Preliminary Discourse on the Study of Natural Philosophy* (London: Longman, Rees, Orme, Brown and Taylor, 1830), p. 12.

would quickly cease, and we should soon congeal into the immobility of China. [50]

Since the theoretical nature of science makes it technologically important, we can easily see the psychological reason why science is widely identified with technology. The average person tends instinctively to oversimplify issues. In particular, he tends to stress the tangible and ignore the abstract. But the technological products of science are most tangible, whereas its theoretical aspects are quite abstract for the person who is not scientifically creative or profoundly reflective.

The current identification of science and technology, however, cannot be properly understood only on the psychological basis mentioned. For it is the result of a long ideological development. To recall it briefly, we must go back to what we have already seen discussing the popular view of science as method. The identification of science and technology in this regard is to be traced to the origins of modern science itself. Francis Bacon is chiefly responsible for it. To him we owe the slogan: "scientific knowledge is power." The angle at which he looked at science as a source of power was that of method (Chapter 1, Section 3a). Bacon's view was taken up and propagated enthusiastically by other philosophers, especially the French encyclopedists and the early positivists. For these people science was meaningful essentially because it supplied man with an infallible method for changing the world according to his wishes. Toward the middle of the 19th century, the situation was ripe for the widespread acceptance of the identity view. The reason was the increasing technological effectiveness of science. Man had the impression that, through science, he could invent whatever he wished (Chapter 4, Section 2a). As Whitehead put it, science was currently thought to have achieved "the invention of the method of invention." [51]

Our rapid survey serves to convey an idea about the nature of the phenomenon we are discussing. People frequently speak in this context not only wrongly but also misleadingly. What is wrong, as has been seen, is to take technology as identical with science. What is even more serious is the superficially confusing way in which the term technology is employed. Technology, as defined above, is a perfectly justified approach to reality, ideologically neutral. The sense in which the term is frequently used today, however, is so ideologically laden as to have become a *cultural ideology* of its own. Contemporary humanistic investigators stress this point more and more. What people often mean by technology

---

[50] H. Poincaré, *The Value of Science,* trans. G. B. Halsted (New York: Dover, 1958), p. 75.

[51] Whitehead (note 37), p. 98.

is not technology at all but rather a worldview that affects the whole attitude of man toward reality and man himself.

We can profitably summarize the insights reached by several independent investigators. Forbes, the noted historian of science and technology, speaks of "technological order." In his view, this is not just one factor of our civilization, but a prime source of the current social outlook. In his own words:

Technology can no longer be viewed as only one of many threads that form the texture of our civilization; with a rush, in less than half a century, it has become the prime source of material change and so determines the pattern of the total social fabric. [52]

The philosopher Skolimowski retains the word "technology," but capitalizes it. He insists on a basic distinction between technologies and Technology. He defines the latter as an ideological embodiment of the technological spirit, something that goes far beyond what has been traditionally understood under the term itself. In brief:

Technology thus is not only and not so much a collection of tools and machines, but *a state of Western mentality.* [53]

The most comprehensive analysis of the phenomenon is attributable to Ellul. He uses the term *technique* to designate it. His definition is enlightening:

The term *technique,* as I use it, does not mean machines, technology, or this or that procedure for attaining an end. In our technological society, technique is the *totality of methods rationally arrived at and having absolute efficiency* (for a given stage of development), in *every* field of human activity. [54]

As Ellul explains in the same context, we have to do here with a truly novel phenomenon, quite different from what was understood under technology in the past. As he put it, "the technique of the present has no common measure with that of the past."

Following Ellul, we can notice two typical traits in the mentality that we may call *technicalism.* The first trait is *rationality.* It arises

---

[52] R. J. Forbes, *The Conquest of Nature: Technology and Its Consequences* (New York: Mentor Books, 1968), p. vii.

[53] H. Skolimowski, "Technology: The Myth behind the Reality," in *Architectural Association Quarterly* (London, 1970), p. 28.

[54] J. Ellul, *The Technological Society,* trans. J. Wilkinson (New York: Vintage Books, 1967), p. xxv. I think I must disagree with two main theses of Ellul's thought: his practical identification of science with technology and his deterministic interpretation of social phenomena. However, as far as a phenomenological analysis of the technical mentality is concerned, I consider his book outstanding and very illuminating.

out of an increasingly exclusive concern for the technical side of problems. The technical or logical-standardized approach to issues is seen as not only essential but as unique and all-powerful. Ellul defines the rationality of technicalism as characterized by a twofold aspect:

...first, the use of "discourse" in every operation; this excludes spontaneity and personal creativity. Second, there is the reduction of method to its logical dimension alone. Every intervention of technique is, in effect, a reduction of facts, forces, phenomena, means, and instruments to the schema of logic. [55]

The second trait of technicalism consists in making the technical approach the source of a new ideal for man as such. This ideal is aptly termed *artificiality* by Ellul. For technical man aspires to revamp the whole of reality according to logical plans. In Ellul's words:

Technique is opposed to nature. Art, artifice, artificial: technique as art is the creation of an artificial system. This is not a matter of opinion. The means man has at his disposal as function of technique are artificial means.... That world that is being created by the accumulation of technical means is an artificial world and hence radically different from the natural world. [56]

In the light of the preceding, we can realize not only why many people confuse science and technology, but also why the confusion in question—as currently prevailing—is most inimical to humanism. The reason is the identification of science with technicalism. For technicalism is more than an error. It is, to speak with Skolimowski, a new and ominous kind of metaphysics. [57] Such a metaphysics rests on the error of taking science for granted. It identifies science with its tangible products and methods. But then it goes so far as to claim that such a pseudoscience is the all-powerful and uniquely acceptable interpretation of reality and man himself.

To conclude, our investigation has led us to two enlightening results concerning the humanistic relevance of science. In the first place, many of the current polemics between science and humanism are beside the point, simply because what is widely attacked in the name of humanism is not science, but rather its technicalist misrepresentation. This is an encouraging result, since it helps to clarify the issue. In the second place, however, we begin to see a humanistically disturbing side of science proper. Indeed, why is science so widely and ominously misinterpreted from the humanistic point of view in the society where it is prevalent? The reason

---

[55] Ellul, *op. cit.*, pp. 78f.
[56] *Ibid.*, p. 79. See the whole chapter "The Characterology of Technique," pp. 61-147.
[57] See note 53.

must be that, somehow, science is incapable of giving rise by itself to a genuine humanism. We must now pursue this important lead further in order to assess adequately the humanistic relevance of the scientific experience as a whole.

3. THE HUMANISTIC INADEQUACY OF SCIENCE

How can we detect—without emotional exaggerations—the humanistic limitations of science? The most promising guideline continues to be the experiential one. Let us ask the creative scientists themselves to present their views, arising from reflection on their own experience. In particular, let them formulate their judgment about science when science has become the principal or even exclusive factor in their lives. To develop this discussion we can begin from a nonphilosophical consideration, namely the psychological impact of science on its practitioners. Then we shall examine more philosophically such sensitive issues as the cognitive and ethical limitations of science itself.

(a) The Psychological Burden

We wish to explore the significance of science for man from the psychological standpoint. In the light of the preceding, one result emerges at once. Science, we have seen, is essentially creativity. But creativity demands the total engagement of man. Hence the first psychological aspect of science: the *consuming concentration* that it demands from its practitioners.

Psychologists are unanimous in stressing the concentration demanded by science. In particular, they point to the isolation of the scientist.[58] *Loneliness* is an unavoidable consequence of the absorption required by scientific creation. Man starts to be scientific when he feels challenged by an aspect of the intelligibility of nature that other people tend to dismiss as meaningless. Thus a researching scientist cannot help being lonely. This is the case, in the first place, because his time and energies are all dominated by his research. But, in the second place, the researcher is also alone because of a social reaction. People feel uneasy about his seemingly odd concerns and tend to ignore him. Thus a sense of isolation tends to mark a scientist throughout his life. For even when a scientist has succeeded in his research, he must be ready to defend his results against the criticisms of other experts and of the public. Finally, if he is truly

---

[58] For data see, for instance, Eiduson (note 4), pp. 46-50.

scientifically minded, the successful scientist cannot rest on his laurels, but must search further, endlessly.

A moving testimony of the consuming character of scientific research is to be found in some of Galileo's letters written toward the end of his life. He was weak with age and psychologically embittered by his confinement. But he was working feverishly on the book that was to be his scientific masterpiece, the *Discourses on Two New Sciences*. The tension of his mind is mirrored in his complaint to Fulgenzio Micanzio, a trusted disciple and friend:

... my restless brain cannot cease from whirling, and with great waste of time. For any thought which last comes to my mind brings along some novelty, and this makes me discard all my preceding results. [59]

A few years later, when the work on the book was finished and he himself was blind, Galileo returns again to complain of the painful tension caused him by his scientific interest. Writing to the same correspondent, he says:

And thus in my darkness I keep fantasying now on this and now on that effect of nature. I cannot give a little peace to my restless brain as desired. This restlessness causes much harm to me, and keeps me in an almost continual watch. [60]

We can characterize more precisely the basic psychological demand made by science on its practitioners. It can be called the *tension of novelty*. This is not an occasional manifestation, but a permanent feature of scientific life. Science is based on novelty and develops through novelty. But novelty exacts a heavy price from the person who pursues it (Chapter 4, Section 2b). For man must be ready to revise his thinking continually —to reorganize his views by taking into account unexpected discoveries. Psychology stresses the heaviness of this requirement. For the person has the impression of running a constant risk. The risk is that of coming apart in one's inmost being. The sociologist Schon has summarized the situation strikingly:

Openness to novelty, then, is openness to experience that is dangerous and potentially destructive. It involves ... "coming apart." Our integration as a person is threatened when we are forced to break a deeply held structuring of experience. The deeper it is, the more intimately tied to other theories, the more threatening it is to break it. [61]

---

[59] Translated from quotation in Galileo Galilei, *Opere*, F. Flora, ed. (Milan: Ricciardi, 1953), p. ix; letter dated Nov. 19, 1634.

[60] Translated from *Opere, op. cit.*, p. xxii; letter dated Jan. 30, 1638.

[61] D. A. Schon, *Invention and the Evolution of Ideas* (formerly *Displacement of Concepts*) (London: Social Science Paperbacks, 1967), p. 98.

The tension of scientific novelty is so serious that the psychologist does not find it inappropriate to establish a comparison between the thinking of the creative scientist and that of a paranoid patient. In both cases there is an urge to break away from what everybody considers normal. The scientist, of course, is not animated by the wish to escape reality. He rather wishes to come to grips with reality in a more objective way. But this is precisely what entails the strain. The tension originates from the discrepancy that occurs between what one has been culturally accustomed to taking for granted and what one wants to perceive through personal investigation. In Eiduson's summary:

The difference between the thinking of the paranoid patient and the scientist comes in the latter's ability and willingness to test out his fantasies or grandiose conceptualizations through the systems of checks and balances science has established—and to give up these schemes that are shown not to be valid on the basis of these scientific checks. [62]

In the light of the strenuous demands that science imposes on man, one should not be surprised, much less scandalized, by the psychological-ethical weaknesses occasionally manifested by the genuine scientists themselves. They are but unmistakable signs of their humanity. Sometimes such weaknesses amount to what can be called *professional deformation*. The scientist may become stunted in some areas of his personality owing to the disproportionate development in some other areas. Darwin, for instance, repeatedly acknowledges in his *Autobiography* that he—to his deep regret—has completely lost in his adult age the great taste for poetry he had as a young man. [63] On the other hand he complains that his mind has become overly absorbed with scientific concerns. As he sadly put it: "My mind seems to have become a kind of machine for grinding general laws out of large collections of facts. . . ." [64]

Often the psychological burden of science manifests itself in the form of *professional defects*. One investigator who gave much attention to this matter is the sociologist Merton. He summarizes the scientists' failings as follows:

... contentiousness, self-assertive claims, secretiveness lest one be forestalled, reporting only the data that support a hypothesis, false charges of plagiarism, even the occasional theft of ideas and, in rare cases, the fabrication of data.... [65]

---

[62] Eiduson (note 4), p. 107.
[63] N. Barlow, ed., *The Autobiography of Charles Darwin* (*with original omissions restored*) (New York: Norton, 1969), pp. 43f and 138f.
[64] *Ibid.*, p. 139.
[65] R. K. Merton, in B. Barber and W. Hirsch, eds. *The Sociology of Science* (New York: Free Press, 1962), p. 485.

The forms of behavior listed are ethically reprehensible and should not be condoned. However, for our purpose it is enough to have pointed out their psychological origin. They are clearly exaggerations of the good qualities that typify the scientist. When people are totally bent on the search for new knowledge, it is only too natural that they may occasionally exaggerate and hence fall into the defects we deplore in them. Eiduson speaks appositely in this connection of what she calls "the seductive graces" of science. These are the attractions that can captivate a scientific person so much as to make him lose the psychological harmony of his personality. Among such "seductive graces" she lists particularly what nonscientists find offensive in scientific workers such as "rebellion against the traditional, the breaking down of what had been fixed, the questioning of the taken-for-granted, the distrust of the obvious." [66] From her own enumeration it is clear that such defects are exaggerations of positive features of the scientific mind.

But the psychological burden of science is not only a source of inner tension and occasional defectiveness. It can cut much deeper into the scientific personality. The most distressing manifestation is the occasional sense of *despondency plaguing the successful scientist*. Kubie, a psychiatrist with much experience of scientific people, stresses the not too infrequent occurrence of such a phenomenon.

The ancient tragedy of human nature, the success which brings no joy with it occurs at least as frequently in the life of the scientific investigator as in art and business. A life of fruitful scientific exploration may end in a feeling of total defeat, precisely because in spite of scientific success the unconscious goals of the search have eluded the searcher. [67]

The psychological explanation of the despondency of the successful scientist is easy to find. Science has attracted him so much that he has enslaved himself to it. Thus he has rushed into and through science, without ever pausing to inquire about the humanizing goal he was pursuing in his work and how he had to proceed in order to attain it effectively. Such a situation accounts for the despondency. For, at a certain moment, the scientist as a human being must realize that science, pursued in this way, is unable to satisfy his profoundest desires; especially, it cannot give him rest and undisturbed happiness. The reason is clear at least because science keeps expanding, and more and more fields of investigation come to the fore while, simultaneously, man's own energy keeps decreasing. Hence science, as a total satisfaction of man's desires, proves to be an ever-receding goal—ultimately, a delusion. Thus, in the end, the scientist

---

[66] Eiduson (note 4), p. 188.
[67] L. S. Kubie, "Some Unsolved Problems of the Scientific Career," in Barber and Hirsch (note 65), pp. 201-229, especially p. 217.

feels that his life has been wasted. He has the impression of having been sacrificed to an impossible dream. Science now appears to him as a tyrant and destroyer.

A pathetic confirmation of the foregoing analysis is found in the writings of Wilhelm Ostwald. He quite clearly speaks from experience. Ostwald was one of the most successful scientists and enthusiastic promoters of science who ever lived. In a book aimed at celebrating the great scientists of the past, he paints a most devastating picture of science itself. To him science appears as a man-devouring goddess, a monster that demands total dedication of her votaries in the prime of their lives and then abandons them to desperation. According to Ostwald, science can be compared to death itself. "Science demands its victims with the same eerie unavoidability as Death." [68] Many scientists—he explains—are drained of their livelihood while still in their youth. But these are the fortunate ones, in his view. For the others must watch their energies gradually disappear and their performance steadily decrease while the expectations of society from them and the responsibilities of their own profession keep mounting unceasingly. Here is the crucial point, the tragedy of the situation: this inherent tension between expectations and hopeless discouragement. In Ostwald's own words:

Here it is where science is indeed wont to offer outward shine to its followers, but by so doing it demands internal sacrifice—human sacrifice unheard-of (*Menschenopfer unerhört*).

The term used by Ostwald is illuminating. *Menschenopfer* is the word usually employed to brand as abominable the killing of people in order to satisfy a bloodthirsty idol. Ostwald insists on this meaning. He speaks exactly of human victims, and does not shrink back from giving an example: "I think that one of such victims was Boltzmann." [69] Ostwald sees destructiveness as an inherent feature of science. This destructiveness operates through the agency of an inescapable psychological mechanism. The man who has first tasted the pleasure of scientific work cannot avoid dedictating his life to it, but science destroys him in the process. In Ostwald's own summary:

He who has tasted the irresistible charm of the work of scientific discovery knows that this excludes any regard for energy and health the

---

[68] W. Ostwald, *Grosse Männer: Studien zur Biologie des Genies* (Leipzig: Akademische Verlagsgesellschaft, 1910), pp. 406f.

[69] L. Boltzmann was the most prominent theoretician of the atomic realm around the turn of the century. He killed himself in 1906 when he was 62. For some time before that date he had been obsessed with an increasing fear of being unable to withstand the pressure of scientific life. He resented particularly Mach's nagging criticism about the meaninglessness of studying atoms. For details see Broda (note 48).

more certainly, the greater the success of the work itself. Thus every important discovery is paid for with a human life. For, if it does make a difference whether the discoverer is totally destroyed or if he survives as a cripple, the difference is all to the worse in the case of the latter.

If the psychological burden of science can be crushing for those who have been successful, it is obviously even more frightening for those whose endeavor has been unsuccessful. Apparently there are many who dedicate themselves to science with the best intentions and total generosity, but actually fail in their intent. The reasons for their failure may be various, but from the experiential standpoint it makes no difference. Science can mean *total disappointment* for those who enslave themselves to it. Kubie rightly speaks in this context of the "expendables of science." He also insists on the frequency of this phenomenon. The stark synthesis he offers is worth pondering:

Every successful scientific career is an unmarked gravestone over the lives of hundreds of equally able and devoted, but oscure and less fortunate, anonymous investigators. Science too has its "expendables".... This is the ultimate gamble which the scientist takes when he stakes his all on professional achievement and recognition, sacrificing to this scientific career recreation, family, and sometimes even instinctual needs, as well as the practical security of money. [70]

To sum up, one should not overemphasize the psychological burden of science. For it is obviously possible for a scientist to become harmoniusly developed in his psychological structure not despite but because of his doing science. Nevertheless, one point becomes increasingly clear. Science does not automatically suffice to make man fully man. Science, therefore, if taken alone, is inadequate as a humanistic factor.

(b) Cognitive Frustration

A second consideration discloses a basic inadequacy of science in the humanistic realm. It regards the promises of knowledge that it holds out for its followers. For the sake of clarity we should notice that we are dealing now with a subject matters different from the preceding. There the emphasis was on the psychological repercussions of science. Here the emphasis is on the cognitive repercussions, where the term is taken in its philosophical rather than psychological connotation. In other words, we are not concerned here with the possibility that science may lead its practitioners to a psychological breakdown but rather with the possibility that it may leave them inclined toward skepticism.

---

[70] Kubie (note 67), p. 228.

The relationship of science to understanding is deceptively simple. A person starts to do science animated by the desire to know. But the more he does science, the more he realizes that science alone will never be able to satisfy his thirst for knowledge. The reason is clear. Understanding or a humanly satisfactory knowledge must, at least in principle, encompass the whole of reality. For man can never feel at rest in his longing for cognition unless he is able somehow to grasp the whole of reality with his mind. But the scientific approach can never satisfy this desire for total understanding. Waiving for the moment any philosophical discussion of the issue, this is such an obvious fact that even the most exuberant of science's promoters cannot fail to point it out. The position of Karl Pearson is a telling example. While writing his *Grammar* to emphasize that science alone is reliable as a source of knowledge, he felt moved to warn the readers against expecting too much from science itself. As he put it:

The goal of science is clear—it is nothing short of the complete interpretation of the universe. But the goal is an ideal one—it marks the *direction* in which we move and strive, but never a stage we shall actually reach. The universe grows ever larger as we learn to understand more of our own corner of it. [71]

Experientially, two principal grounds prove the inadequacy of science as a source of knowledge. The first is hinted at by Pearson in the words cited. The more science develops, the more it uncovers new problems. The frustrating nature of this endless development is expressed by the scientists themselves in frequently mentioned similes. Some employ the "door and room" image. They have the impression that a discovery is something like opening a door to a room. When you think you have arrived, you notice that the room you are in is nothing but a passage way. Other doors and other rooms wait for you, in endless succession. Some other scientists speak of light and darkness. The greater the circle of the knowledge you have attained, the larger the surrounding area of ignorance you become aware of. The increase of light seems only to point out the unconquerableness of darkness. From this experiential point of view one is almost forced to agree with G. B. Shaw's cruel gibe: "Science is always wrong. It solves problems only to replace them with others."

The second experiential ground of the cognitive inadequacy of science stems from the preceding. Not only does science proliferate endlessly, but this proliferation seems to threaten the validity of the results achieved in the past. New discoveries leave man baffled about the very nature of scientific truth. To express this in a simile, it is not just that science confronts man with a merely "ideal" goal, as Pearson put it in the words

---

[71] K. Pearson, *The Grammar of Science* (London: Dent, 1937), p. 18.

quoted. Science seems to confront man with a mirage. The more man strives after knowledge by doing science, the more he has the impression that knowledge itself recedes and dissipates. At any event, knowledge becomes increasingly fragmented and unsurveyable. Even such a sympathetic historian of the scientific mentality as Merz cannot help putting it bluntly:

The encyclopedic treatment of knowledge ... has shown that the extension and application of learning leads to the disintegration, not to the unification, of knowledge and thought. [72]

The topics discussed are convincing evidence of the cognitive inadequacy of science. They become still more convincing when one considers that they are acknowledged as valid also by creative scientists for whom science has definitely a humanistic significance. A telling testimony in this regard is that of William Thomson, Lord Kelvin. He had been one of the most brilliant, successful, and creative physicists of all time. He had to his credit no less than the average of one scientific paper a month throughout his scientific career. Many of his contributions were rightly regarded of enduring validity. And yet, in his seventies, he felt compelled to characterize his entire scientific effort with the dismal word of "failure." Here is his own statement:

One word characterizes the most strenuous of the efforts for the advancement of science that I have made perseveringly during fifty-five years, and the word is *failure*. I know no more of electric and magnetic forces or of the relation between ether, electricity and ponderable matter, or of chemical affinity than I knew and tried to teach to my students of natural philosophy fifty years ago in my first session as professor. [73]

Of course, one can easily extenuate testimony such as this. One may point to psychological considerations. Thomson perhaps felt dejected because of his old age, which impeded the continuation of his creative research. Also one may say that Thomson probably had expected too much from science owing to the positivistic mentality of the times. Be it as it may, the words cited indicate that science tends not to fulfil the cognitive expectations aroused in its followers. This conclusion is even clearer when we examine statements by scientists to whom the extenuating circumstances mentioned do not apply. One such scientist is Heisenberg.

---

[72] J. T. Merz, *A History of European Scientific Thought in the Nineteenth Century* (New York: Dover, 1965), vol. I, p. 37.

[73] Quoted in J. J. Thomson, *Recollections and Reflections* (New York: Macmillan, 1937), p. 425.

From all his writings it is obvious that Heisenberg does not belittle the humanistic importance of science. Nor does he overrate it in a positivistic way. And yet, when commenting on his own experience as a creator of quantum physics, he cannot refrain from pointing to a discouraging aspect of science. He stresses that quantum physics demanded a complete change in the scientist's frame of mind when compared with the hitherto established mentality of classical mechanics. This Heisenberg sees as an inherent requirement of science as such—the continual necessity of changing one's outlook without ever hoping to attain a perfect and unchangeable knowledge. His words emphasize the disquieting character of such a requirement:

Yes, the necessity is to break away those things which seem to be obvious and which actually are the basis on which you stand. One always, in such a situation, is forced to cut the branch on which one is sitting. That can't be helped, because after all, one never can rest. There is no solid bottom. One is always somewhere in the middle and one can get some clarity around oneself but one can never hope that these fundamentals will rest forever. [74]

To clarify his point, Heisenberg mentions in the same context the famous theoretical physicist Paul Ehrenfest, whose attitude toward science he designates as "despair." Continuing his own comments, Heisenberg himself seems to agree with the basic reasons for such an attitude.

He knew that we must, so to say, always be in despair about our situation. Out of this despair we must do something which leads a little bit further, and that's all we can do. We certainly can't hope to solve the problems once and for all. That's just out of the question.

To close, we should not overstress the cognitive frustration produced by science. For science is genuine knowledge, and deserves to be openly acknowledged as such. Nonetheless, the fact remains: science proves itself inadequate as a source of knowledge. This is the more obvious the more man places all his trust in science. Jean Rostand, a scientist for whom science is the only trustworthy form of knowing, put it chillingly:

In proportion as science increases her power, science feels less assured of her knowledge. Like a person afflicted with a "failure neurosis," science finds occasion for her gravest doubts in her greatest triumphs. [75]

(c) Ethical Disgust

Another area in which science proves itself humanistically inadequate is that of ethics. This statement, of course, does not come as a surprise

---

[74] In Kuhn *et al.* (note 25); interview February 27, 1963, pp. 20f.
[75] Rostand (note 1), p. 85.

today when so much criticism is voiced by the public against the negative aspects of our scientific civilization. However, it raises an issue that must be analyzed here, in view of our specific goal. For we are concerned not with the reactions of the public, who frequently misunderstands the nature of science. Rather, we want to know how scientists themselves come to evaluate the ethical relevance of science by reflecting on their own experience.

An attitude of ethical repugnance toward science has become increasingly noticeable in recent years among some of the best scientists. The reasons are multiple. Chief, of course, is the shock produced by the use of the atom bomb, and the constant threat that the bomb poses to society. In this connection, possibly no more poignant case can be found than that of Max Born. This great theoretician was an acknowledged leader among the brilliant people who developed quantum physics. He did not participate in any way, even indirectly, in the production of the atom bomb. Thus one would have expected of him a sense of serenity concerning science. And yet, in his old age, Born could not help declaring that science was causing him a deep sense of foreboding and dissatisfaction. In his own words:

We, the atom and I, have been on friendly terms, until recently. I saw in it the key to the deepest secrets of nature, and it revealed to me the greatness of creation and the Creator. It supplied me with satisfactory work, in research and teaching, and thus provided me with a livelihood. But now it has become the source of deep sorrow and apprehension, to myself as well as to everyone else. [76]

One could easily try to lessen the impact of the words cited by explaining them in purely psychological terms. Born could be seen as the typical disgruntled old theoretician who is naïvely surprised at the harshness of real life. But such an interpretation would simply miss the point that Born is trying to make. He himself in another passage of the same work indicates the profound reason that motivates his stand. This reason goes well beyond any contingent historical manifestation and seems to indicate a weakness inherent in science itself. To quote him again:

I am haunted by the idea that this break in human civilization, caused by the discovery of the scientific method, may be irreparable. Though I love science I have the feeling that it is so greatly opposed to history and tradition that it cannot be absorbed by our civilization. The political and military horrors and the complete breakdown of ethics which I have witnessed during my lifetime may be not a symptom of an ephemeral

---

[76] M. Born, *My Life and Views* (New York: Scribner, 1968), p. 88.

social weakness but a necessary consequence of the rise of science—which in itself is among the highest intellectual achievements of man. [77]

One could still object that any criticism of science on ethical grounds is misplaced and hence pointless. For genuine science is, after all, essentially knowledge or theory. Thus everything that can be harmful to man in practice, even if derived from science, is no longer science but rather technicalism. The rebuttal is justified, yet the nagging question remains: why does science allow itself to be so easily technicalized? This is the central problem that worries all those who love and respect science. And this is the reason why scientists become wary in regard to science when ethical issues are involved. Einstein, for one, while sending his wishes to a society dedicated to ethical culture, could not refrain from stressing that in the ethical realm "no science can save us." [78] Then, going on to explain his thought, he diagnosed a certain public loss of ethical sense as due to the scientific tendency to insist on a purely intellectual attitude. This, he took care to note, is not really a matter of technical spirit but something even more basic. As he put it:

I believe, indeed, that overemphasis on the purely intellectual attitude, often directed solely to the practical and factual, in our education, has led directly to the impairment of ethical values. I am not thinking so much of the dangers with which technical progress has directly confronted mankind, as of the stifling of mutual human considerations by a "matter-of-fact" habit of thought which has come to lie like a killing frost upon human relations.

Another distressing consequence of science that bears on ethics is pointed out by Polanyi. He shows how science can lend itself to such a misinterpretation as to become a source of pseudoethics and pseudoideal. In brief:

Alleged scientific assertions, which are accepted as such because they satisfy moral passions, will excite these passions further, and thus lend increased convincing power to the scientific affirmations in question—and so on, indefinitely. [79]

The sociopsychological mechanism Polanyi refers to here is well known. Science is frequently seen by many as a revolutionary mentality that, for the sake of progress, rejects and destroys traditional ethical values. As Born put it:

The scientific attitude is apt to create doubt and skepticism toward tradi-

[77] *Ibid.*, p. 58.
[78] Einstein (note 8), p. 53.
[79] Polanyi (note 16), p. 230.

tional, unscientific knowledge and even toward natural, unsophisticated actions on which human society depends. [80]

The consequence of the scientific questioning of traditional ethics is obvious. Since man cannot live without ethics, it is natural that science itself be considered the source of a new ethics. This self-styled scientific ethics is the more readily accepted the more science is deemed superior to the prescientific mentality it displaces. Hence we have the paradox of an ethics that becomes more dogmatic and oppressive as it increasingly claims to free man from the unenlightenment of the prescientific mentality. In short, science may easily lead to ethical perversion. There is no need to insist on the seriousness of such a situation. We have all witnessed in recent years the catastrophic effects of such a mentality when it is adopted and ruthlessly exploited by totalitarian regimes. But the threatening danger of such an attitude is growing rather than diminishing, even in the democratic countries. In Polanyi's own summary:

This explains not only the deliberate unscrupulousness of modern totalitarianism, but also the moral appeal of its declared resolve to act unscrupulously. For this resolve is taken to certify that its power embodies righteousness, and may therefore acknowledge no higher obligation than that of defending its own supremacy which it must do at all costs. Those who rule in its name are entitled to scorn mercy and honesty, not simply for reasons of expediency... but on account of their moral superiority over the emotionalism, hypocrisy, and general woolliness of their moralizing opponents. [81]

The result of the situation we have analyzed is that science tends to be seen with growing ethical disgust even by those for whom science is the most positive achievement of modern civilization. A synthesis of this attitude is to be found in a very bitter passage by Bertrand Russell. [82] He who had been a most enthusiastic promoter of the scientific mentality came to think of science as the tyrant of man and nature.

Russell starts out by acknowledging that science arose as an attitude of love:

Science in its beginnings was due to men who were in love with the world. They perceived the beauty of the stars and the sea, of the winds and the mountains. Because they loved them their thoughts dwelt upon them, and they wished to understand them more intimately than a mere outward contemplation made possible.

But, he goes on to explain, in the process of time: "The lover of nature has been baffled, the tyrant over nature has been rewarded." The reason

---

[80] Born (note 76), p. 54.
[81] Polanyi (note 16), p. 231.
[82] B. Russell, *The Scientific Outlook* (New York: Norton, 1959), pp. 262-264.

for this development, as Russell sees it, is twofold. First, science has been unable to satisfy the aspirations of the lover of nature in that it has reduced nature itself to pure mechanism. Clearly, mechanism cannot attract the love of anyone because reality comes to appear as "a skeleton of rattling bones, cold and dreadful, and perhaps a mere phantasm." The second reason for the transition from love to tyranny is the fact that science has proved to be a growing source of power for man over nature. Hence man has tended increasingly to take advantage of science in order to exploit nature and also damage himself. Russell concludes by forecasting a scientific society of the future completely dominated by tyranny and sadism.

Disappointed as the lover of nature, the man of science is becoming its tyrant.... Thus science has more and more substituted power-knowledge for love-knowledge, and as this substitution becomes completed science tends more and more to become sadistic. The scientific society of the future as we have been imagining it is one in which the power impulse has completely overwhelmed the impulse of love and this is the psychological source of the cruelties which it is in danger of exhibiting.

To close, it is clear again that we should not overemphasize our findings. For science, having a definite humanistic significance, cannot avoid having also a positive ethical relevance. Nonetheless, a point is established. Science is not so humanistically significant as to be adequate in the most vital area of humanism, namely the realm of ethics. Hence the humanistic inadequacy of science becomes more and more evident.

(d) Agnosticism and Meaninglessness

A final area in which the scientific approach manifests its humanistic inadequacy regards the absolute. As has been seen (Chapter 3, Section 3), science can lead man to a personal experience of the absolute. However, when man tries to reach an assurance about ultimates and absolutes by relying solely on the scientific approach, he is likely to feel perplexed and even desperate. This is what constitutes the phenomenon of scientific agnosticism—the benumbing suspicion that, as a result of science, one must resign oneself to ultimate meaninglessness.

Historically speaking, it is illuminating that the very term *agnosticism* came into vogue at the height of the public enthusiasm for science during the 19th century. The coiner of the word and its indefatigable propagandist was Thomas Henry Huxley. It may appear odd that this man, a distinguished biologist and enthusiastic promoter of science, was so much in favor of agnosticism. But he did not see any inconsistency in his position; rather the opposite. We can gain a valuable insight by briefly studying his views.

According to Huxley, agnosticism is the necessary product of the scientific mentality. Recalling the reason why he came to call himself an agnostic, Huxley mentions the social custom of labeling oneself according to the ultimate certainties that one entertains because of religious or philosophical motives. In his own conviction, he differed from most other people in that he was certain that nothing ultimate could be known with certainty. Hence he opted for agnosticism.

They were quite sure they had attained a certain "gnosis"—had, more or less successfully, solved the problem of existence; while I was quite sure I had not, and had a pretty strong conviction that the problem was insoluble.... [83]

As is intimated in the passage cited, Huxley define the term "agnosticism" following its etymology. He takes it as the exact opposite of gnosis or knowledge. However, according to him, this term does not stand for something truly negative. Rather, it expresses a conviction that is eminently positive and constitutes a conquest of the mind. Thus he endeavors to describe agnosticism as a critical attitude, as a method leading to certainty:

Agnosticism, in fact, is not a creed, but a method, the essence of which lies in the rigorous application of a single principle. That principle is of great antiquity, it is as old as Socrates; as old as the writer who said: 'Try all things, hold fast by that which is good'... [84]

In particular, Huxley connects agnosticism to the scientific approach. Thus, in fact, be goes on to explain in the text just cited:

...it is the fundamental axiom of modern science. Positively the principle may be expressed: In matters of the intellect, follow your reason as far as it will take you, without regard to any consideration. And negatively: In matters of the intellect do not pretend that conclusions are certain which are not demonstrated or demonstrable.

Above all, Huxley wants to vindicate agnosticism as a great advance in thought. He opposed it to what he considers the continual failure of philosophy to achieve ultimate certainties. Thus, even if agnosticism were to be deemed something merely negative, he still contends that it expresses a valuable insight that makes the mind more realistic. In his own terms:

I do not very much care to speak of anything as "unknowable." What I am sure about is that there are many topics about which I know nothing; and which, so far as I can see, are out of reach of my faculties.... It is getting on for twenty-five centuries, at least, since mankind began

---

[83] In Bibby (note 15), p. 19.
[84] *Ibid.*, p. 111.

seriously to give their minds to these topics. Generation after generation, philosophy has been doomed to roll the stone uphill; and, just as all the world swore it was at the top, down it has rolled to the bottom again.... [85]

Agnosticism as propounded by Huxley is a beguiling mixture of self-assurance and skepticism, of critical attitude and absolute self-confidence. But apparently Huxley excluded the possibility that science itself could be engulfed by the agnostic mentality. This is, however, the position adopted by another prominent interpreter of science who flourished at the same time as Huxley. Emil du Bois-Reymond, a leading physiologist, defended the view that science itself should be interpreted agnostically.

Contemporary biologists, especially in Germany, were increasingly convinced that biology definitively proved the validity of mechanistic materialism. The argument was the familiar one: life is nothing but a series of chemical-physical phenomena that are perfectly understood in terms of the mechanistic conceptions of matter and energy. In some famous lectures—particularly one tellingly entitled "The Seven World Enigmas" (1880)—du Bois-Reymond shocked his colleagues by flatly denying their basic assumption of total knowledge through science. His view was diametrically opposite. He maintained that even though it could be proved that the problems of life are reducible to those of physics and chemistry, the physical concepts of matter and force will themselves remain unexplained and not understood. For these concepts are but abstractions derived from phenomena, and man cannot know the ultimate reasons why phenomena present the features that actually characterize them. Therefore, du Bois-Reymond inferred, we have to admit that reality remains essentially unknown and unknowable, forever.

Does du Bois-Reymond's agnostic interpretation of science entail that man had better give up scientific research? Quite the contrary, as he saw it. With great emphasis, he insisted that one should rather increase one's own commitment to science. His four Latin slogans have become classic. According to du Bois-Reymond, scientific man should first of all be realistic, and openly acknowledge his complete ignorance—*ignoramus*. Furthermore, he should avow his radical incapability of attaining to knowledge even in the future—*ignorabimus*. But this situation should not induce him to abandon research. Rather, it should spur him toward adopting an increasingly questioning attitude—*dubitemus*. And this inquisitive attitude should express itself into a determined effort to work hard—*laboremus*. [86]

It is not our task here to examine critically the doctrine of agnosticism.

---

[85] *Ibid.*, p. 114.
[86] For details see, for instance, Merz (note 72), p. 53, note 2.

Thus we may ignore the contention that agnosticism is the inherent consequence of science. We may also overlook the doubtful consistency of the two agnostic positions outlined. But we must acknowledge the great humanistic significance of agnosticism. It is truly a view that affects the whole of man—all of his values and ideals. What is then the message of agnosticism to man as such? We have seen that it can be interpreted in many ways. For some, like Huxley, to be agnostic is something one can boast about. For others, like du Bois-Reymond, agnosticism is more of a sobering challenge. We must pursue this issue further. In particular, we must see whether such optimistic assessments respond to the experience of the scientist who has probed agnosticism in depth.

We find an illuminating analysis of the corrosive nature of agnosticism in Darwin's frank and detailed confession of his religious travails. Darwin stands out through the restraint and profundity with which he faces the issue. He does not flaunt his agnosticism. He just acknowledges it modestly, with obvious distress. We can follow the whole arc of Darwin's anguished groping for ultimate certainties. At the beginning, there seemed to be no problem. He recalls that he was extremely sensitive to religious experience in his youth. He writes that "whilst standing in the midst of the grandeur of a Brazilian forest" he could not help sensing "wonder, admiration, and devotion which fill and elevate the mind." [87] As he remarks immediately afterward, this experience gave to him also a sense of the unique dignity of the human person: "I well remember my conviction that there is more in man than the mere breath of his body." And yet, in his old age and after so much scientific success, he feels sadly moved to confess that he cannot experience anything of the kind any longer.

But now the grandest scenes would not cause any such convictions and feelings to rise in my mind. It may be truly said that I am like a man who has become colour-blind, and the universal belief by men of the existence of redness makes my present loss of perception of not the least value as evidence.

Despite his experiential crisis, Darwin did not abandon the search for ultimate certainties. While acknowledging his insensitivity to religious experience, he struggled to find other approaches that could assure him an unshakable certitude and enlighten him about the overall meaning of reality. This he did by resorting to reasoning. Reflecting on the significance of natural orderliness he wrote:

Another source of conviction in the existence of God, connected with the reason and not with the feelings, impresses me as having much more weight. This follows from the extreme difficulty or rather impossibility

---

[87] In Barlow, (note 63), pp. 91-94.

of conceiving this immense and wonderful universe, including man with his capacity of looking far backwards and far into futurity, as the result of blind chance or necessity. When thus reflecting I feel compelled to look to a First Cause having an intelligent mind in some degree analogous to that of man; and I deserve to be called a Theist.

Darwin notes that the attitude described was strong in his mind at the time when he wrote *The Origin of Species*. But then he adds that he has had "many fluctuations" on this point. The overall consequence is that Darwin came to be thoroughly dissatisfied and anguished in his search for ultimates and absolutes. He felt overwhelmed by the complexity of the issues involved and resignedly accepted to call himself an agnostic. In his own summary:

I cannot pretend to throw the least light on such abstruse problems. The mystery of the beginning of all things is insoluble by us; and I for one must be content to remain an Agnostic.

To Darwin, the agnostic condition was a far cry from the optimistic situation depicted by Huxley and du Bois-Reymond. It was a purely negative result, the position of a mind that had become so completely uncertain that it did not dare to take any firm stand—except perhaps to exclude the possibility of any certainty. In particular, Darwin always refused to admit he was certain about the nonexistence of God. As he wrote to a friend toward the end of his life:

In my most extreme fluctuations I have never been an atheist in the sense of denying the existence of a God. I think that generally (and more and more as I grow older), but not always, that an agnostic would be the more correct description of my state of mind. [88]

To Darwin agnosticism was no less than a source of complete disconsolateness bordering on desperation. Thus he kept speculating, despite the apparent hopelessness of his endless thinking and the profound grief he derived from it. He confesses as much to a friend in his late years:

Your conclusion that all speculation about preordination is idle waste of time is the only wise one; but how difficult it is not to speculate! My theology is a simple muddle; I cannot look at the universe as the result of blind chance, yet I can see no evidence of beneficent design; or indeed of design of any kind, in the details. [89]

From the analysis of Darwin's case it is clear that science can have a completely negative humanistic significance. Far from leading man to

---

[88] Letter to J. Fordyce; quoted in G. de Beer, *Charles Darwin: A Scientific Biography* (New York: Doubleday Anchor Books, 1965), pp. 268f.
[89] Letter to J. D. Hooker, 1870, quoted in Barlow (note 63), p. 162.

complete certainty, it can lead him to ultimate meaninglessness. This is the eventual outcome for many persons who had expected every possible certainty from science. Science leads them to agnosticism, and agnosticism breeds desperation. An example of such desperation is contained in a text by Fred Hoyle, the leading contemporary astrophysicist. After having discussed the present-day knowledge of the universe, he concludes by pointing to the situation in which modern man comes to find himself. He compares man to a climber perched precariously on the brink of a precipice.

Here we are, in this wholly phantastic universe, without scarcely a clue as to whether our existence has any real significance.... We are in rather the situation of a man in a desperate, difficult position on a steep mountain. [90]

As Hoyle goes on to explain, there is something absurd in the situation of scientific man. He is able to penetrate the universe with his mind, and yet does not understand the meaning of his own life.

Perhaps the most majestic feature of our whole existence is that while our intelligences are powerful enough to penetrate deeply into the evolution of this quite incredible universe, we still have not the smallest clue to our own fate. [91]

Science can depress man to the limit. An illuminating example, because of the reasons given, is the discussion of the human import of science undertaken by Jean Rostand, the distinguished biologist and restless philosophical inquirer. Rostand analyzes the message of science from the standpoint of evolution interpreted as struggle for existence and survival of the fittest. His conclusion is a most dreary one. To him man, as a result of the light thrown on him by science, appears a totally lonesome being, wandering aimlessly in a completely meaningless universe. In his own words:

All that he [man] holds dear, all that he believes in, all that matters in his eyes, began in him and will end with him. He is alone—foreign to all else. There is nowhere any slightest echo to his spiritual needs. And all that the surrounding world offers him is the spectacle of a somber and sterile charnel house where the triumph of brute force is staged in disdain for suffering, indifference toward the individual, the group, the species, and toward life itself. [92]

Rostand is very convinced of his view. To him this distressing picture

---

[90] F. Hoyle, *The Nature of the Universe* (New York: Harper & Row, 1960), p. 138.
[91] *Ibid.,* p. 141.
[92] Rostand (note 1), pp. 61f.

is all that science can give to man—a message of desperation. Thus in fact he concludes the passage cited:

Such, it seems, is science's message. It is arid. Up to now, one must admit, science has done no more than to give man a clearer conscious-ness of the tragic strangeness of his condition, while awakening him, as it were, to the nightmare that grips him.

We can close our investigation about the humanistic limitations of science. There is no denying that—for all the genuine humanism of science—the scientific approach does not suffice in itself to make man more human. From step to step, beginning from psychological influences down to implications concerning absolutes and values, it seems that science takes a perverse pleasure in consternating man and threatening him with destruction. To be sure, this is not a necessary consequence of science as such. For many persons have become more richly human through their doing science, and all others are called to imitate them. Nevertheless, it is disturbingly instructive to notice how so many people who have generously given themselves to science come in the end to feel bewildered, confused, discouraged, prostrated, and utterly disgusted. It is always distressing to watch man wandering about in desperation. It is even more distressing to watch the desperation threatening scientific man, since science is one of the greatest achievements of man. Hence the need arises urgently to find a remedy to the situation. For the corrosive process goes on and tends to engulf everything dear to man, including science itself. Rostand, for instance, in the consistency of his position, does not refrain from seeing the very wonder of science as just an absurd astonishment. As he put it:

Our way of judging the works of nature must be rather like that of the layman judging a work of art. Full of absurd astonishments. If there is anything to be amazed at in nature, it is surely not where we think. [93]

The humanistic inadequacy of science should arouse the concern of every reflective person, scientist and nonscientist alike. For it is not merely an academic question that can safely be ignored by the average, pragmatically inclined man. Rather, given the unavoidable consistency of the human mind, and the growing social influence of science itself, it becomes increasingly a challenge to man as such—even a destructive threat. Rostand, with his customary penetration, has pointed out the complex relationship of trust and hatred engendered in contemporary man by science. "One denies more than one should when one cannot believe as much as one would like to." [94] But man, to live as such, needs both believing

---

[93] *Ibid.*, p. 48.
[94] *Ibid.*, p. 273.

and denying. Hence, if his reaction to science is just emotional and not harmoniously reflective, he may well end up by either believing too much or denying too much—and thus destroying himself in the process.

4. THE DRAMA OF SCIENTIFIC MAN

In this first part of our investigation we have examined systematically the experience of scientific man. We must now try to give a preliminary answer to our central question. What is the general significance of science to man? On the basis of our results the answer is twofold, negative and positive.

Negatively speaking, it is clear that science is not what many people hold it to be. The popular misconceptions of science are many. Some border on caricature. To begin with the least sophisticated of such views, science is not a hybrid mixture of practical recipes and abstruse esoterism. The scientist is not a combination of magician and charlatan, inebriated by his own success, cocksure in his approach to all kinds of problems. Naturally enough, scientists can be found who act like, and also think of themselves as, infallible priests of an almighty goddess. But such a mythologization of science, if it exists, clamors for debunking. Intelligent man, including the scientist, enjoys a good chuckle at his own expense. [95] Nor can science be considered an arrogant, ruthless and naïve self-projection of *homo faber*. In other words, science is not to be confused with technology nor, much less, with technicalism. To be sure, many a scientist may present the defects of the technicalist mentality. But the humanistic investigator should know better than to ascribe to science the failings of some of its practitioners. In particular, he should not perpetuate such a disgusting misinterpretation of science as a matter of principle. [96]

A more refined form of misunderstanding considers science as a sort of expensive hobby, an evasion of the roughness and the demands of real life, an elitist contempt for what cannot be fitted into neat theoretical schemes. This view, unfortunately, keeps recurring under the pen of humanistic writers. Thus, for instance, Miguel de Unamuno is reported to have said that science not only does not deal with the problems of real man, but it "turns against those who refuse to submit to its orthodoxy

---

[95] For a witty satire of the pompous scientist see, for instance, A. Standen, *Science Is a Sacred Cow* (New York: Dutton, 1950). The author is a chemist and editor of *Encyclopedia of Chemical Technology*.

[96] For a notable example for such misinterpretation of science see L. S. Feuer, *The Scientific Intellectual: The Psychological and Sociological Origins of Modern Science* (New York: Basic Books, 1963).

the weapons of ridicule and contempt." [97]   Another writer goes so far as
to dub science "the glorious entertainment." [98]   No safer way could be
found to perpetuate the cultural split of our civilization than by adopting
such a position, which is objectively false and subjectively offensive.   The
falsity of interpreting science as escapism and play is clear in the light of
our investigation and ought to be known to seriously concerned scholars. [99]
The accusation of unrealism is especially offensive for the scientists because
their central concern, to which they dedicate their time and even their
lives, is nothing but respectful attention to reality itself.

A positive characterization of the experiential significance of science
for man is difficult to formulate.   The reason is a seemingly inherent
double-sidedness of science as a man-affecting factor.   Reflective scientists
themselves are puzzled and alarmed by the situation.   They frequently
point it out.   A distinguished physiologist writes, for instance, that science
is essentially dilemmatic:

Some kind of dilemma seems to be inherent in science, for every achievement, by its limitations, implies also a failure.   The answer to a question generally raises a host of new questions, scientific, political, philosophical or ethical. [100]

Another great scientist concurs that science seems to be eminently ambiguous: to one person it is a source of joy, to another a source of sad-
ness.   In his own words:

Natural knowledge we commonly account a joyous thing; but it is clearly
a saddening thing as well.   Nature is like a music to which two friends
can listen and both be moved and yet each by a different train of
thought. [101]

Two main reasons make scientific man feel embarrassed about his
science.   One is the easiness with which the best attainments of science
can be abused by man-oppressing technicalism in the realm of politics,
economy, and various forms of human engineering.   Einstein sees this
as the almost inescapable tragedy of the scientific researcher.   As he put it:

Thus the man of science, as we can observe with our own eyes, suffers
a truly tragic fate.   Striving in great sincerity for clarity and inner in-

---

[97] For de Unamuno's quotation and additional comments see R. Dubos, *The Dreams
of Reason: Science and Utopias* (New York: Columbia University Press, 1961), pp. 148f.
    [98] J. Barzun, *Science: The Glorious Entertainment* (New York: Harper & Row, 1964).
    [99] An implicit confutation of Barzun's central thesis can be found in J. Huizinga's
Homo Ludens: A Study of the Play Element in Culture (Boston: Beacon, 1955), especially
pp. 203f.
    [100] Hill (note 30), p. 3.
    [101] C. Sherrington, *Man on His Nature* (New York: Doubleday Anchor Books, 1955),
p. 294.

dependence, he himself, through his sheer superhuman efforts, has fashioned the tools which are being used to make him a slave and to destroy him also from within. [102]

The second reason why the scientist feels wary about science is the intimate dissatisfaction it is likely to leave. Science appears to be brutally intellectualistic. As a consequence, man has the impression of being exposed to a mercilessly cold climate in which nothing but the rigor of the mind dominates. To summarize with Rostand:

The reign of science has opened a sort of glacial era in the mental history of our species: we have as yet no clear proof that the shivering human soul is able to stand the rigorous climate of reason. [103]

What then is science, so we can give an answer to many of our contemporaries who feel torn by the issue? Numerous observers arrive at a catastrophic conclusion. Science to them appears to be an unmitigated, if inescapable, tragedy. As we have just seen, even Einstein seems to share this opinion. Another philosopher-scientist who comes to the same conclusion is Rostand. He writes:

Science explains nothing, I agree, yet those who have tried to provide something better might as well have kept quiet. Science or silence. [104]

I am all the more afraid of science because I believe in nothing else. [105]

Rostand's words just cited—for all their apparent gloomy finality—contain a precious clue concerning our question. They reveal the reason why many people are driven to desperation by science. The reason is simple. People all too frequently take for granted that man, living in the scientific age, is obliged to choose for or against science. Under this either/or assumption it is clear that science must lead to tragedy. It must do so because, while offering man the brightest promises, it proves unable to fulfil any of them. Indeed, we have seen that science is truly a humanistic factor, but an inherently inadequate one. Accordingly, to choose sides for or against science amounts to trying to build a humanism on a basis that fails to take into account man as a whole. Needless to add, however, the either/or attitude relative to science is not founded on any convincing grounds. It is merely the result of psychological fascination. Since science is obviously important for man, people tend to see it as all-important.

If we try to avoid the pitfall just indicated, we can finally give a direct answer to our central question. What is science as far as of relevance

---

[102] Einstein (note 8), p. 358.
[103] Rostand (note 1), p. 64.
[104] *Ibid.*, p. 293.
[105] *Ibid.*, p. 296.

to man living in the scientific age?   Science is the *drama* of contemporary man.   The term must be taken rigorously.   According to the dictionary, drama is a dynamic human situation the outcome of which hangs in a precarious balance.   In other words, drama is a process involving persons and is characterized by the tension between two possible outcomes: human success or failure.   But the condition of man living in the scientific age fits exactly the definition of drama given.   That is to say, since science has become a predominant factor of our civilization, contemporary man is faced with the threat of social and cultural failure unless he succeeds in becoming more human through science.

To sum up, the insights we have achieved by examining science as an experience of the whole man confront man himself with an intellectual challenge.   If science is humanistic but not self-evidently nor adequately so—up to the point of producing a dramatic situation for those who live under its influence—it is necessary to investigate how a satisfactory scientific humanism can be developed.   That is to say, it is necessary to explore the new humanistic perspectives disclosed by science and integrate them into a comprehensive understanding of man.   This is therefore the goal we set ourselves in the second part of the present investigation.   We intend to inquire systematically into the main features of an adequate doctrine of man that really suits the scientific age.

PART TWO

THE HUMANISTIC PERSPECTIVES OF SCIENCE

CHAPTER 5

EXPERIENCE AND REFLECTION:
THE EPISTEMOLOGICAL PERSPECTIVE OF SCIENTIFIC HUMANISM

As we begin the second part of our investigation, let us recall briefly the methodological guidelines we are following. Our aim is to understand the significance of science for man as such. The approach adopted is inspired by philosophy. In the first part we attempted a kind of pre-philosophical or phenomenological analysis of the scientific experience. In this second part we shall discuss science in a strictly philosophical, albeit introductory, manner.

The issue of the relationships between science and philosophy is a very touchy one. To avoid misunderstandings, this is the position I have adopted on the subject. To begin with, I am well aware that scientists avoid philosophical discussions, and rightly so, when pursuing science. The reason that justifies their attitude is the intrinsic distinction between science and philosophy. However, if this attitude is justified, it is nonetheless clear that it involves a negative decision, a basic renunciation. This inherent limitation of the scientific approach has, for instance, been pointed out by the psychologist Jean Piaget as follows:

It consists simply in renouncing certain discussions which divide the spirits and in pledging oneself—by convention or gentleman's agreement—not to speak of anything but those questions which are accessible to the exclusive use of certain methods which are common and communicable. There is, therefore, in the constitution of a science, a necessary renunciation, a determination not to mix those preoccupations for which one perhaps cares most as a person to results obtained or to explanations attempted....[1]

From the situation described we can infer that philosophy, because it is a different type of knowledge from science, may usefully attempt to investigate those questions that scientists leave unexamined. The characteristic approach of philosophy is personal reflection. Our aim here is to discover, by means of reflection, what constitutes the essence and the

---

[1] Translated from J. Piaget, *Introduction à l'Epistémologie Génétique* (Paris: Presses Universitaires de France), vol. I, p. 9.

implications of scientific knowledge. Needless to add, when advocating the right and duty of philosophy to study science, one should not claim any superiority of philosophy over science. For philosophy is purely a reflective form of knowing. Hence philosophy of science can operate only by depending on science and thus should not claim any superiority over science. However, philosophy can and should strive to bring its contribution to the understanding of science itself.

By adopting a respectful attitude toward science, we can easily find a clue for our overall philosophical investigation. Philosophy is reflective exploration of knowledge. But knowledge presents three main aspects: that of an activity of the subject, that of an information about the object, and that of a stimulus to action. Accordingly, we begin our philosophical analysis of science by studying knowledge as an activity of the subject. This is then the plan of discussion in the present chapter. First we are going to analyze the specific feature that distinguishes scientific knowledge, namely observability. Then we shall examine some basic insights about the nature of knowledge that come to the fore through the success of science. Afterwards we shall consider the problematic nature of two of the main features of science, namely its objectivity and its progressivity. To close, we shall attempt to outline the traits that should characterize a doctrine of knowledge or epistemology really satisfactory for man living in the scientific age.

1. Observability: The Epistemological Characteristic of Science

The reason for starting our epistemological discussion of science by considering observability is obvious. Science claims to be the knowledge of reality as observable. To study the topic methodically, we shall start out by surveying the terminology currently employed. Then we shall discuss the influence exerted by observability on the development of science as such. Finally we shall try to formulate the principle of observability as characteristic of the scientific way of knowing.

(a) Scientific Observability: A Terminological Analysis

To arrive at the notion of scientific observability we can profitably begin from ordinary parlance. In everyday speech, to observe means to perceive through the senses, in particular through the sense of sight. The meaning of the term in science is basically the same. The essential difference is science's characteristic concern for interpersonally testable objectivity. Examination of this point offers the fundamental clue to realizing the specificity of scientific knowing.

The ordinary person, when observing, is not centrally concerned with interpersonally testable objectivity. To him it is enough to notice how things present themselves to his senses. The scientist, on the contrary, is not satisfied with that. He wants to know not only how things appear to his senses, but also how things are in themselves. Thus the scientist does whatever possible to remove the subjectivism inherent in his unaided senses. Two main sources of subjectivism can be found in the human senses. One is their selective range of sensitivity. For instance, the human ear cannot perceive vibrations that are above or below two well-defined thresholds of frequency. The other source of subjectivism is the oscillation in sensitivity to which the senses are exposed according to circumstances or the history of the person involved. Thus the human eye perceives the same source of light as having greater or smaller brightness as a result of the situation preceding the perception. One may be dazzled when entering an illuminated room, but after a while one finds the light there quite tolerable. To remove the subjectivism of the senses the scientist devises a way to objectify the senses themselves. This objectification is obtained by means of *instruments or observation apparatus.* The devices designated by this name are, fundamentally, nothing but sense aids. What makes them important is their ability to overcome the two forms of subjectivism mentioned. Instruments extend man's sensitivity beyond the narrow range that is normally accessible to it. Also instruments standardize the sensitivity of man by making it independent of the idiosyncrasies of individual observers.

On the basis of the preceding, we can briefly define the principal terms bearing on our topic. *Observable,* for science, is anything that can be perceived through the senses, either directly or by means of instruments. From this fundamental point of view, there is no essential difference between scientific observation and experiment. The terms are psychologically and methodologically distinct, in that observation is the first step in the scientific approach to the study of nature, while experiment is the last stage in the same. Or, if one likes, observation is the seeking stage of science and experiment is its checking stage. But, epistemologically speaking, observation and experiment bear the same characteristics. Both are aimed at perceiving through the senses what is objectively there. *Observation,* however, is in strict sense a perception of what nature spontaneously offers to man. *Experiment,* on the contrary, is a public, schematic, and repeatable observation.

To forestall possible confusion, a few remarks should be added concerning the terms frequently employed in connection with observation. First, a word about *experience.* Properly speaking (see Preliminaries, Section 2), the term "experience" far exceeds the range of scientific observation. For experience means any act or process through which man becomes aware of something else, without specifying the type of perception involved. Thus we can experience entities that we cannot see or touch through the senses—for instance, beauty and goodness. Observation, on the contrary, is precisely defined as perception through the senses. However, since observation is obviously a kind of experience, we can define its object, the observable, in terms of experience itself. We can then say that an *observable is anything that can be experienced publicly.* That is to say, something is observable if it can be detected through the senses, either directly or indirectly, by anyone who takes sufficient pains to become aware of it. Second, a remark should be made concerning *reality*. Reality, properly speaking, is the most comprehensive term of all—it stands for whatever exists, or even can exist. Frequently, however, when speaking in connection with science, there is the tendency to employ the word loosely. To avoid misunderstandings, it is necessary to be more accurate. Thus one should speak of *nature or observable reality* when referring to that segment of reality that is scientifically observable. As for other nonobservable segments of reality, if any exist, it is not the business of science —by definition—to decide whether they are real or not.

(b) The Role of Observability in Science

As understood by the scientist, observability is a philosophical attitude that marks the transition between ordinary and scientific knowledge. To clarify this pivotal point we must turn to history. A few examples will illustrate how science is born and develops. In particular, they will show why scientists attach such importance to observability.

Concerning the origin of science, history shows that the decisive factor was not better instrumentation but a new philosophical awareness about the intelligibility of nature (see Chapter 1, Section 1a). This awareness constitutes the essence of scientific observability. A revealing example is offered by the history of optics. Optical lenses began to circulate in the West around the year 1300. If mere seeing with the eyes—observation in the ordinary sense—could be sufficient to make science, much science would have been made during the three centuries that elapsed between the invention of the lenses and their use by Galileo. Actually no science was made. The lenses were indeed used, but only for practical purposes. They were made into spectacles for those whose eyesight was defective or they were combined to produce gadgets that served to entertain. Even the telescope—a combination of lenses—was invented. But it was regarded solely as an instrument of amusement. The reason why all this optical instrumentation remained scientifically sterile was an epistemological principle widely shared in contemporary learned circles. The slogan was:

"Valid knowledge cannot be attained through sight alone" (*Non potest fieri scientia per visum solum*). [2]

What was Galileo's contribution in the optical field? A manifold one, of course. But essentially it was a new philosophical awareness, which we have called scientific observability. He grasped at once that the telescope was not a plaything but an instrument "of inestimable usefulness" as he put it. Animated by this conviction, he proceeded forthwith to manufacture good telescopes, even though he himself did not understand the scientific theory of the instrument. Similarly he produced the first microscopes. Then he used the instruments extensively to study nature carefully. As has been rightly said, Galileo by so doing gave new eyes to mankind. The new sense of sight consisted in taking seriously and objectively nature as accessible to the senses of man. His contemporaries regarded sense observation as inherently subjective, and thus distrusted sight as a source of knowledge. Galileo realized that observation could be objectively reliable, being capable of interpersonal testing. Even historical dates stress the accelerated tempo in the cognition of nature brought about by this new philosophical attitude. Before Galileo the lenses had been sterile for about three centuries. Beginning with Galileo, they produced a rapidly increasing harvest of information. Galileo published his first telescopic discoveries in his *Starry Messenger* in 1610. Shortly afterwards, in 1632, Anthony van Leeuwenhoek, the great microscopist, was born. In a matter of a few decades the lenses disclosed to man an immense amount of the intelligibility of nature.

Observability is essential in science not only at its beginnings, but throughout its development. It would be fascinating to follow in some detail the role played by observability inside a single science. I have tried to do this elsewhere in regard to the study of the atomic structure of matter. [3] For the purpose of the present discussion, it may be enough to summarize the main data here. Even at the atomic level, it is observation that is paramount. The renowed crystallographer Fedorov, for instance, could assert that through X-ray diffraction the atoms were made visible. Or, more accurately, "if not the atoms themselves, then the photographic images caused by them." [4]

Observation is essential in science because, in the first place, through

[2] The historical and philosophical problems of optics have been discussed by V. Ronchi in several of his books. See especially his *The Nature of Light,* trans. V. Barocas (Cambridge, Mass.: Harvard University Press, 1970).

[3] For details see, for instance, Part I of E. Cantore, *Atomic Order: An Introduction to the Philosophy of Microphysics* (Cambridge, Mass.: MIT Press, 1969).

[4] E. S. Fedorov to N. A. Morozov, letter dated Oct. 2, 1912. Quoted in P. P. Ewald, ed. *Fifty Years of X-Ray Diffraction* (Utrecht: Oosthoek's Uitgereversmaatschappij, 1962; International Union of Crystallography), pp. 346f.

it, speculation ceases and genuine knowledge begins.   The history of atomism offers a particularly convincing example because the doctrine was debated by philosophers for over two millennia without gaining any decisive support among investigators of nature.   Then, very rapidly, atomism started to gain more and more supporters until it became universally recognized as the satisfactory interpretation of the structure of matter. The reason for the transition?   The observations made by scientific investigators, beginning with John Dalton at the start of the 19th century. In the second place, observation is essential in science because of the increasing attainments that it enables science itself to achieve.   Once again, atomism convincingly illustrates this statement through all the steps that mark its history as a scientific theory.   In particular, it is important to emphasize that observation accounts for all the conquests of atomic physics, even those that appear most recondite and inaccessible to the senses.   These conquests, as is well known, are especially those embodied in the theory of quantum or wave mechanics.   For all its complexity and abstractness, this theory is the result of observation.   Max Born, one of the most creative quantum physicists, summarized the situation aptly when he wrote:

I think that not a single step in wave mechanics would have been possible if some necessary foothold in facts had been missing.   To deny this would mean to maintain that Planck's discovery of the quantum and Einstein's theory of relativity were products of pure thinking.   They were interpretations of facts of observation, solution of riddles given by nature—difficult riddles indeed, which only great thinkers could solve. [5]

Sometimes, to be sure, the validity of the example we have just brought is impugned by critics.   They object that, at least in quantum mechanics, observability does not play a decisive role.   The reason normally given is that numerous physicists, notably Einstein, dispute its permanent validity.   The implication is that, if quantum mechanics were observational, all physicists would accept it without exception.   We cannot discuss this complex issue here.   Nevertheless, the essential point stands firm.   Einstein himself, the great opponent, is far from critical of the observational foundation of the quantum theory.   On the contrary, he openly acknowledges that this foundation is solid and indisputable.   In his own words:

It must be admitted that the new theoretical conception owes its origin not to any flight of fancy but to the compelling force of the facts of experience. [6]   [This theory] even now presents a system which, in its

---

[5] M. Born, *Natural Philosophy of Cause and Chance* (New York: Dover, 1964), p. 90.
[6] A. Einstein, *Ideas and Opinions,* trans. S. Bargmann (New York: Crown, 1954), p. 334.

closed character, correctly describes the empirical relations between statable phenomena as they were theoretically to be expected. . . . [7]

To sum up, we can see why those who appreciate science rate observability so highly. It is a vital matter. Science stands and falls with observability. Thus, while scientists insist much on idea or theory as indispensable to science (Chapter 1, Section 1b), they are equally insistent on observability. The reason is that the theory cannot constitute genuine understanding unless it passes the test of observation. The theory, in other words, has to be an idea, but of a special kind. Claude Bernard, the great experimentalist, speaks of the experimental idea:

. . . the experimental idea is by no means arbitrary or purely imaginative; it must always have a support in observed reality, that is to say, in nature . . . essential to any hypothesis is that it must be as probable as may be and must be experimentally verifiable. [8]

(c) The Principle of Scientific Observability

Given the central importance of observability for science, it may be advantageous to express the attitude involved by means of a philosophical principle. Scientists themselves often attempt such a philosophical formulation. An example is, for instance, Dirac's statement:

Only questions about the result of experiments have a real significance, and it is only such questions that theoretical physics has to consider. [9]

The essential function of the principle of observability is that of defining what is frequently called *scientific meaningfulness*. It must determine what is meaningful or not meaningful for science as a specialized form of knowledge. To fulfil this function it seems that the principle can be expressed in a twofold manner. (a) *No statement should be considered scientific unless it refers to an observable entity.* (b) *Any statement must be considered scientific if it refers to an observable entity.*

The first formulation stresses the critical attitude taken by science toward any proposition that claims to have scientific relevance. To science as a specific form of knowledge the only object of interest is observable reality. Hence nothing can be asserted to be relevant scientifically unless it proves to be observable. To be sure, it may happen that something is not observable for a while, then it becomes such. An example, as we have seen, is the atomic structure of matter. In any case, if something

---

[7] In P. A. Schilpp, ed., *Albert Einstein, Philosopher-Scientist* (New York: Harper Torchbooks, 1959), p. 666.

[8] C. Bernard, *An Introduction to the Study of Experimental Medicine,* trans. H. C. Green (New York: Dover, 1957), p. 33.

[9] P. A. M. Dirac, *The Principles of Quantum Mechanics* (New York: Oxford, 1947), p. 5.

proves to be unobservable in principle, science rules it out of consideration as totally meaningless. A well-known example of this type is the simultaneity of events for different observers, irrespective of their state of motion or rest relative to the events themselves. Such a simultaneity had always been taken for granted. Einstein proved that it could not be observed in principle, owing to the constancy of the velocity of light. As a result, simultaneity was dropped from the scientific conception of nature —notwithstanding the instinctive repugnance that arises in the human mind when something that appears so obvious is denied. [10]

The second formulation of the principle stresses the complete openness and continual developability of science. The realm of observability is by itself limitless. Correspondingly, the reach of science is limitless, too. It extends to anything and everything that is susceptible of being observed.

To avoid misunderstandings about the epistemological essence of science we must add some additional remarks. The observability principle, as formulated, should be recognized as having a threefold character. (1) The principle is *specific*. It determines what falls within the realm of science as a specialized form of knowledge. By the same token, it clarifies what science leaves out. Thus it defines science proper, as distinct from any other form of knowledge. In brief, the principle of observability serves to avoid confusing science with other possible ways of knowing reality. (2) The principle is *positive, not exclusive*. It stakes out a certain segment of reality as susceptible of study by means of science. But it does not exclude that other segments of reality may exist besides the observable one. Nor does it assert that observable reality itself can be known exclusively and exhaustively by science. In brief, the principle of observability serves to assert the originality of science, but not at the expenses of other possible forms of knowledge. (3) The principle is *methodological, not dogmatic*. It declares that the observational approach is an important avenue to the attainment of knowledge. But by no means does it prove that observation has a self-evident epistemological validity— nor does it exclude the validity of other forms of knowledge. In brief, the principle is not a dogmatic statement, but a heuristic or methodological one.

To conclude our discussion, we should assess scientific observability properly. It cannot be said to be coextensive with science as such. For science, if taken in its entirety, is an experience of the whole man. Nevertheless, observability is pivotal because it constitutes the epistemological characteristic of science. Science, as an original form of knowledge, is

---

[10] For a simple and nontechnical, yet profound, presentation of the foundations of relativity see Einstein's own *Relativity: The Special and General Theory—A Popular Exposition* (New York: Crown, 1961).

defined by observability. Consequently, to pursue our aim of realizing the humanistic meaning of science, we must now examine the results of observability. We shall first analyze the new epistemological insights it has brought to light. Then we shall explore the problems that arise from the success of observability.

2. EPISTEMOLOGICAL INSIGHTS OF SCIENCE: KNOWLEDGE AS PUBLIC EXPERIENCE

The insights about the nature of knowledge disclosed by the consistent application of the principle of observability in science can be summarized under three main headings. Science has given man a new sense of the objectivity of his knowing. It has manifested that he can understand nature intrinsically. It has revealed that his knowing is not static, but inherently dynamical. We shall survey in some depth these important contributions of science to the self-understanding of man as a knower.

(a) Objectivity

Objectivity is the most prominent feature of science. Scientists to a man insist that, if their undertaking has any meaning at all, it is because of its objectivity. As Heisenberg, for instance, put it:

Every scientist who does research work feels that he is looking for something that is objectively true. His statements are not meant to depend upon the conditions under which they can be verified. Especially in physics the fact that we can explain nature by simple mathematical laws tells us that here we have met some genuine feature of reality, not something that we have—in any meaning of the word—invented ourselves. [11]

But, if science is so concerned with objectivity, what is new in this situation? Is not any genuine knowledge objective? This is the first question we must examine if we are to realize the epistemological novelty of science.

To begin very rapidly from the beginning, prescientific man was inclined to think that he really knew something only when he grasped—by abstraction or intuition—the essence of that thing. This is the *essentialist view of knowledge*. From history we are informed that the Aristotelian version of such a view was long dominant in the West. It consisted of a very simple position. Each species of bodies was defined by a nature or essence—such as lightness, heaviness, animality, spirituality, and so

---

[11] W. Heisenberg, *Physics and Philosophy: The Revolution in Modern Science* (New York: Harper Torchbooks, 1962), p. 82.

forth.   The process by which the mind was supposed to perceive the essence was called abstraction.   But once the mind had abstracted the essence of things, the acquisition of knowledge was virtually closed.   The reasons why things were light or heavy, the structure of animality, and the like were considered out of bounds for the human mind.   Hence the Aristotelian notion of occult qualities or causes.   A radicalization of the essentialist view was proposed by Descartes in the 17th century.   He ridiculed Aristotle for having proclaimed that something in nature could remain hidden from the human mind.   His own position was that the mind was able to intuit the essence of things in such a way that nothing at all was left to be known.

The first epistemological insight contributed by science was the exposure of the essentialist view.   Science arose with the conviction that knowledge was objective and reliable, but not essentialist.   The texts are clear.   Galileo, though proud of his knowledge, repeatedly stated his conviction of not knowing the essence of any of those bodies that contemporary Aristotelian philosophers claimed to know, such as water, fire, earth, moon, and sun.   He declared that man would never succeed in knowing such essences during his lifetime. [12]   Concerning light, for instance, he disclaimed any ascription of a personal opinion about its essence, stating that "he had always naïvely avowed that he did not know what light was." [13]   Newton, the other giant of early science, took exactly the same position.   We read in fact in the Scholium Generale of his *Principia*:

...what the real substance of anything is we know not.   In bodies we see only their figures and colours, we hear only the sounds, we touch only their outward surfaces, we smell only the smells, and taste the savours; but their inward substances are not to be known either by our senses, or by any reflex act of our minds. [14]

What is the epistemological import of the nonessentialist view of knowledge?   People frequently imply that it amounts to a kind of positivist phenomenism.   The truth is quite different.   Scientists are all for the objectivity of knowledge, but dissociated from its essentialist interpretation.   We can realize the new way of thinking by studying the polemic between Newton and the Cartesians.   The Cartesian philosophy was dominant on the Continent.   Thus, when Newton's *Principia* appeared, everybody in continental Europe was full of praise for his clarity of mathematical

---

[12] Quoted in Galileo Galilei, *Opere*, F. Flora, ed. (Milan: Ricciardi, 1953), note to pp. 596f.

[13] Letter to Fortunio Liceti, Sept. 15, 1640, in *Opere, op. cit.*, p. 1075.

[14] Quoted in A. Koyré, *Newtonian Studies* (Chicago: University of Chicago Press, 1968), p. 159.   See the whole study, "Gravity an Essential Property of Matter?," pp. 149-163.

formulation and logical rigor of demonstration. However, the leading scientific periodical, *Le Journal des Savants,* in its review of the book considered the *Principia* devoid of any scientific value because "they do not fulfil the necessary requirement of rendering the universe intelligible." [15] Even more, some leading scientists and philosophers—such as Huygens and Leibniz—went so far as to accuse Newton of having reintroduced occult qualities in his interpretation of the world. The reason was that Newton had explained mechanical motion by reducing it to universal gravitational attraction. He had claimed to know it as a fact, but had also disclaimed any knowledge of its essence or substance. This position was preposterous to the Cartesians: "... a relapse into medieval conceptions that had been thought exploded, and ... a kind of treason against the good cause of natural science." [16] As a response to the chorus of protests, Newton had his disciple Roger Cotes write a preface to the second edition of the *Principia*, and he himself added to the book the Scholium Generale. In it, as we have just seen, he explicitly underlined his non-essentialist conception of knowledge. In particular, to justify his position, he uttered his famous statement: *"Hypotheses non fingo."*

To understand Newton's mind it is necessary to study what he meant by the frequently quoted but seldom understood statement. These words have been often hailed by empiricists and positivists as a vindication of their doctrinal standpoints. Such an interpretation is wrong, as shown by historical research. [17] To summarize the data currently available, we must begin by challenging the ordinary English translation due to Motte in his standard version of the *Principia*. There Newton is made to say: "I do not frame hypotheses." The actual meaning of his words, on the contrary, is: "I do not *feign* hypotheses." The translation makes a great difference here. For, as Koyré has rightly pointed out, "'Feign' implies falsehood, and 'frame' does not, or at least not necessarily." [18] As a consequence, instead of supposing that Newton intended to reject hypothesis outright, we are led to think that he might have meant to reject "feigned"

[15] Quoted by R. Dugas and P. Costabel in R. Taton, ed., *A General History of Science,* trans. A. J. Pomerans (London: Thames and Hudson), vol. II, p. 265.

[16] E. J. Dijksterhuis, *The Mechanization of the World Picture,* trans. C. Dikshoorn (Oxford: Clarendon Press, 1961), pp. 479f.

[17] Probably the most comprehensive study of the subject is to be found in I. B. Cohen, *Franklin and Newton: An Inquiry into Speculative Newtonian Experimental Science and Franklin's Work in Electricity as an Example Thereof* (Philadelphia: The American Philosophical Society, 1956), pp. 113-201 and 575-589. Another important study is R. M. Blake, "Isaac Newton and the Hypothetico-Deductive Method," in R. M. Blake, C. J. Ducasse, and E. H. Madden, eds., *Theories of Scientific Methods: The Renaissance through the Nineteenth Century* (Seattle: University of Washington Press, 1960), pp. 119-143.

[18] Koyré (note 14), p. 35. See the whole study, "Hypothesis and Experience in Newton," pp. 25-52.

or false hypotheses only. But, what did Newton mean by the use of this term?

The best definition of the Newtonian *hypothesis* was given by Newton himself. It is found in a manuscript note which was intended as an addition to the four Regulae Philosophandi of the *Principia* but was not published during Newton's lifetime. In this document we read:

Whatever is not derived from things themselves, whether by the external senses or by the sensation of internal thoughts, is to be taken for a hypothesis. Thus I sense that I am thinking, which could not happen unless at the same time I were to sense that I am. But I do not sense that any idea whatever may be innate.... And those things which neither can be demonstrated from the phenomenon nor follow from it by argument of induction, I hold as hypotheses. [19]

In the passage cited, the polemical reference to Descartes is unmistakable. Hence we can gather Newton's meaning in his celebrated phrase. He had no intention of rejecting what is ordinarily called hypothesis, namely a theorizing effort of the mind that is based on observation and tested by experimentation. Newton could not exclude hypothesis thus understood because he, as a creative scientist, could not dispense with hypothesis and theory in his own work (Chapter 1, Section 1b). Moreover, Newton himself indulged in much speculation which he left precisely in the form of tentative views in the Queries of his *Opticks*. However, Newton condemned with all his might *pure* speculation in the Cartesian sense of the term. Thus, by rejecting "hypothesis" as something "feigned," Newton intended to brand as absurd the Cartesian claim that the mind can intuit the essences of things by means of personal reflection and intellectual clarity. In other words, Newton intended to proscribe a view that, to him, bore a resemblance to fiction or falsity. For, according to Newton's conviction, Descartes' conception of knowledge—being wholly speculative—led exactly to fiction and falsity. In short, the "hypothesis" Newton rejected was for him

...simply identical with fiction, and even a gratuitous and necessarily false fiction... the substitution of a fictitious reality for the given one, or at least of a reality in itself inaccessible to perception and knowledge, a pseudoreality endowed with properties imagined or fancied in an arbitrary manner for that very purpose. [20]

In the light of the preceding, we can see why Newton stoutly defended the objectivity of his scientific discovery of gravitation while, at the same time, determinedly rejecting any essentialist conception of knowledge.

---

[19] Quoted in Koyré, *op. cit.*, p. 272.
[20] *Ibid.*, p. 36.

Newton's defense of the objectivity of science was passionate. To avoid personal involvement, he first had Cotes deny explicitly the accusation that gravity could be considered an occult cause. Cotes stated in the Preface:

... those are indeed occult causes whose existence is occult, and imagined, but not proved; but not those whose real existence is clearly demonstrated by observations. Therefore gravity can by no means be called an occult cause of the celestial motions, because it is plain from the phenomena that such a power does really exist. [21]

Since the offensive view persisted, Newton himself entered the fray. In an essay he published unsigned in *Philosophical Transactions* about the Leibniz-Clarke correspondence (*Commercium Epistolicum*), Newton contrasts his own position with that of his adversaries and refutes their accusation of occultism:

Mr. Leibnitz hath accused him [Newton] of making Gravity a natural or essential Property of Bodies, and an occult Quality and Miracle.... It must be allowed that these two Gentlemen differ very much in Philosophy. The one proceeds upon the Evidence arising from Experiments and Phenomena, and stops where such Evidence is wanting; the other is taken up with Hypotheses, and propounds them, not to be examined by Experiments, but to be believed without Examination. The one for want of Experiments to decide the Question, doth not affirm whether the Cause of Gravity be Mechanical or not Mechanical; the other that it is a perpetual Miracle if it be not Mechanical. [22]

To sum up, we can realize a first epistemological insight stemming from science. It is a new view of objectivity. As against the conviction of prescientific man, science leads to a conception of knowledge that is unmistakably objective but, by the same token, nonessentialist. Scientific knowledge is objective because it is founded on observation and tested by experiment. Scientific knowledge is nonessentialist because it never indulges in fiction—that is, it never purports to give a literally adequate representation of the objects it studies. It is satisfied to state what it has been able to gather so far through observation and the effort of theorization.

(b) Understanding

A second major feature of knowledge is understanding. Objectivity refers to the reliability of the information obtained. Understanding refers to the penetration of the mind into the structure of the object one has been

---

[21] *Ibid.*, p. 142.
[22] *Ibid.*, pp. 146f. See the whole study, "Attraction an Occult Quality?," pp. 139-148.

informed about.   The aspiration after understanding is of course—as
Aristotle put it aphoristically at the beginning of his *Metaphysics*—the
central desire of the human mind.   What is new in the conception of
understanding brought about by science?

Historically, the novelty of scientific understanding is a most conspicuous phenomenon.   To state it in brief with Koyré, we have here:

... one of the profoundest, if not the most profound, revolutions of
human thought since the invention of the Cosmos by Greek thought: a
revolution which implies a radical intellectual "mutation".... [23]

To investigate our theme we begin by summarizing the prescientific conception of understanding.   Then we shall explore the characteristic new
features of scientific understanding.

The conception of prescientific understanding, at least in the West,
was embodied in the Aristotelian synthesis.   The reason for the universal
acceptance of this synthesis is easy to detect.   The Aristotelian view coin-
cided with that of so-called common sense.   The historian Koyré put it
aptly: "Common sense, indeed, is—as it always was—medieval and
Aristotelian." [24]   What was the conception entertained by such a mentality
in regard to understanding?   In principle it was just an application of the
essentialist interpretation of knowledge.   Man was thought to really understand when he succeeded in grasping the essences of things and in deriv-
ing, from them, a systematically comprehensive view of reality.   The comprehensiveness of the view was paramount.   For to understand, of course,
means to see many as forming one, to perceive a universal orderliness
embodied in many component parts.   But reality does constitute a whole;
everything is interconnected with everything else.   Hence the prescientific
conception of understanding demanded a total synthesis of rational type.
Man was supposed to grasp with his mind the essences of things and
derive therefrom, by logical deduction, a comprehensive system of the
world.

In the practice of his life, prescientific man realized only too clearly
his inability to understand nature according to the essentialist epistemological criteria of the times.   Hence the conviction arose that nature was
intrinsically unknowable to man.   This is the source of the view that
ascribed occult qualities or causes to nature itself.   As a consequence of
this opinion, prescientific man developed a conception of the knowledge
of nature that was frankly man-centered and subjective.   All the possible
understanding of nature was supposed to have been attained when man

---

[23] A. Koyré, *Metaphysics and Measurement: Essays in Scientific Revolution* (London:
Chapman & Hall, 1968), p. 16.

[24] *Ibid.*, p. 5.

was able to realize how things were in relation to him and succeeded in constructing a rationally unified structure out of this phenomenistic information. Such a conception was embodied in the then accepted phrase that stated that the understanding of nature amounts to the "saving of the appearances." The phrase can be traced all the way back from the times of Galileo to those of Plato. [25] It was universally taken for granted that it embodied the necessary limitation of the human mind, which no amount of effort could expect to overcome.

In the light of the preceding it is clear what a hard battle the pioneers of science had to wage to have science accepted as genuine understanding. They had to do nothing less than destroy one conception of understanding and replace it with another. With Koyré, the situation can be put synthetically:

...they had to destroy one world and to replace it by another. They had to reshape the framework of our intellect itself, to restate and to reform its concepts, to evolve a new approach to being, a new concept of knowledge, a new concept of science—and even to replace a pretty natural approach, that of common sense, by another which is not natural at all. [26]

The novelty and improvement in the conception of understanding brought about by science emerge when we contrast the goal, the means, and the achievements of science with those of the prescientific mentality. Beginning with the goal, the contrast is sharp. Prescientific man was essentially satisfied with a phenomenistic view of nature. To scientific man, phenomenism is no knowledge at all. Rather, his goal is to penetrate beyond appearances in order to reach the intrinsic intelligibility of things.

The means used by science in its pursuit of understanding further distinguish the two mentalities. Prescientific man contented himself with a very superficial observation of nature, coupled occasionally with anthropomorphic and anthropocentric speculations such as the geocentric system, with all its crystalline spheres, epicycles, and the like. He did not even suspect the possibility of creating with his mind an ideal representation of the intelligible structure of nature as objectively existing. But this is what science continually strives to do by means of all its endeavors, which go under the name of abstraction, schematization, hypothesis, and theory. To be sure, scientific man is not bent on idealization as an end in itself. As Max Weber, for instance, put it: "The construction of abstract idealtypes recommends itself not as an end but as a means." [27] Nonetheless,

---

[25] See the important historical study of this conception by P. Duhem, *Sozein Ta Phainomena: Essai sur la Notion de Théorie Physique de Platon à Galilée* (Paris: Hermann, 1908).

[26] Koyré (note 23), pp. 20f.

[27] M. Weber, *The Methodology of the Social Sciences*, trans. E. A. Shils and H. A. Finch (New York: Free Press, 1949), p. 92.

the idealizing approach of science stresses the new conception of understanding that science has disclosed. For only idealization enables man to
overcome the apparent obviousness of sense appearances in order to reach
the intrinsic intelligibility of things.

The effective achievements of science stress the greatness of the
epistemological realization brought about by science itself concerning understanding. The attitude of prescientific man was ultimately one of
resigned ignorance. When confronted with the complexity and variety of
nature, he was convinced that he could do nothing more than accept
information as it presented itself to him more or less phenomenistically.
Scientific man is convinced that he can penetrate with his mind the intrinsic
structure of nature—that he can understand it by means of an explanation
that is objective and also lasting.

In the scientists' conviction, science really makes man understand
because, in the first place, it formulates complete theories of natural phenomena. A theory is said to be complete or explanatory when it embraces all
the observable regularities of a given type, and traces them back rigorously
to a basic observable property of nature. An outstanding example of such
an explanation is the Newtonian gravitational synthesis. This theory is
complete because it encompasses all the types of local motion. It is explanatory because it proves that all such motions are the necessary consequence of the experimentally detectable property of nature called gravitational attraction (Chapter 2, Section 1b).

In the second place, scientists are convinced that science really makes
man understand nature because scientific theories, once they have been
accepted as complete by the community of researchers, prove to have lasting
validity. They are not reversed or called into question by subsequent
discoveries, no matter how novel or surprising these may turn out to be.
What constitutes the finality of scientific theories has been described clearly
and forcefully by Heisenberg:

...in the exact sciences the word "final" obviously means that there are
always self-contained, mathematically representable, systems of concepts
and laws applicable to certain realms of experience, in which realms they
are always valid for the entire cosmos and cannot be changed or improved. [28]

An example of final scientific theory is, again, the Newtonian synthesis.
The finality of this theory, as shown by Heisenberg in the context cited,
consists in its perennial validity. For it is indeed true that motion has
been studied more precisely by subsequent theories, namely relativity and

---

[28] W. Heisenberg, *The Physicist's Conception of Nature,* trans. A. J. Pomerans (New
York: Harcourt, 1958), p. 27.

quantum mechanics. Nonetheless the Newtonian results remain adequate even in our own time when the bodies in motion are of the same macroscopic type as those that Newton himself investigated.

To sum up, the epistemological significance of science for man becomes increasingly clear. Man is by nature a knower. His basic aspiration is to understand. But science has thrown surprisingly new light on understanding. It has especially freed man from the bondage of sense appearances and enabled him to attain the intrinsic intelligibility of nature. Of course, these new humanistic insights can be difficult to assess in their profound import. For science is complex and its epistemological analysis is demanding. However, the central point stands out. Science is epistemologically meaningful because it leads man to understand better his own understanding.

(c) Dynamism

We have still to consider a third major epistemological insight brought to the fore by science. Knowledge, as shown by science, is not only objective without being essentialist, nor just able to attain intrinsic understanding without stopping at appearances. Over and above everything else, knowledge manifests itself as inherently dynamical. To appreciate the importance of this insight we must once again start by surveying the epistemological views of prescientific man.

The prescientific conception of knowledge, as embodied in the Aristotelian synthesis, was definitely a static one. As has been said rightly: "Aristotle considered scientific thought as thought at rest, fixed and immovable." [29] This statement, of course, should not be interpreted as meaning that Aristotle was unaware of a certain changeability and progressivity of knowledge. He was so much aware of it that Werner Jaeger calls him "the inventor of the notion of intellectual development in time." [30] However, what Aristotle and many thinkers after him were unable to realize was the inherent developmental character of human knowledge. The insight about this point came only gradually to light as a consequence of the development of science itself.

We have already seen that the prescientific conception of knowledge was an essentialist one. The static view of knowledge resulted from such an essentialist basis. The connection between the two views is readily detected. Indeed, what was for countless centuries the paradigmatic embodiment of ideally perfect knowledge? The answer is familiar. The ideal form of knowledge was the one embodied in the axiomatic-deductive

---

[29] L. Robin, in R. Taton, ed., *A General History of Science,* trans. A. J. Pomerans (London: Thames and Hudson, 1963), vol. I, p. 228.

[30] W. Jaeger, *Aristotle* (New York: Oxford, 1962), pp. 3f.

structure of Euclid's famous geometric treatise. Why such a preeminence given to the geometric model? The answer is not difficult to find following the Aristotelian argumentation.

Aristotle divided knowledge into two main categories: cognition of the "what" and cognition of the "why." In the medieval formulation *cognitio propter quia* and *cognitio propter quid*. The first category of cognition was capable of novelty—but only because it was empirical or merely factual, without attaining to the essence of things . Thus, in Aristotle's opinion, the first kind did not deserve to be called knowledge in the genuine sense of the term. The reason was precisely its inability to go beyond the factual—its limitation to the "what" of things, instead of reaching the "why" or the intrinsic reasons why things were as they were. The second kind of cognition, on the contrary, was postulated to reach the "why" of things. This it was supposed to do by grasping the essences of things themselves.

It is now clear why Euclid's axiomatic-deductive system was seen for many centuries as the paradigmatic pattern of genuine knowledge. The axioms were supposed to express the essences of the objects known. The deductive formulation ensured that the statements of this knowledge were all intrinsically necessary, never merely factual. But, of course, a form of knowledge like Euclidian geometry is not capable of genuine novelty. For, in principle, it must limit itself to developing through intrinsic logical deduction. That is, its only possibility of growth consists in making explicit by means of conclusions the content of truth, which is already implicitly present in the premises. As an overall result, the prescientific person thinking along Aristotelian lines had no choice. If he wanted a form of knowledge that he considered genuine—that is, providing intrinsic certainty and not merely factual information of an empirical kind—he had to opt for a static type of cognition.

Science, in its historical beginnings, benefited much from Aristotle's epistemological systematization. But it also gradually outgrew its stifling limitations. This can be seen clearly by surveying rapidly the history of mechanics. As the pioneer of mechanics, Galileo attempted nothing less than to implement the Aristotelian-Euclidean ideal as regarded the study of motion. Galileo, to be sure, rejected vehemently the essentialist conception of knowledge (Chapter 5, Section 2a). But he was just as vehemently opposed to the mere factualness of scientific statements (Chapter 1, Section 1bc). Thus Galileo condemned those scientists whom he called "mathematical"—that is, the ones who were satisfied with factual observations. In opposition to them, he ranged himself among the "philosophical" scientists—namely those who studied "the true constitution of the universe." [31]   About the possibility of discovering such a constitution of the universe Galileo was emphatic.   He wrote in the context just cited:

For such a constitution [of the universe] exists; it is unique, true, real, and could not possibly be otherwise; and the greatness and nobility of this problem entitle it to be placed foremost among all questions capable of theoretical solution.

The axiomatic-deductive ideal of knowledge continued to be influential throughout the development of what is usually called classical mechanics, that is the mechanics systematized by Newton and his followers.   Newton's great work was significantly entitled *Principia Mathematica Philosophiae Naturalis* and was clearly patterned on the Euclidian model.   The other great mechanicians followed the same pattern with the increasingly clear intention of making mechanics a strictly rational science, the counterpart of geometry.   To cite only the most prominent ones, we can begin with Euler.   He wrote two fundamental treatises: one the dynamics of the point and the other on the dynamics of solid bodies (1736 and 1760).   In both of these works his guiding principle is clear.   René Dugas, the historian of mechanics, puts it as follows:

Euler's set purpose is to develop dynamics as a rational science, starting from definitions and logically ordered propositions.   He intends to demonstrate the laws of mechanics in such a way that one may understand that they are not only certain, but also necessary truths. [32]

The attempt to make mechanics the geometry of the moving bodies was pursued systematically, especially by the French school.   D'Alembert published a treatise on dynamics in 1743.   In its *Discours Préliminaire* he rejected one of Euler's axioms as not being intrinsically necessary.   This lack of intrinsic necessity he found very objectionable because, as he put it, "it would ruin the certainty of mechanics and would reduce it to be nothing more than an experimental science. . . . " [33]   Lagrange published his great treatise in 1788, under the title *Mécanique Analytique*.   The very title of this work is revealing because analysis, as contemporarily understood, meant precisely an axiomatic-deductive treatment, with all the rigor of mathematical systematization.   That Lagrange meant to attain the analytical ideal can be seen especially from his own declarations in the *Avertissement*. [34]   Finally, the geometrization of mechanics was carried to

---

[31] In S. Drake, ed. and trans., *Discoveries and Opinions of Galileo* (New York: Doubleday Anchor Books, 1957), p. 97.
[32] Translated from R. Dugas, *Histoire de la Mécanique* (Neuchâtel: Griffon, 1950), p. 229.
[33] Translated from citation in Dugas, *op. cit.*, pp. 236f.
[34] See citation in Dugas, *op. cit.*, p. 319.

completion in Laplace's *Système du Monde* (1796). In this monumental work Laplace gives the impression that mechanics is really *the* rational —uniquely, precisely and rigorously deductive—cognition of nature. For instance, he confides to the reader:

Of course, one should not measure the simplicity of the laws of nature by means of our ability to conceive them. But when those laws which seem to us the most simple agree perfectly with all the phenomena, then we are well justified to consider them as rigorous, too. [35]

Should one infer from the preceding that the Aristotelian idea of static knowledge had proved itself true? Many people tended to take it for granted, especially during the course of the 19th century. But, as events were to disclose soon, such an interpretation was far from justified. Heisenberg, for one, severely condemns the opinion in question as "*an inherently uncritical* philosophy." [36] The main reason he gives for his judgment illuminates why scientists came to adopt a strictly dynamical conception of knowledge. To cite him directly:

The scientific concept of the universe of the nineteenth century... is called rational, since its centre, classical physics, can be built up from a small number of axioms capable of rational analysis, and since it rests on belief in the possibility of a rational analysis of all reality. It must, however, be stressed that the hope of gaining an understanding of the whole world from a small part of it can never be supported rationally.... Thus the hope of understanding all aspects of intellectual life on the principles of classical physics is no more justified than the hope of the traveller who believes he will have obtained the answer to all problems once he has journeyed to the end of the world. [37]

Following Heisenberg's lead, we can understand how science brought to the consciousness of man the dynamical character of his knowledge. The static or axiomatic-deductive conception of knowledge was exposed as a subjective illusion. Man, striving after it, was in the situation of a traveler striving to reach the end of the world. Before Columbus and Magellan, people had normally accepted the view that there was such a thing as the end of the world. But these two explorers proved that the idea of the end of the world was without foundation and even objectively absurd. In an analogous way, science—just by developing normally, that is, by being faithful to its observational character—proved eventually that knowledge could never be imprisoned within the rationalistic scheme of an axiomatic-deductive system.

---

[35] Translated from citation in Dugas, *op. cit.*, p. 344.

[36] Heisenberg, *Conception* (note 28), p. 128; cf. pp. 121-151.

[37] Heisenberg, *Philosophic Problems of Nuclear Science,* trans. F. C. Hayes (London: Faber, 1952), pp. 23f.

Heisenberg is in an especially authoritative position to interpret the epistemological lesson offered by postclassical mechanics. For he was an early supporter of the theory of relativity and contributed chiefly to the creation of quantum theory. But both these mechanical systems are radically new when compared with their classical counterpart. They disclose totally unexpected properties of motion. Heisenberg therefore is entitled to condemn so sternly the static-rationalistic conception of knowledge as embodied in classical mechanics. However, one should take care not to misunderstand the dynamical nature of knowledge as brought to light by the new mechanics. To this end, Heisenberg's own remarks on the subject are particularly illuminating.

The basic point is that scientific knowledge does not develop by revolution—that is, by simply discarding the past and replacing it with something else. Rather, science develops through internal consistency, namely by adhering unswervingly to what we have called the observability approach to nature. Heisenberg stresses this point.

Modern theories did not arise from revolutionary ideas which have been, so to speak, introduced into the exact sciences from without. On the contrary, they have forced their way into research which was attempting consistently to carry out the program of classical physics—they arise out of its very nature. [38]

In particular, Heisenberg wants to rebut the widespread idea that quantum theory is due to a revolutionary attitude in the study of nature. Thus he insists that this theory is nothing but the outcome of the observational attitude, which is as old as science itself.

In quantum theory, too, the turning away from the principles of the classical description of nature was not effected by the penetration into our science of ideas new and alien to the spirit of earlier physics. On the contrary, science was forced step by step to yield the ground of classical physics as a result of a succession of the most memorable experimental discoveries. [39]

From the example of the historical development of mechanics it is now clear how science has enabled man to arrive at a new conception of his knowledge. The static conception had presented itself so convincingly to prescientific man as to become a kind of ideal. Science, simply by being faithful to its observational vocation, manifested the illusory character of such an ideal. The genuine ideal for man as a knower is rather that of having the courage to strive for literally new discoveries. This is the dynamical conception of knowledge: never attempt to imprison the

---

[38] *Ibid.*, p. 13.
[39] *Ibid.*, p. 14.

inexhaustible intelligibility of reality within the narrow bounds of a man-made axiomatic-deductive system. On the contrary, keep seeking for ever-new truth—unprecedented and unexpected, but all the more rewarding. This is the reason why Columbus fittingly exemplifies the spirit of the scientific person. For he, as Heisenberg put it, "possessed the courage to leave the known world in the almost insane hope of finding land again beyond the sea." [40]   Genuine dedication to truth cannot remain static but must be continually dynamical.

To sum up, the humanistic significance of science in the epistemological sense of the term is by now clear. However, since we have covered con-siderable ground in the discussion of this theme, it may be advantageous to complete our investigation by means of a systematic exposition of the results obtained.

(d) The Epistemological Originality and Message of Science: A Synthesis

Two main points emerge from the foregoing epistemological analysis. The first is that science constitutes an original form of knowledge, with typical characteristics and autonomy of development. The second point con-sists in a series of disclosures that science, through its own success, brings to the fore about the nature of knowledge as such. We are going to survey both of these points systematically.

(da) The Epistemological Originality of Science

The originality of the epistemological nature of science can best be synthesized by speaking of a consistent openness to experience; here, how-ever, the term is taken in its public or interpersonal sense. With this qualification, the position of Ferdinand Gonseth, the indefatigable promoter of the so-called open philosophy, is undoubtedly right. In his own words:

Now, the perspective changes completely (du tout au tout) when one becomes aware that scientific research begins, establishes itself, and un-folds as an actualization of an option—the option consisting in openness to experience. [41]

What distinguishes science from any other form of knowledge? Preeminently its *interpersonal* character. As we have seen, science is based upon and thrives on observability. But what is observable is pre-cisely interpersonal: it can be perceived, studied, and tested by many persons. Various designations are normally employed to express such a

---

[40] *Ibid.*, p. 25.
[41] Translated from F. Gonseth, *Le Problème du Temps: Essai sur la Méthodologie de la Recherche* (Neuchâtel: Griffon, 1964), p. 10.

character of science. One is *publicity*. This word stresses that science is by its nature a form of knowledge that can be shared, in principle, with all people—not just by teaching, but by making other people observe what constitutes the foundation of any scientific statement. Another term designates science as *consensible* knowledge.[42] This is meant to emphasize that science makes its results inspectable to anyone who is interested in testing their validity.

To avoid misunderstanding the epistemological originality of science, one should be careful in interpreting the terms used to designate it. Thus, for instance, scientific *consensus* should not be taken in an empiricist acceptation. For what makes scientists agree on a given statement is not, properly, their sensing things in the same way. Rather, their source of agreement is the mental realization that things are perceivable, objectively, in a given manner. Another equivocal designation is that of *impersonality*. It is frequently used by scientists themselves. Bridgman, for instance, closes an epistemological discussion by saying: "The ideal of science as at present understood is thus an impersonal one."[43] These words are ambiguous because they give the impression that science is done at the expense of the personality of its practitioners. But such an interpretation would amount to an essential misunderstanding of science itself. For —as we have abundantly seen—science is the product of human creativity, and creativity is one of the highest manifestations of the human personality. Nonetheless, one can speak of impersonality if he opposes science to other forms of knowledge that are more specifically personal or internally reflective. A third term that lends itself to equivocation is scientific *objectivity*. Objectivity, as understood by the scientist, is just another way of stressing the observability or interpersonalness of science. Thus it should not be interpreted as a polemical designation aimed at excluding the possible objectivity of other forms of knowledge. Nor should objectivity be equated to objectivism—namely, the tenet that reality is nothing but what can be observed about it. Furthermore, the emphasis laid by science on objectivity should not be construed as a position that tends to exclude as meaningless any consideration of the human subject as such. Rather, it should be viewed as a critical requirement that opposes false subjectivity or subjectivism—that is, wishful thinking, phenomenism and the like.

In particular, the epistemological originality of science can be brought to a focus by discussing an episode that the popular mind has transformed into a myth. Galileo is standing in front of a telescope and challenging

---

[42] J. M. Ziman, *Public Knowledge: An Essay Concerning the Social Dimension of Science* (London: Cambridge University Press, 1968), *passim*.

[43] P. W. Bridgman, *The Way Things Are* (New York: Viking Press, 1961), p. 129.

his adversaries to look at the skies through it. If only they could be persuaded to follow Galileo's example (so the widespread interpretation goes), science would triumph and Aristotelianism would cease to dominate public opinion. This interpretation is mythical because it misconceives the epistemological character of science. It assumes that science is basically a matter of skepticism and empiricism. Skepticism: one should never accept anything as convincingly proved about nature by relying on the words of other people. Empiricism: it suffices to look with one's own eyes to accept the statements of science as evident. History takes care to dispel this illusion. We know at least the name of one of the persons whom Galileo never succeeded in persuading to use his telescope: Cesare Cremonino. [44] This man was a personal friend of Galileo's. As for skepticism, he was abundantly supplied with it—so much so that he had repeatedly incurred the suspicions of the Inquisition. Why did he never accept science? Because he could not convince himself that the images produced by the telescope were worthy of trust as representing the real world.

The preceding episode shows that science is not essentially a matter of looking, but a matter of thinking in a new way about nature and the manner of knowing it. In other words, science comes to be, not through sense impressions but through an original philosophical insight. The insight is that interpersonal experience or public observation supplies trustworthy knowledge. This is the original contribution of science to epistemology—the disclosure that man has a cognitive power whose potentialities remained hidden to the prescientific person.

Two main consequences follow from the recognition of science as an unprecedented form of knowing. One is the autonomy of science with relation to philosophy. Science is autonomous basically because it has been able to discover an approach to truth of which philosophers were unaware. Hence science should be openly acknowledged the right to proceed in its own way, without interference on the part of philosophers.

The second consequence regards the duty for both philosophers and scientists to collaborate to bring to light the epistemological significance of science as well as its implications. For science is indeed independent of philosophy in its proper realm. But, on the other hand, science has a great philosophical relevance. It is therefore necessary to make explicit what this relevance is so that the humanistic significance of science can be properly understood. We shall revert to this question shortly. But, before that, we must synthesize the new information about the nature of knowledge that emerges from the success of science.

---

[44] For details see V. Ronchi, *Storia del Cannocchiale* (Vatican City: Pontifical Academy of Sciences, 1964), pp. 827-829.

(db) The Epistemological Message of Science

What insights about the nature of human knowledge does science offer to the philosopher who reflects on its successful originality? Without attempting to discuss the subject in all its complexity, we can summarize the epistemological message of science by means of several systematic propositions.

Knowledge is experience, not intuition.   As we have abundantly seen, science originates from an encounter between the whole man and the concrete object of his investigation.   Abstract formulations about the object may be important, indeed necessary.   But science is essentially a contact between man and nature.   This is the profound significance of the observational character of science.   Scientific observation is essentially experience: it is the experience of the whole man insofar as it is testable on an interpersonal level.   The epistemological message that derives from this fact consists in the extension of the experiential interpretation to knowledge in general.

It is remarkable that science, through its experiential attitude, has been able to rediscover an age-old epistemological insight that philosophy largely lost sight of in the course of history.   The ancient adage put it plainly: nothing can be present in the mind that has not passed through the senses.   This solid foundation of knowledge, although more or less continuously adhered to by philosophy in words, was largely ignored in fact.   Already Aristotle saw knowledge mainly as a matter of intuition. His essentialist doctrine can hardly be called any other way.   But the intuitionist conception of knowledge became predominant in modern times. Beginning with Descartes and culminating with Kant and his followers, modern philosophy has constantly put a prize on intuition or some sort of aprioristic view of cognition. Philosophers, even of positivist and empiricist persuasion, have always aspired after a kind of cognition that would enable man to grasp things the way they must necessarily or self-evidently be.

Science makes man recover the sense of modesty and concreteness that should characterize genuinely human knowing.   Man, the dependent knower, has not the power to grasp how things must be a priori.   Hence his conduct is not justified—either when he strives for a superhuman intuition of essences or when he lapses into skepticism because such an intuition proves unattainable to him.   This is therefore the basic epistemological lesson of science to reflexive man: knowledge is essentially a lived experience.   That is, to know means to encounter something concretely existing by contacting it with one's whole being.   The awareness that results from such an encounter constitutes the experience called knowledge.   In brief, through science man is invited to overcome the seduction of intuitionist knowing while rediscovering the humaneness of experiential knowing.

Knowledge is intellection, not rationality. As emerges from our entire discussion, the core of the scientific experience lies in discovery. The scientist aims at discovery and feels rewarded by it. But discovery is a question not of rationality, but of intellection. To discover, in fact, means to become intellectually aware of something as objectively true. Rationality, on the contrary, is a matter of logical systematization obtained by way of reasoning.

Once again we notice that science, through its experiential approach, has rediscovered and made more precise an old epistemological insight. Already the medievals used to distinguish between intellect and reason. Intellect was the ability of the mind to perceive the truth contained in the data supplied by the senses. Reason was the ability of the mind to systematize, in a logical fashion, the truth it had perceived. Beginning with Descartes, however, modern philosophy insisted so heavily on reason that rationality and logic became the predominant features of knowledge. To know and to reason came to be taken as practically synonymous. As a consequence, modern Western culture is preeminently a rationalistic one.

Science seemed for a while to be the vindication of rationalism. The enormous successes of classical mechanics made people forget that its foundations were not intuitions but simply intellectual perceptions of some of the properties of motion. But science itself, in its historical development, took care to dissipate the error. The establishment of the new kinds of mechanics, relativistic and quantistic, has broken the spell of the rationalistic view. Contemporary man, as a consequence, is now more enlightened about the genuine nature of his knowing. To know is not to reason, but to see with the mind or intellect. Thus man can now rejoice in his intellectual seeing without seeking a delusive ideal in logic and rationality.

Knowledge is openness and service, not possession. Since science is preeminently discovery or intellectual perception, it is clear that the attitude it inspires toward the object known is basically one of openness rather than possession. The scientist is moved to be open to his object throughout his research, without ever thinking of possessing with his mind all the intelligibility of the object itself. Openness marks the scientific enterprise from beginning to end; scientific man must be animated uniquely by the desire to perceive reality as it is, without reading anything into or out of it. Openness instead of possession is the final result of science because experience tells the reflective scientist that the object he has investigated is effectively inexhaustible in its intelligibility.

Still, however, we notice that science, by simply being faithful to its experiential character, enables man to rediscover an ancient epistemological insight that was frequently disregarded in subsequent philosophical developments.   To primitive man it is obvious that to know means essentially to entertain an attitude of personal openness toward the object, as it exists in itself, without ever presuming to possess with one's own mind all the intelligibility of the object itself.   Such an attitude is spontaneous to primitive man because he is aware of his own mental limitations when confronted with the overpowering greatness and variety of the universe. But advances in intellectual sophistication only too often tend to make man forget his inherent mental limitations.   He rather inclines to think that he can encompass with his mind all the intelligibility of reality—that he can exhaust the overall structure of the world by expressing it in clear and distinct ideas formulated with logical rigor.   This, as we have seen repeatedly, was the great temptation of intuitionism with its attendant manifestation, rationalism.   Man inclined to think that he was able to possess truth as his own.

By dispelling the illusions of both intuitionism and rationalism, science enables man not only to understand his knowledge better, but also to practice it in a more humanizing way.   The intuitionist-rationalist conception of knowledge is ultimately dehumanizing.   For it confronts man with an impossible task, hence it leads him to dejection and obscurantism once he realizes that he cannot master it.   As for those who mistakenly think they can effectively attain a rationalistic intuition of the universe, there is no end to the mischief they tend to cause.   For there is no more arrogant and inhuman person than the one who thinks he owns the truth.   As a consequence, the epistemological illumination given by science is humanizing at least because it frees man from the dangers of intuitionism and rationalism.   But the contribution of science can be even more meaningful from the humanistic standpoint if man strives to practice consistently, in the manifold manifestations of his life, the spirit of openness and service of truth, which is typical of science itself.

The spirit of science is essentially one of service toward truth.   The reason is obvious.   The openness that characterizes the scientific enterprise is not ultimately directed toward the various material objects on which research is carried out, but is directed toward truth itself.   The genuine scientist, in fact, is the person who aims at nothing less than making himself more and more open to truth.   Thus, in brief, one can say that the genuine scientist is the servant of truth.   But if such a spirit of service is effectively heeded by scientific man in the various manifestations of his life, it is obvious that he cannot fail to become more human as a result of his being scientific.   This, of course, is the highest reason that justifies speaking of science as a humanistic undertaking.

3. THE PROBLEM OF SCIENTIFIC OBJECTIVITY

So far in this chapter we have been concerned only with the epistemological originality of science and its humanistic implications. We have just concluded with a summary that may easily seem overoptimistic although it is no more than an indication of what the success of science discloses to the philosopher reflecting on the nature of knowledge. The time has now come to face the epistemological problems raised by the very success of science.

In the light of the preceding, the discussion we are about to begin may perhaps present itself as superfluous. Indeed, if science is so epistemologically successful while developing autonomously of philosophy, what need is there to discuss the philosophical issues raised by science? The answer is that the need for such a discussion is twofold.

The first reason for studying science philosophically arises from the very real possibility of misunderstanding the spirit of science itself. We have already examined some forms of this misunderstanding (Chapter 4, Section 2). Such a misunderstanding, as we have seen, is not just a matter of academic interest but entails quite serious humanistic consequences (Chapter 4, Section 3). As a result, it is obviously necessary to investigate directly the philosophical nature of science. This we must do at least in order to prevent science from becoming harmful to man and, ultimately, destroying itself. But, of course, the negative motivation for the philosophical study of science is not the chief one. The second reason for this study arises from the quite positive significance of science for man. We have already outlined the main traits of such a humanistic significance. But we have not yet discussed the problems that are connected with them. This is what we must do now.

The approach we are going to follow will remain the exploratory one we have adopted so far. We do not intend to call into question science and its achievements. Rather, we start out by accepting them in their fullness. Our only intention is to explore and make explicit what is actually present in science but is left out of consideration by the scientists acting as such.

The basic epistemological problem that presents itself to our consideration is that of scientific objectivity. In accordance with our methodological principles, we start out by trying to make explicit the reasons why scientists are convinced of the objectivity of their endeavor. Then we shall probe deeper into the issue, beginning with the fact that science is made by man and all the consequences that this entails.

(a) The Man-Independence of Science

Scientists ordinarily do science without bothering at all about possible epistemological questions affecting the objectivity of their work. What is the reason that justifies such an attitude? It may be called the obvious man-independence of scientific statements. It is a conviction arising from several epistemological criteria, tacitly shared and more or less clearly realized. The main criterion can be formulated explicitly as follows.

The overriding motive for scientists' confidence of attaining objective truth through their science is the *interpersonal approach of science itself*. There is something very persuasive in the fact that many people—quite independent in their thinking, critically minded, and seriously concerned with truth—consent to accept as true a given interpretation of nature. It is this convergence of many minds on one proposition that stresses the convincingness of the interpersonal approach of science. Indeed, how could the many converge if not because a given proposition is objectively true; that is, it leaves to those who understand its import no alternative but to accept it? One can perhaps in this connection speak of the experience of *natural truth*, meaning thereby a statement that appears to reveal, all of a sudden, a profound and universal intelligibility of nature. A clear case of such an experience is contained in a letter written to Darwin by a distinguished botanist shortly after the publication of *The Origin of Species*. Referring to the principle of natural selection Watson wrote:

Your leading idea will assuredly become recognized as an established truth in science.... It has the characteristics of all great natural truths, clarifying what was obscure, simplifying what was intricate, adding greatly to previous knowledge.... Now these novel views are brought fairly before the scientific public, it seems truly remarkable how so many of them could have failed to see their right road sooner. [45]

Another famous example that shows the irresistible power of the scientific truth is the reaction of Lothar Meyer to Cannizzaro's pamphlet explaining the method for measuring the relative weights of atoms. The German researcher had been long anguished by the apparently unsolvable complexities of the problem. While he was reading the pamphlet, the situation of his mind changed suddenly. As he was to write later, referring to the episode: "... the scales fell from my eyes, doubts vanished, and the feeling of calm certainty came in their place." [46]

A second important motive that convinces the scientists that science is man-independent, hence objective, is the *fecundity, cognitive and prac-*

---

[45] Letter of H. C. Watson to C. Darwin, quoted in J. T. Merz, *A History of European Scientific Thought in the Nineteenth Century* (New York: Dover, 1965), vol. II, p. 335, note.

[46] Quoted in J. R. Partington, *A Short History of Chemistry* (New York: Harper Torchbooks, 1960), p. 256.

tical, of scientific theories.   There is again something irresistible in the fact that a scientific theory manifests a validity that extends far beyond the empirical basis from which the theory itself was derived in the first place.   One of the main reasons for instance, why the Newtonian gravitational synthesis exerted such a powerful influence on researchers for centuries was exactly its seemingly inexhaustible universality.   Newton, of course, had formulated it by starting from a limited number of observable regularities, especially those studied by Galileo and Kepler.   And yet, the theory itself proved capable of explaining *all* types of mechanical motions of macroscopic bodies.   Thus the scientific mind could not help feeling that the theory had really hit upon the intrinsic intelligibility of nature itself.   In other words, the cognitive fecundity of a scientific theory is convincing because, if the theory itself were just a convenient man-made summary of empirical data, such a fecundity would be utterly impossible. As a consequence the theory in question is accepted as truly objective.

But, if the cognitive fecundity of scientific theories is convincing, there should be no objection of principle to recognizing a similar convincing power in the practical fecundity of such theories.   For, indeed, in both cases man is enabled to achieve results that he could not attain were the theories involved not expressions of objective insights into nature.   In this sense one can prize the practical effectiveness of science as a sign of its truth.   As Rostand put it incisively:

Practical value, sign of truth.   I admire useful learning because it attests connivance with the real. [47]

In this connection, however, a word of caution must be added.   The convincingness of science arising from its practical effectiveness should not be misinterpreted in a pragmatic vein.   The rank-and-file scientist is frequently heard to say: "I accept this theory because it works."   Thus the impression is conveyed that science, after all, is nothing but a trial-and-error guessing, purely aimed at practical results and whose concern for objective truth is merely incidental.   Such an interpretation would be most misleading.   The creative scientist—the person to whom theories are due in the first place—argues quite the opposite way.   His central concern is for truth.   Hence he does not adopt a theory because of its practical effectiveness but rather because of its ability to express the content of truth contained in observational data.   When, then, he realizes the practical fecundity of a theory, he rejoices but, once again, because of his concern for truth.   He is glad because he finds in such a practical fecundity a confirmation of the objective truth of the theory he has labored to formulate.

---

[47] J. Rostand, *The Substance of Man,* trans. I. Brandeis (New York: Doubleday, 1962), p. 273.

(b) The Man-Dependence of Science

Although the man-independence of science is an obvious and undeniable datum, it is not the only such feature of science itself. In fact we may say just as rightly that science is obviously and unavoidably a man-dependent form of knowledge. A certain subjectivity in science is inescapable. Already Galileo was aware of this puzzling situation as we can gather from a revealing remark of his spokesman Salviati. When discussing the trajectory of projectiles, Salviati admits the validity of the objections raised by his two interlocutors against the actual verification in practice of the precise results obtained in theory. In his own words:

I concede that the conclusions thus demonstrated in the abstract alter themselves in the concrete and falsify themselves to such a point that neither the horizontal motion is uniform, nor the acceleration of the vertical motion increases with the rate supposed, nor the trajectory of the projectile is parabolic, and the like. [48]

To be sure, Galileo in this passage manages somehow to extricate himself from the difficulty. He refers to the constant custom of the scientists and points out that there could be no science at all if the researcher were not allowed to make use of abstractions. However, the issue is raised, and cannot be brushed aside if one is seriously concerned with the epistemological significance of science.

The subjectivity or man-dependence of science becomes strikingly evident if one adverts to the plain—although frequently overlooked—fact that *science is made by man*. Science, indeed, is the product of man's own creativity (Chapter 1, Section 1); it is due to the effort of the mind, which strives to penetrate beyond appearances and attain genuine understanding (Chapter 5, Section 2b). But if this is so, how could one hope to really understand science without discussing this fundamental issue?

To clarify how momentous or, perhaps, embarrassing this feature of science can be, we may survey rapidly the successive phases of the scientific endeavor. Beginning with observation, we have already discussed the indispensable heuristic role of theory (Chapter 1, Section 1bb). In simple terms, scientific research cannot get started unless the scientist adopts some man-made interpretation about the structure of nature. This interpretation alone, in fact, enables him to distinguish between relevant and irrelevant observable data. But, then, how can he know that his interpretation is truly objective? To move a step further in the process of science making, we encounter the complex process of theorization. It is obvious that here, if anywhere, the role of man is paramount. Hence the question

---

[48] Translated from G. Galilei, *Discorsi e Dimostrazioni Matematiche intorno a Due Nuove Scienze,* A. Carugo and L. Geymonat, eds. (Turin: Boringhieri, 1958), pp. 301f.

about objectivity becomes more urgent. Not to delay our survey, a word will be said only about the so-called scientific models. These are visualizable representations of observable objects that scientists use as starting points for their theoretical work. What is the assurance that models are objective? For instance Bohr, the famous theoretician, had no qualms about admitting that models could somehow be compared to poetical metaphors or images. Speaking out of his experience in atomic research, he formulated his position to Heisenberg as follows:

These images (*Bilder*) are derived or, if you like, guessed from experiences.... We must realize clearly that language can be employed here only in a way similar to that of poetry. In poetry, too, there is no intention of representing situations of fact with precision, but only of arousing images in the consciousness of the hearer and thus bringing about intellectual connections. [49]

If models are made by man, how can one be certain that the entire structure of science is not founded on a purely subjective interpretation of nature? The standard answer is that one can be certain because of experimentation. But, is this answer really to the point? No one who has reflected on the nature of experiment can avoid having misgivings. For experiment is dependent on man at least on one count. This consists in the obvious fact that experimental data remain meaningless for science until they are interpreted by means of a theory (Chapter 1, Section 1c). But, again, this theory is necessarily dependent on man. Hence, also at the experimental stage, it is not possible to do away with the intervention of man in science except at the absurd condition of doing away with science itself. As Schrödinger pithily summarized it:

...the observer is never entirely replaced by instruments; for, if he were, he could obviously obtain no knowledge whatsoever. [50]

To conclude, science—precisely because it is knowledge *by* man— unavoidably requires the intervention *of* man as a contributor to such knowledge. In other words, there is no escape from subjectivity in science. To be sure, one could object by saying that, in the end, the universal consensus of scientists suffices to take away any subjectivity from science. But this rejoinder is not convincing. For the history of science shows that the more science advances in objectivity, the more it also comes to depend on the active intervention of man doing science. To realize this, it is enough to recall how much more relativity and quantum physics require

---

[49] Translated from W. Heisenberg, *Der Teil und Das Ganze: Gespräche im Umkreis der Atomphysik* (Munich: Piper, 1969), p. 63; cf. p. 57.

[50] E. Schrödinger, *What is Life? The Physical Aspect of the Living Cell* and *Mind and Matter* (London: Cambridge University Press, 1967), p. 176.

the creativity of man than does classical mechanics. But, as a result, the mere consensus of the scientists on the objective value of a theory cannot by itself remove all objections. Who, in fact, should have the last decisive word in such a controversial matter? The scientist who develops the theory? But he would be the one most easily exposed to the danger of subjectivism. Should the scientific community as a whole be the decisive arbiter? But who speaks for it? The vast majority of scientists do not discuss epistemological questions, but simply adopt a theory because it works. As for the relatively few scientists who discuss such issues, their consensus, if any, can be meaningful only in view of the philosophical reasons they give for their positions. In the end, then, we have really no escape from the problem posed by the apparent contradiction: science is both man-independent and man-dependent. As a consequence, unless we are disposed to reduce science to pure pragmatism—hence deny its humanistic significance—we must face the problem of scientific objectivity squarely.

(c) The Empiricist Opinion

As has been seen, the reflective person is led to the realization that the objectivity of science is not a self-evident proposition. However, various philosophical opinions on this issue are possible. One of the most widespread and appealing is empiricism. Supporters of this opinion agree that the objectivity of science presents a philosophical problem but claim that the solution can be found in sense experience alone. Since a position of the kind is widely assumed to be typical of the scientific mentality, we must analyze it in some detail.

(ca) The Structure of the Opinion

Empiricism is, by itself, only a vaguely defined position. To realize its import more clearly we must examine several of its major forms currently influential among philosophers of science.

Instinctivism. The most radical form of scientific empiricism is credited to Ernst Mach. It can be called instinctivism. The reason for such a designation stems from Mach's own conception of science. Science, to him, is nothing but instinct. Or, more precisely, it is an instinctive form of knowledge expressed in abstract terms. Thus, for instance, he describes the two fundamental traits of the scientist:

Indeed, it is perfectly certain, that the union of the strongest instinct with the greatest power of abstract formulation alone constitutes the great natural inquirer. [51]

If one accepts Mach's position, it is clear that the problem of the objectivity of science can be solved easily enough. No special philosophical investigation is needed, but psychology suffices. Science should be considered objective simply because it is the product of instinct. In fact, what is instinctive is obviously objective being the necessary consequence of nature's own laws. Instinctivism, therefore, is what Mach considers the adequate explanation of scientific objectivity. As he put it:

...instinctive knowledge enjoys our exceptional confidence. No longer knowing *how* we have acquired it, we cannot criticize the logic by which it was inferred. We have personally contributed nothing to its production. It confronts us with a force and irresistibleness foreign to the products of voluntary reflective experience. It appears to us as something free from subjectivity, and extraneous to us, although we have it constantly at hand so that it is more ours than are the individual facts of nature. [52]

If science is instinctive, how should one account for its origin and development? Mach is consistent: empiricism, taken in its most undiluted form, supplies the answer. That is, science arises simply because man is passively receptive to sense impressions. These, by imprinting themselves on him as an imitation of the processes of nature, constitute science. In his own words:

How does instinctive knowledge originate and what are its contents? Everything which we observe in nature imprints itself *uncomprehended* and *unanalyzed* in our percepts and ideas, which then, in their turn, mimic the processes of nature in their most general and most striking features. [53]

In the light of the preceding, one can understand Mach's epistemological doctrine contained in the so-called *principle of mental economy*. Mach took great pride in the principle as expressing an original philosophical insight. [54] According to this principle, science is nothing but an organization of sense impressions, a kind of shorthand summary of past information. In other words, the significance of science is reduced by Mach to its practical usefulness. Science, in his opinion, is that procedure which enables man to economize mental energy when he has either to

---

[51] E. Mach, *The Science of Mechanics: A Critical and Historical Account of Its Development*, trans. T. J. McCormack (La Salle, Ill.: Open Court, 1960), p. 35.
[52] *Ibid.*, p. 94.
[53] *Ibid.*, p. 36.
[54] For detailed discussion of the principle see Mach, *op. cit.*, pp. 577-595.

apply or to transmit information about the world. This, in particular, is the way Mach interprets the essence of natural or scientific laws:

The communication of scientific knowledge always involves description, that is, a mimetic reproduction of facts in thought, the object of which is to replace and save the trouble of new experience. Again, to save the labor of instruction and acquisition, concise, abridged description is sought. This is really all that natural laws are. [55]

To sum up, Mach's doctrine is the most rigorous formulation of epistemological instinctivism concerning science. This is the reason for the immense influence it has constantly exerted among philosophers of science. However, the diffusion of the instinctivist mentality is not due to Mach alone. It stems rather from a widespread inclination to view science as a self-justifying form of knowledge. Already during Mach's lifetime, but independently of him, views similar to his were propounded by philosophers like Richard Avenarius and scientists like Gustav Robert Kirchhoff. We shall revert to this point when trying to assess the philosophical import of the doctrine in question. But for the moment we must continue our analysis of other forms of empiricism.

Operationalism. A second form of empiricism, probably derived from that of Mach at least indirectly, is the doctrine of operationalism, proposed by Percy William Bridgman. Bridgman starts out by claiming an absolute empiricist position:

The attitude of the physicist must therefore be one of pure empiricism.... Experience is determined only by experience. [56]

In Bridgman's terminology, experience is to be identified with the totality of sense impressions produced by instrumental observation. Actually, the instrumental interpretation of knowledge constitutes the originality of operationalism itself. The reasons underlying operationalism are obvious. Bridgman contends that genuine knowledge demands total clarity and unambiguousness of concepts. But he is convinced that such concepts can only be formulated on the basis of instrumental observation. Hence, in Bridgman's view, each concept—for instance, length—is unambiguously and adequately defined only by a set of instrumental operations. He goes so far in this contention that he identifies a concept with the set of operations used to determine it.

In general, we mean by any concept nothing more than a set of opera-

---

[55] E. Mach, *Popular Scientific Lectures,* trans. T. J. McCormack (La Salle, Ill.: Open Court, 1943), pp. 192f. See the whole essay, "The Economical Nature of Physical Inquiry," pp. 186-213.

[56] P. W. Bridgman, *The Logic of Modern Physics* (New York: Macmillan, 1946), p. 3.

tions; *the concept is synonymous with the corresponding set of operations.* [57]

Bridgman does not admit exceptions of principle to his operationalism. Thus, in particular, as he himself takes care to remark, one should not speak of one concept when more than one kind of operation is used to define it. Rather, one should speak of just as many concepts as there are kinds of operation. This he discusses in detail while considering the case of length. In summary:

In *principle* the operations by which length is measured should be *uniquely* specified. If we have more than one set of operations, we have more than one concept, and strictly there should be a separate name to correspond to each different set of operations. [58]

As a consequence of his operationalist rigor, Bridgman claims to have achieved total clarification of language. Meaningful concepts are those that are defined by means of operations. All other concepts are rejected as meaningless. [59]

What is the ultimate foundation of Bridgman's position? He does not attempt to justify it in strictly philosophical terms. He just seems to assume that science, as a set of operations, demonstrates convincingly the obviousness of his ideas. The root of his doctrine, then, appears to be a psychological conviction. He starts from the assumption that he knows how the mind works. This is in fact Bridgman's own admission—thrown in almost as an afterthought—at the end of his celebrated *Logic*:

...all the discussion of this essay has been subject to one explicit assumption, namely, that the working of our minds is understood.... [60]

Conventionalism. The two doctrines just reviewed are the most consistent expressions of empiricism. However, many people find them objectionable precisely because of their consistency. Still a third doctrine, therefore, should be considered here—one that only indirectly can be called empiricist. It was proposed by Henri Poincaré under the name of conventionalism.

Poincaré's epistemological interpretation stemmed from the increasing evidence, at the turn of the century, that classical mechanics was not the perfect form of science as hitherto assumed. Poincaré starts out by remarking that people have for too long overrated the cognitive significance of science. The reason, he explains, lies in a mistaken view that science

---

[57] *Ibid.*, p. 5.
[58] *Ibid.*, p. 10.
[59] For details see Bridgman, *op. cit.*, pp. 28-31.
[60] *Ibid.*, p. 197.

is based exclusively on facts and does not contain any hypotheses. Against this position, Poincaré asserts that science is permeated by hypotheses. He lists several kinds of them. Some of these hypotheses, he contends, are obviously conventions or definitions in disguise. In his own words:

> ...some [hypotheses] are verifiable, and when once confirmed by experiment become truths of great fertility... others may be useful to us in fixing our ideas... others are hypotheses only in appearance, and reduce to definition or to conventions in disguise. The latter are to be met with especially in mathematics and in the sciences to which it is applied. [61]

What is Poincaré's justification of conventionalism? At first sight he seems to imply that none is needed. Indeed, he claims that his doctrine simply accounts for the rigorousness of science. Thus, in fact, he goes on to explain in the text cited:

> From them [conventions], indeed, the sciences derive their rigor; such conventions are the result of the unrestricted activity of the mind, which in this domain recognizes no obstacle. For here the mind may affirm because it lays down its own laws....

As for the the possible objection that scientific conventions may be arbitrary, Poincaré makes short shrift of it. He continues in the same text: "Are they arbitrary? No; for if they were, they would not be fertile."

So far, no doubt, it appears that there is no connection between conventionalism and empiricism. And yet, when we read other texts by Poincaré, the connection becomes obvious, even though only by inference. It is surprising, in fact, how Poincaré defines the role of theory in science. It seems that for him theory is nothing but a matter of practical usefulness. The similarity to Mach's position is striking. Here are Poincaré's own words:

> In physics... [theoretical] principles... are only introduced when it is of advantage. Now they are advantageous precisely because they are few, since each of them very nearly replaces a great number of laws. Therefore it is not of interest to multiply them. [62]

Even more illuminating is Poincaré's notion of scientific objectivity. Despite his disclaimer that science is not artificial, he goes on to contend that science cannot know the external world. What science calls objective is nothing but a subjective view of reality typical of man as such. As he put it:

---

[61] H. Poincaré, *Science and Hypothesis,* trans. W. J. G. (New York: Dover, 1952), pp. xxiif.

[62] H. Poincaré, *The Value of Science,* trans. G. B. Halsted (New York: Dover, 1958), p. 126.

... scientific laws are not artificial creations; we have no reason to regard them as accidental, though it be impossible to prove they are not. Does the harmony the human intelligence thinks it discovers in nature exist outside of this intelligence? No, beyond doubt, a reality completely independent of the mind which conceives it, sees or feels it, is an impossibility. A world as exterior as that, even if it existed, would for us be forever inaccessible. But what we call objective reality is, in the last analysis, what is common to many thinking beings, and could be common to all.... [63]

To sum up, also for conventionalism—as for the two other doctrines—science seems to be basically a matter of sense experience alone. Instinctivism and operationalism take this position openly. Conventionalism takes a similar position, but only indirectly. The objectivity of science, in any case, is reduced to an attitude of subjective conviction. This conviction is founded—more or less persuasively—on sense impressions alone. The overall assumption is that, besides sense impressions and the psychological consensus that arises from them, there is no way of justifying the objectivity of science.

(cb) Scientific Critiques

As already indicated, empiricism—especially in its Machian formulation—is widely assumed to be the proper interpretation of science in epistemological terms. We must now address ourselves directly to this question. Is the widespread assumption justified or not? In order to find an answer without having to resort to lengthy philosophical investigations, let us ask creative scientists themselves to comment on this issue. Since they know science from within, they can give us valuable indications.

Scientists are usually very much in favor of empiricism's professed goal of making science genuinely positive. They agree that science should be based on observational data and thus be free of any dogmatism. Hence the frequent praise of Mach that one hears from their lips. Einstein, for instance, says admiringly: "I see Mach's greatness in his incorruptible skepticism and independence...." [64] This Machian attitude, as Einstein goes on to explain, contributed much to prepare his own spirit toward the development of the relativity theory. For it freed him from the apparently unconquerable domination exerted by classical mechanics. Nevertheless, Einstein himself was one of the most persistent critics of empiricism in general, and Mach in particular. This disagreement, despite the admiration, can do much to illumine the question under discussion.

The first reason creative scientists usually reject empiricism is plain: *Empiricism ignores scientific creativity.* This is for instance the basic

---

[63] *Ibid.*, p. 14.
[64] In Schilpp (note 7), p. 21.

objection raised by Einstein against Mach in the text just cited. He castigates him for failing to realize the importance of theory in the development of science. As he put it:

Mach's epistemological position... today appears to me to be essentially untenable. For he did not place in the correct light the essentially constructive and speculative nature of thought and more especially of scientific thought, in consequence of which he condemned theory on precisely those points where its constructive-speculative character unconcealably comes to light, as for example in the kinetic atomic theory.

Rejection of empiricism because of its failure to take into account the complexity of scientific creativity is to be found in various forms in the scientists' sayings. Many attack the boast of empiricism to be only concerned with clarity of ideas. Pasteur's scornful comment is worth quoting: " I pity people who have nothing but clear ideas." [65] The reason is obvious. If science is essentially creation, one must agree with Meyerson that "Obscure ideas are indispensable to the scientist." [66] Other scientists reject the current tendency among empiricist philosophers to reduce science to logical schemes. Ziman, for instance, writes out of his experience as a theoretical physicist that such an interpretation of science is unsatisfactory at least because it "does not leave enough room for genuine scientific error. It is too black and white." [67]

The second reason why scientists are dissatisfied with empiricism reaches more deeply. *Empiricism dehumanizes science.* This view comes to the fore in various ways, especially in the common condemnation of Mach's economy principle. Einstein finds that the principle has "a suspiciously commercial character." [68] Born terms the principle "a materialistic expression." [69] He vehemently denounces it as a complete misunderstanding of science, whose genuine nature, in his mind, is "hunger for knowledge and understanding, a sister of art, philosophy, and religion." [70] Historical research into scientific creation underscores further the superficiality of the empiricist interpretation. Two contemporary historians, for instance, condemn the cavalier attitude frequently adopted toward the complex philosophical and theological aspects of Newton's creative thought. In their view, empiricist interpreters "have reduced the complexity of Newton's thought to absurd simplicity." [71]

---

[65] Translated from quotation in E. Meyerson, *De l'Explication dans les Sciences* (Paris: Payot, 1921), p. 623.

[66] Meyerson, *op. cit.,* p. 630.

[67] Ziman (note 42), pp. 5f.

[68] Translated from Heisenberg, *Der Teil* (note 49), p. 95.

[69] M. Born, *Natural Philosophy of Cause and Chance* (New York: Dover, 1964), p. 207.

[70] *Ibid.,* p. 128.

[71] Dugas and Costabel (note 15), p. 263.

Since Mach is the chief representative of empricism, he is some-times singled out for attack on this score of ignoring the humanism of science. Einstein is particularly bitter. Summing up his views in front of the French Philosophical Society (1922), he uttered the verdict:

The system of Mach studies the relations that exist among the data of experience; the totality of such relations is, for Mach, what constitutes science. That is a bad viewpoint. In sum, what Mach has done is a catalogue, not a system. Just as much as Mach was a good mechanician, he was a deplorable philosopher. [72]

Another condemnation of Mach because of his failure to understand the human significance of science was uttered by Heisenberg. Heisenberg seems to see Mach's position as making science ultimately irrelevant for man as such. As he put it:

I was never much impressed by Mach.... It was too—I would say not too negative, but too modest in what he wanted. It was, perhaps I should say, too little poetical. [73]

Still a third reason can be found in scientists' utterances as a motiva-tion to condemn empiricism. *Empiricism sterilizes science.* It is not for nothing, in fact, that Mach himself became the living symbol of the opposi-tion to scientific progress. His critical power was immense, his experi-mental ability great. Nevertheless he did his best, by ridicule and insinua-tion, to impede research in two of the most fruitful fields of physics, the science he had dedicated himself to. These two fields, as is well known, were atomism and relativity. Why did Mach act so unenlightenedly? Einstein sees no other ground but Mach's own empiricist enmity toward theory in the making of science. Speaking of Mach and of Ostwald, Einstein accuses them of philosophical prejudice with regard to atomism:

The antipathy of these scholars towards atomic theory can indubitably be traced back to their positivistic philosophical attitude. This is an interesting example of the fact that even scholars of audacious spirit and fine instinct can be obstructed in the interpretation of facts by philosophical prejudices. The prejudice—which has by no means died out in the meantime—consists in the faith that facts by themselves can and should yield scientific knowledge without free conceptual construc-tion. [74]

Another example of the damaging influence exerted by the empiricist attitude regards Poincaré. Poincaré, as we have seen, was not an empiricist

---

[72] Translated from quotation in F. Herneck, *Physikalische Blätter* **15** (1959), 564.

[73] T. S. Kuhn, J. L. Heilbron, P. Forman, L. Allen, eds. *Sources for History of Quantum Physics* (copyright: American Philosophical Society); interview November 30, 1962, p. 4.

[74] In Schilpp (note 7), p. 49.

in the strict sense of the term. Nevertheless his position bordered on empiricism because, in his view, scientific theory was but a convenient man-made summary. Louis de Broglie, the great theoretical physicist, discusses in one of his works the reason why Poincaré—the leading mathematical physicist of the times—failed to discover relativity. Certainly it was not for lack of information nor of intellectual ability, de Broglie notes. After all, the question was hotly debated in physical circles, and Poincaré was exceptionally gifted as a theoretical physicist. According to de Broglie, the reason of Poincaré's failure was of an epistemological type. He could not see the issue as scientifically meaningful because of his "philosophical tendency toward nominalistic convenience." [75]

To sum up, it is clear that empiricism—especially instinctivism—is far from justified in its claim of being able to provide an adequate epistemological interpretation of science. Rather, it proves to be just one among various possible philosophical opinions that stand or fall according to the validity of the arguments mustered to support them. To complete our investigation, therefore, we must now proceed further and explore directly the philosophical issue raised by the objectivity of science.

(d) The Epistemological Problem of Scientific Objectivity

When we try to realize the structure of the philosophical problem raised by scientific objectivity, it may be useful to recall once again the motives of our investigation. We do not intend to challenge the existence of scientific objectivity. Nor do we want to claim that philosophy is superior to science because the former takes over where the latter leaves off. Our only motivation is respect for science and consistency with ourselves as knowers. Scientific objectivity is a fact, but this fact puzzles reflective man. Hence we feel it our wish and duty to contribute to the clarification of the issue involved. Needless to add, such a clarification is also in keeping with the genuine spirit of science itself. For science is essentially a search for knowledge. Hence any new question that presents itself to the scientifically minded person should stimulate him to seek further, in order to achieve better knowledge. In particular, of course, one should strive to understand what makes the fundamental greatness of science itself—its very objectivity.

The structure of the epistemological problem raised by scientific objectivity comes to the fore if we begin from the obvious fact that science is knowledge. Knowledge needs to be formulated by means of concepts and propositions. Accordingly, the problem we must investigate involves the objectivity of scientific concepts and scientific propositions.

---

[75] For details see L. de Broglie, *Sur les Sentiers de la Science* (Paris: A. Michel, 1960), p. 368.

To consider *scientific concepts* first, their problematic nature becomes apparent at once if we dare to ask a seemingly preposterous question. The question reads: what difference is there between scientific concepts and the objects designated by them?   In other words: how far do scientific concepts represent reality and how far do they differ from it?   Admittedly, the question looks at first blush absurd.   Indeed, does it not appear self-evident that scientific concepts stands for objectively existing things as these can be observed by man?   Thus, how on earth could one come upon the idea of distinguishing between the two?   And yet, the question is not preposterous at all.   For an example, let us consider the situation prevailing at the time when the teen-aged Einstein was beginning to work on his theory of relativity.   Who would not have felt nonplussed and irritated had he been confronted with a question of the kind enunciated?   Everybody was taking for granted that the nature of time was obvious, and so people spoke of absolute time and simultaneity as though they were self-evident concepts that accurately represented a situation of fact.   But Einstein, in his genius, knew better.   He knew that a concept does not necessarily represent all the intelligibility of an observable entity, and thus he kept searching for a better concept of time.   Hence his theory of relativity was born.   Hence, too, is the philosopher justified when he poses the basic question concerning the difference between scientific concepts and the things they stand for.   For only if one is able to see a distinction between the two can one begin to realize the complexity and richness of scientific knowledge.

The issue of the significance of scientific concepts should be taken quite seriously by anyone who respects science.   The example of the great scientists is illuminating in this regard.   They do not feel ashamed to engage in long, and often heated, philosophical debates precisely on this subject.   The purpose of their debates is that of pinning down with precision the meaning and limitations of the concepts they employ in their science.   Instances of such discussions are numerous.   To restrict ourselves to quantum physics, we can mention the long-drawn-out controversy between Einstein and Bohr, the controversy between Schrödinger and Born, Louis de Broglie's criticism of the so-called Copenhagen interpretation of quantum physics, and the like.   Why all of this activity, which appears pointless to the many persons who are tranquilly satisfied in taking the results of quantum physics as self-evident?   Born gives an apposite answer while discussing the nature of his exchanges with Schrödinger. He admits that the issue is not a scientific but a philosophical one.   But he hastens to add that such discussions about the precise meaning of words are just as important as the mathematical formalism of science itself.   In his own terms:

The whole discrepancy is not so much an internal matter of physics, as one of its relation to philosophy and human knowledge in general.... The difference of opinion appears only if a philosopher comes along and asks us: Now what do you really mean by your words, how can you speak about electrons to be sometimes particles, sometimes waves, and so on? Such questions about the real meaning of our words are just as important as the mathematical formalism. [76]

In brief, the philosophical investigation of scientific concepts and their content of truth is but a necessary consequence of the respect that one entertains for science as knowledge. If one is convinced that science is genuinely cognitive, one cannot help striving to realize what kind of cognition science is and how far it goes.

As regards the epistemological research to be undertaken on scientific concepts, the philosopher should strive for two goals, one positive and one negative. The negative goal consists in not ascribing to scientific concepts a content that they do not have. In other words, the first concern of the philosopher should be not to misunderstand science. The second and positive goal set before the philosopher consists in becoming reflectively aware of the richness of the content of scientific concepts. The promising approach to this issue is the genetic one. Thus, in the first place, the philosophical investigator should strive to become explicitly and concretely aware of what is actually implied by the process of scientific abstraction. Then he should explore the nature of scientific models, and why scientists employ them. Above all, he should try to understand how and why and to what limit scientific concepts express something that exists in itself and does not depend for its existence on the mind of the knowing man. Only at this point will one be in a position to assess adequately the objectivity of the concepts science employs, and thus avoid misleading interpretations of either the dogmatic or skeptical type.

The other major epistemological realm concerning scientific objectivity deals with *scientific propositions or laws*. We are faced with a problem here because all scientific statements present themselves as generalizations, with the form: all As are Bs. That is to say, scientific statements claim to be valid for all objects of a given type, independently of any limitation of space and time. But this universality of scientific propositions is very much surprising; and there is no way of suppressing the surprise by simply referring to observation or induction. Observation does not suffice in itself to justify scientific statements simply because observation is necessarily always something particular, whereas scientific statements are universal. Thus, as Born for instance remarks, a scientific law is of necessity something that exceeds observation. His words are worth quoting:

---

[76] M. Born, *Physics in My Generation: A Selection of Papers* (New York: Pergamon, 1956), pp. 140f.

Observation and experiment are crafts which are systematically taught. Sometimes, by a genius, they are raised to the level of an art.... So it looks as if science has a methodical way of finding causal relations without referring to any metaphysical principle. But this is a deception. For no observation or experiment, however extended, can give more than a finite number of repetitions, and the statement of a law—*B* depends on *A*—always transcends experience. Yet this kind of statement is made everywhere and all the time, and sometimes from scanty material. [77]

As regards induction, a similar point is made by Whitehead, himself a competent scientist and well-known philosopher. He argues against the widespread tendency to take induction as the self-evident justification of the validity of scientific laws. Induction is the generalization of individual observations, each one made at a different point in time and space. But scientific laws claim to have a universal validity, one that is independent of particular circumstances of time and space. Whitehead points out that such a situation is truly surprising:

Induction is not based on anything which can be observed as inherent in nature ... the order of nature cannot be justified by mere observation of nature. For there is nothing in the present fact which inherently refers either to the past or the future. [78]

If the validity of scientific laws is surprising, it is obviously dutiful to investigate reflectively what the reasons are that justify speaking of scientific laws at all. In particular, one should philosophically examine in what precise sense scientific laws are universal. Likewise one should inquire what is effectively meant by the time-independence of scientific statements and under what conditions such statements can be expected to have a perennial validity. All of these issues must be faced straightforwardly if one esteems science. Otherwise the danger is great. Scientific statements run the risk of being seen as mere results of psychological-social conditioning or the product of tacit conventions. Science, then, would be effectively destroyed in what counts most, its cognitive objectivity.

To sum up, recognition of the epistemological problem of scientific objectivity does not stem from the pressure of outside interests that want to force their way into science. Rather, such a recognition is but the consequence of taking the cognitive significance of science seriously. If one is convinced that science is genuine knowledge, one must explore why this is so. Of course, such an exploration has to be conducted by means of an approach that is different from the scientific one. But no one should be surprised that this is the case. For philosophy, too, is a form of knowledge—a different but complementary one to science.

---

[77] Born (note 69), p. 6.

[78] A. N. Whitehead, *Science and the Modern World* (New York: Mentor Books, 1948), p. 52.

4. THE PROBLEM OF SCIENTIFIC DYNAMISM

The second major aspect of scientific knowing, besides objectivity, is dynamism. We have just discussed the problematic character of scientific objectivity. We must now examine the epistemological situation of scientific dynamism. As usual, our goal in this study will simply be that of understanding more deeply the nature of scientific knowledge and thus perceiving its humanistic significance better. The procedure to adopt in order to attain our goal is marked out for us by the general trend of our investigation. We shall begin by studying the reasons why scientists are convinced of a genuine progressivity of science. Then we shall investigate whether progressivity is a self-evident property of science or not. Finally we shall draw the epistemological conclusions that derive from the preceding analysis.

(a) Scientific Progressivity

As has been seen, one of the major epistemological insights due to science consists in the realization of the dynamical character of human knowledge (Chapter 5, Section 2c). We must now return to this insight and examine it critically. How is the dynamical property of knowledge—notably, science—to be interpreted in epistemological terms? Should one think that the mind of man is subject to continuous change, endless and aimless? Or should one think that the mind moves closer and closer to a perfect knowledge to be reached eventually by way of successive approximations? Or, perhaps, neither of these hypotheses applies and the dynamism of knowledge should be interpreted in still another way? Since our issue is very complex, we are going to face it by degrees. We start out by analyzing how reflective scientists assess the dynamism of science in philosophically meaningful terms.

To begin with, it is worthwhile to notice that scientists are convinced that knowledge is *inherently developmental*. That is to say, scientists as a rule reject the hypothesis of a perfect knowledge to be attained eventually by way of successive approximations. Rather, they are persuaded that man will never be able to know nature fully and exhaustively. As Heisenberg put it: "We will never find an end to our attempt to understand nature." [79] This is a very important point because it expresses most clearly the epistemological originality of science as experienced by its practitioners. The sources of the conviction outlined are both experimental and theoretical. Experiment is intrinsically capable of endless refinements and improvements. Accordingly, it is not surprising that experimental circles

---

[79] In Kuhn *et al.* (note 73); interview February 28, 1963, p. 2.

were the first to stress the developmental character of human knowing. [80] Moreover, reflection on theory leads to the same conclusion. Scientific theories may be true, but they are inescapably limited—hence changeable and never wholly adequate. A strong expression of the scientists' conviction on this matter is expressed by Claude Bernard, who writes:

When we propound a general theory in our sciences, we are sure only that, literally speaking, all such theories are false. They are only partial and provisional truths which are necessary to us, as steps on which we rest, so as to go on with investigation; they embody only the present state of our knowledge, and consequently they must change with the growth of science, and all the more often when sciences are less advanced in their evolution. [81]

If scientists reject the idea of a perfect knowledge to be attained through successive approximations, they are also very opposed to the conception of knowledge as endless change. This conviction is very strong. As against the widespread cliché that sees science as a revolutionary endeavor, continually destroying its past in order to make place for the future, reflective scientists insist on both conservatism and novelty as typical of scientific knowing. For want of a more specific word, this conviction can be designated as one of *evolution*. As in the case of a living being, scientists think that science develops in time by conserving the past and improving upon it. Poincaré has made this point forcefully:

The advance of science is not comparable to the changes of a city, where old edifices are pitilessly torn down to give place to new, but to the continuous evolution of zoologic types which develop ceaselessly and end by becoming unrecognizable to the common sight, but where an expert eye finds always traces of the prior work of the centuries past. One must not think then that the old-fashioned theories have been sterile and vain. [82]

What are the motives that, in the eyes of the scientists, justify the evolutionary interpretation of human knowing? They are of several kinds. A fundamental one is a *psychological realization* about the genesis of knowledge in general. Piaget, the master of genetic psychology, synthesized this realization clearly. It is a fact that the human individual, especially during his evolving years, finds himself in a "state of unstable equilibrium; every new acquisition modifies previous ideas or risks involving a contradiction." [83] Nevertheless it is also a fact that the well-balanced individual

[80] See E. Zilsel, "The Genesis of the Concept of Scientific Progress," *Journal of the History of Ideas* **6** (1945), pp. 325-349.

[81] Bernard (note 8), pp. 35f.

[82] Poincaré (note 62), p. 14.

[83] J. Piaget, *The Psychology of Intelligence*, trans. M. Piercy and D. E. Berlyne (Totowa, N. J.: Littlefield Adams, 1966), pp. 39f.

manages to assimilate the new information without breaking down psychologically under the weight of the continual novelties to which he is exposed. Just the same, Piaget concludes, this is what happens in the historical development of science:

An exact science, despite the "crises" and reforms on which it prides itself to prove its vitality, constitutes a body of ideas whose detailed relationships are preserved and even strengthened with every new addition of fact or principle; for new principles, however revolutionary they may be, justify old ones as first approximations drafted to a certain scale; the continuous and unpredictable work of creation to which science testifies is thus ceaselessly integrated with its own past.

Another motive frequently adduced by scientists to support the evolutionary interpretation of science is alluded to by Piaget in the passage cited. It is a kind of *quantitative criterion*. Subsequent discoveries inside one science do not wipe out preceding results, but incorporate them as rough approximations. Louis de Broglie discusses this point in an important philosophical text. He starts out by asserting the finality of scientific law:

Each time a law has been verified in an incontestable manner to a certain degree of approximation (all verification carries with it a certain degree of approximation), we have a definitely acquired result which no later speculation is able to undo. If it were not thus, no science would be possible. [84]

Although a scientific law is final in the sense described, de Broglie goes on to explain, new research in the field usually shows that the law is not precise but only approximate. He refers to the history of optics. At first people thought that the laws of geometrical optics were rigorously precise, but subsequent research proved that they were just rough approximations of the laws discovered by wave optics. The discovery of wave optics, thus, did not invalidate what had been affirmed by geometrical optics, but simply proved its approximate character. The same considerations apply, according to de Broglie, to the development of science in general.

It is just by this process of successive approximations that science is capable of progressing without contradicting itself. The structures that it has solidly built are not overthrown by subsequent progress, but rather incorporated into a broader structure.

In particular, de Broglie applies his criterion of continual approximations to the relationships between quantum and classical mechanics. He

---

[84] L. de Broglie, *The Revolution in Physics,* trans. R. W. Niemeyer (New York: Noonday, 1953), pp. 20f.

notes that the former is different from the latter. Nevertheless, when quantum mechanics is applied to the study of macroscopic bodies, for which classical mechanics was first devised, the two theories lead to the same results. Accordingly, he sees here another case of scientific progressivity.

Thus we meet again the customary process of scientific progress: well-established principles, well-verified laws are conserved, but they can be considered valid only as approximations of certain categories of facts.

To put it synthetically, the dynamical character of science is frequently assessed by reflective scientists in a twofold manner, negatively and positively. Negatively speaking, scientists reject the widespread view that science develops by a series of revolutions. Particularly when discussing the relations between quantum mechanics and its classical counterpart, its foremost creators insist that the former did not destroy the latter, but just corrected its unwarranted extrapolations. Heisenberg, for instance, explains the situations by means of a geographical comparison. Pre-Columbian people were accustomed to extrapolate their contemporary knowledge of the earth to regions hitherto undiscovered. So they felt upset when Columbus achieved his famous discoveries. Objectively, however, it would be wrong to speak of revolution in the case of Columbus' achievements. Likewise, Heisenberg concludes, one should not speak of revolution in physics. In his own words:

It is equally wrong to speak today of a revolution in physics.... Only the conception of hitherto unexplored regions, formed prematurely from a knowledge of only certain parts of the world, has undergone a decisive transformation. [85]

This view is shared by other experts in both forms of mechanics. Born, for instance, writes:

Quantum mechanics is no more and no less revolutionary than any other newly propounded theory. Once again, it is really a conquest of new territory. [86]

To put it in positive terms, the normal scientific attitude amounts to a deep-seated conviction that science is characterized by genuine progressivity. That is to say, science is seen as a continual advance, preserving the conquests of the past and constantly adding to them. A famous passage by Planck expresses this conviction in an almost lyrical way. Although meant originally to celebrate the significance of relativity vis-à-vis classical mechanics, it describes well how scientists feel about the developmental character of science in general.

---

[85] Heisenberg (note 37), p. 18.
[86] Born (note 76), p. 35.

[Modern physics] does not merely disorganize and destroy, but in a much higher degree organizes and constructs—it simply discards a form which, owing to the advance of science, was already out of date. In place of the old confined structure, it erects a new one, more comprehensive and more lasting, and includes all the treasures of the former ... in a different, clearer grouping, and yet has room for discoveries still to be made. It shuts out of the physical universe those unessential factors introduced by the contingencies of human views and customs, and purges physics of its anthropomorphic elements, which arose from individual peculiarities of physicists, and the complete exclusion of which, as I have endeavoured to show elsewhere, is the real end of all physical knowledge. It opens to the progressive mind a perspective of almost immeasurable breadth and height, and leads him to coordinate results in a way unthought of in former periods. [87]

(b) The Consensualist Criterion

The widespread conviction among scientists, as has been seen, is that science is essentially a developmental-progressive enterprise. It is developmental in that it can never come to a stop in its investigation of nature. It is progressive in that its development is a continual improvement. At this point, the critical question comes due. How can one be sure that the development of science, which is supposed to be endless, is effectively progressive instead of being merely a continual change? The instinctive answer is that science is progressive because scientists normally agree about its being so. This is the *consensualist criterion*: the principle that claims that science is progressive because its practitioners hold it to be so. We must now scrutinize the validity of the criterion and consider the consequences that follow from relying upon it.

(ba) The Circularity of Consensualism

The consensualist position is appealingly simple. To probe its structure and consequences we can perhaps find no better guide than Jean Piaget, who upholds the principle and tries to vindicate it in a systematic way. [88] His analysis is illuminating because of its profundity and consistency.

Piaget begins by stressing that the developmental-progressive interpretation of science, as usually admitted, is far from self-evident, but presents a problem. The problem arises because science claims to be both inherently developmental and truly progressive. If science were

[87] M. Planck, *A Survey of Physical Theory* (formerly: *A Survey of Physics*), trans. R. Jones and D. H. Williams (New York: Dover, 1960), p. 43.

[88] Piaget (note 1), vol. I, pp. 38-51 and vol. III, pp. 306-319. The passages quoted in the text were translated from pp. 40f.

not inherently developmental, it would be easy to validate the progressivity of science itself. One would have in fact a "fixed frame of reference" against which to plot the march of science. This frame of reference would consist in the perfect and definitive cognition that one could expect to achieve by means of scientific investigation. The march of science, therefore, could be conceived as a series of successive steps, moving closer and closer to an unchangeable goal. The goal would consist in the definitive knowledge that one could expect science to supply. The inherently developmental character of scientific knowledge, however, makes it impossible to think of such a fixed frame of reference or unchangeable goal. For the developmental character of science implies, precisely, that man will never be able, through science, to achieve perfect and definitive cognition of anything. As a consequence, the view that science is truly progressive cannot be seen as obvious but rather quite problematic.

The problem of scientific progressivity, as analyzed by Piaget, arises from the impossibility of finding a fixed frame of reference against which to evaluate the march of science. Since Piaget is convinced that the progressivity of science must be upheld, he opts at this point for the consensualist criterion. Then he goes on to examine the consequences that follow from its adoption. Let us suppose, with him, that we take the consensus of scientific experts as the ultimate frame of reference to assess the progress of scientific statements. What consequences follow from this option? The basic consequence is clear from history. Scientific consensus is not something that remains fixed and permanent. It is rather something that keeps changing with time. As a result, if one adopts the consensualist interpretation of science, it is possible to try to vindicate the progressive character of science. But this can only be done by studying genetically the contents of scientific consensus along the course of time. In other words, one must rely on a changeable frame of reference—the scientific consensus—to evaluate the continual changeability of science. This situation is upsetting, as Piaget himself points out. Indeed, how can one know whether science effectively progresses if the significance of its changes is to be judged against a frame of reference that itself keeps changing?

If one adopts the consensualist criterion—Piaget makes it clear—the problem remains to justify the validity of the criterion itself. But this, he goes on to explain, cannot be done satisfactorily if one continues to adhere to the criterion in question. For then one is faced with an inescapable dilemma from the epistemological point of view. The dilemma consists in the fact that one has really no choice but either to take the criterion as self-evident or to admit that it needs a proper justification. But the first alternative is epistemologically unsatisfactory because a genetical-historical analysis of the sciences shows that the scientific consensus changes over the times. Nor is it self-evident that such changes need necessarily be progressive. As for the second alternative, it is unsatisfactory simply because it is based upon a circular procedure. To summarize with Piaget himself:

If genetical analysis relies necessarily on the reference frame made up by the sciences that are established at a given epoch, it is this reference frame, then, that one should explain in order to generalize the genetical explanation to knowledge as a whole. But, at this point, one finds oneself in the presence of the following dilemma: either genetical analysis will not succeed in giving an account of its own frame of reference, and thus fail to constitute a general epistemology—or it will succeed, but only at the price of an obvious circularity (*cercle évident*), since genetical analysis relies, in this second case, on a system of reference which depends itself on it.

Piaget is rightfully upset about the possible circularity of scientific knowledge. For, if science could be reduced to a circular form of knowing, science itself would be no better than an illusion. Hence Piaget, after having conjured the specter of circularity, tries manfully to lay it to rest. First, he denies that the circularity involved is a vicious one. Or, at least, he intimates that people should accept it as a necessity of nature. Thus, in fact, he goes on to say in the text cited:

However this circularity, no matter how factual, is not vicious or, at least, it is imposed by the very nature of things.

Apparently dissatisfied with his own conclusion, Piaget attempts subsequently to extenuate it by means of a comparison. He likens the system of the sciences that, when taken together, constitute scientific consensus to a system of scientific instruments that mutually check each other. He gives as an example the measurement of time. When the duration of an event has to be measured accurately, the experimentalist employs numerous clocks. Each is used as a control of the others. In the same way, Piaget infers, the convergence of the system of the sciences shows that the circularity under discussion should not cause any serious concern. In his own words:

Only, if it is ineluctable, such a circularity is susceptible of successive enlargements, comparable in this regard to certain well-known circles of science, such as the one adopted in the measurement of time.... One can then extend the chain [of the sciences] endlessly without leaving the circularity. But the more this is enlarged, the more the convergences observed enable one to find in such a growing coherence the assurance that the circularity is not vicious.

Needless to say, Piaget's reasoning can hardly satisfy the critical thinker. In particular, the example he adduces seems to lead to the opposite conclusion of what he means to prove. For, if it is true that numerous instruments of measurement can serve as mutual controls, they can do this only under the condition that they be mutually independent in their operations. But exactly opposite is the situation that obtains in the case of the scientific consensus. For the various sciences, which are supposed to check each other, are not mutually independent. Rather, they influence each other, continually and necessarily. This happens, of course, because the same human mind is at the origin of them all. As a consequence, the convergence of the sciences or scientific consensus cannot provide an adequate criterion to prove that scientific knowledge genuinely progresses instead of merely changing with time.

(bb) Dangers of Consensualism

The consensualist interpretation of scientific progress, unsatisfactory because of its circularity, is attended by further evils that threaten to invalidate the entire cognitive significance of science. We must still pursue these ramifications of our problem because of the serious implications they present.

The deleterious character of consensualism comes immediately to the fore when we realize that, after all, this criterion is nothing but another form of instinctivism. It is most natural or instinctive to admit that the convergence of opinions of the experts of a given time is a sign of truth. In particular, it is instinctive to assume that science progresses continually because scientists agree that this is the case. However, if there is nothing better than instinct to prove it, how can one be sure that the progress of science is no more than mere appearance—an aimless wandering of the mind instead of an advance toward more objective knowledge?

Ernst Mach, the champion of instinctivism, discusses the dynamism of science with great consistency on the basis of his epistemological postulates. Science is for him uniquely and exclusively instinctive knowledge made up of sense impressions (Chapter 5, Section 3ca). As a consequence, Mach does by no means rule out the possibility that future sense impressions may overthrow what is presently held for certain and replace it with other views, totally incompatible with it. He is firm on this point as an inescapable corollary of his instinctivism. In his own words:

Instinctive knowledge is, after all, only experimental knowledge, and as such is liable ... to prove itself utterly insufficient and powerless, when some new region of experience is suddenly opened up. [89]

What is then the connection between different scientific views or principles that prevail at different periods of history? Mach is forthright

---

[89] Mach (note 51), pp. 94f.

in his answer. There is no connection but a historical one. Thus he goes on to explain in the text cited:

The *true* relation and connection of the different principles is the *historical* one. The one extends farther in this domain, the other farther in that.

Finally, what is the overall significance of the continual development of science? Mach concludes logically that all that is in question here is just a zigzag meandering, arbitrary and endless. This is his view in the text under examination:

All principles single out, more or less arbitrarily, now this aspect now that aspect of the same facts, and contain an abstract summarized rule for the refigurement of the facts in thought. We can never assert that this process has been definitely completed.

Following Mach's lead, we can easily see the main implication of consensualism when it is taken as a self-evident criterion of truth. Consensualism can hardly avoid becoming *historicism*. That is to say, the temporal unfolding of science tends to be seen not as genuine progress or a continually increasing advance in objective knowledge. On the contrary, science seems to develop by means of successive revolution. More precisely, there appear to be just as many sciences as there are different consensuses of experts in different periods of history. Scientists, to be sure, vehemently reject such an interpretation (see Chapter 5, Section 4a). But logical necessity compels consensualist philosophers to override scientific objections and defend historicism as the only acceptable interpretation.

The view of science as a series of successive revolutions keeps recurring. For instance, an early work by Adam Smith, composed toward the middle of the 18th century, defends the thesis that science changes radically from time to time, according to the mentality prevailing in a given epoch. [90] Smith finds evidence for such a view in the history of astronomy. He contends that both the Ptolemaic and the Copernican astronomies deserve to be called scientific because both were upheld by the experts of different historical epochs.

As was to be expected, the historicist-revolutionary interpretation of science gained wide currency especially after the far-reaching changes of perspective introduced into physics by the rise of relativity and quantum theory. An entire philosophical school is at present defending this interpretation. Thomas Kuhn, the leader of the school, is rigorously consistent in his historicism. For instance, he finds it perfectly justified to assert

---

[90] The work was entitled *The Principles Which Lead and Direct Philosophical Inquiries, Illustrated by the History of Astronomy.* For a short discussion see O. H. Taylor, *A History of Economic Thought: Social Ideals and Economic Theories from Quesnay to Keynes* (New York: McGraw-Hill, 1960), pp. 50-56.

that the Einsteinian and Newtonian mechanics are just as mutually incompatible as the Copernican and the Ptolemaic astronomies were. In his own words:

From the viewpoint of this essay these two theories are fundamentally incompatible in the sense illustrated by the relation of Copernican to Ptolemaic astronomy: Einstein's theory can be accepted only with the recognition that Newton's was wrong. [91]

In the light of the preceding, there is not much need for insisting that consensualism is dangerous for science. It is so, obviously, because its attendant historicism tends to reduce scientific convictions to passing fashions. If one is justified in speaking of different sciences on the basis of different consensuses of experts, no science, in the genuine sense of the term, will remain. Science ceases to be objective knowledge and becomes subjective conviction. In particular, one cannot retain the notion of scientific progress, except in a very equivocal sense. Thus consensualism, though claiming to vindicate the progressivity of science by relying only on science itself, leads in the end to the virtual destruction of science and its progressive character.

An even greater danger is likely to arise from consensualism. It consists in the sense of utter meaninglessness which can hardly fail to embitter the life of the reflective scientist. Science comes to be experienced as a source of cognitive frustration, leading man absolutely nowhere in his passionate quest for knowledge (Chapter 4, Sections 3b and 3d). We have here a nihilist conclusion that threatens to engulf the whole personality of the scientist by making science appear to him completely irrelevant from the human point of view. Especially serious is the fact that even the greatest scientists are not shielded against such an utterly negative interpretation of what constitutes the central motivation of their lives. A telling example is a testimony of Heisenberg. He, the convinced champion of scientific progressivity, cannot help wondering whether man will ever be able to understand anything by means of science. For the very meaning of understanding appears to be changing with the times. As he put it:

We will never find an end to our attempt to understand nature. The trouble is that only by doing so we learn again and again and afresh what the word "understanding" means. The word "understanding" means in the sixteenth century something which is quite different from its meaning in the twentieth century and probably in the twenty-first century it will again mean something quite different from now. [92]

[91] T. Kuhn, *The Structure of Scientific Revolutions* (Chicago: University of Chicago Press, 1962), p. 97.

[92] W. Heisenberg, in Kuhn *et al.* (note 73) interview February 28, 1963, p. 2.

To sum up, it is clear that, if one really esteems science, one cannot refuse to admit the existence of the epistemological problem of scientific dynamism, as a typically philosophical question, to be solved by philosophical procedure. This is so because the refusal to acknowledge such a problem as strictly philosophical implies historicism, and historicism amounts to a virtual destruction of science. To be sure, the historicist interpretation of science is self-evidently absurd. Who, in fact, could seriously claim that—for instance—the Ptolemaic and the Copernican world systems have the same cognitive status? Who could dismiss the dedication and sacrifices of genuine scientists throughout the centuries as pure waste? Briefly, who in his right mind could defend the view that, ultimately, science and nonscience are cognitively the same—namely, different forms of passing intellectual fads? And yet, unless one is willing to acknowledge the philosophical investigation of scientific dynamism as a necessary form of inquiry, there cannot be any escape from historicism. The reason is, as we have just seen, that there cannot be any escape from logic. Logic, therefore, compels us now to face the epistemological problem of scientific dynamism directly.

(c) The Epistemological Problem of Scientific Dynamism

The structure of the epistemological problem of scientific dynamism can easily be detected by going back to fundamentals. Science is knowledge. But the scientific kind of knowledge presents two principal peculiarities. One is novelty. The scientific information is something new when compared with the information available to nonscientific man. The other peculiarity is the continual changeability of the scientific information itself. It follows that the philosopher who intends to understand the epistemological significance of the dynamism of science is faced with two main issues.

In the first place, the philosopher should explore the import of the *cognitive break* that takes place in the transition between ordinary and scientific knowledge. A question arises here because the discontinuity between the two kinds of knowledge is obvious. And yet, it cannot be simply taken for granted that science replaces nonscientific or ordinary knowledge entirely and absolutely. In fact, science is itself a specialized form of knowing, restricted to reality inasmuch as observable. Hence one would not be justified to assume a priori that science makes all other forms of knowing superfluous and meaningless. Accordingly, the first issue to be explored by the epistemological investigator regards the relationships of science with ordinary knowledge. His task is to make explicit the original contribution of science to human knowing in general. But, at the same time, he should try to clarify the specific limitations of scientific knowledge itself.

In the second place, the philosopher should explore the *continuity and unending improvement* that are typical of science. As has been seen, this constitutes the central issue of scientific dynamism. A question arises here because the development of science is characterized by two features that seem to contradict each other. On the one hand, science is continuous; that is, it somehow remains unchanged. On the other hand, science improves constantly; that is, it keeps changing all the time. The task of the epistemologist is to investigate, by means of concrete examples, in what sense science can be said to be changeable as well as unchangeable in its historical development. He should try to find out what new content of truth later discoveries add to earlier ones in a given field of scientific research. In particular, the epistemologist should try to formulate a clear-cut criterion to determine why people are justified to speak of progress in science according to the strict sense of the term.

Once the philosopher has succeeded in understanding the progressive character of scientific knowing, he should still tackle a more general question—the issue of the *dynamical character of knowledge in general*. This point is pivotal if one can ever hope to integrate the new epistemological perspectives disclosed by science into a satisfactorily comprehensive doctrine of knowledge as such. Indeed, why does the progressivity of science upset people if not because the currently prevailing doctrine of knowledge is much too static and not dynamic enough? Many persons, for instance, feel compelled to choose between the validity of the Newtonian mechanics, on the one hand, and that of the Einsteinian mechanics, on the other. Such an attitude is not justified except on the assumption that a statement can only be completely and unchangeably true or completely and unchangeably false. But such a static interpretation of knowledge is certainly neither self-evident nor, probably, convincing. The hypothesis, at least, can be advanced that knowledge may have a dynamical or progressive nature. That is to say, it may be quite possible for various statements about the same object to be different from each other, and yet be all true.

In brief, the central challenge presented by scientific dynamism is that of formulating a systematic doctrine of knowledge that takes into due account a major experiential feature of *knowledge as a living activity of man*. Experientially, it is a fact that knowledge can be true and valid without being static. For, indeed, experience shows that knowledge is essentially a contact between subject and object. In other terms, knowledge is awareness of a presence. But all affirmations about something that is present are true, even though they differ from each other, provided they do not deny that something is really present. Thus, experientially, the dynamism of knowledge does not present any difficulty, but it is rather an unavoidable manifestation of genuinely human and living cognition. Indeed, how could one be in living contact with an object for a length of time without coming to perceive the object itself in different and increasingly more adequate ways? The difficulty arises when one tries to express in explicit and systematic terms such an inherent progressivity of experiential knowing. This is then the humanistic task of the epistemologist who takes seriously the epistemological challenge of scientific dynamism. He has to help thoughtful man to understand reflectively the nature of his own living knowledge.

5. KNOWLEDGE AS EXPERIENCE AND REFLECTION: THE EPISTEMOLOGICAL PERSPECTIVE OF SCIENTIFIC HUMANISM

In the present chapter we have analyzed the epistemological structure of science: its originality and main problems. For the purpose of this book we must now study the synthetical features of the new epistemological perspective disclosed by science. In other words, we must bring together the various threads of our discussion and weave them into a unitary pattern. The pattern must befit a humanism that is truly scientific.

To attain this aim we begin by exploring the epistemological root of the current split between science and philosophy. Then we shall discuss the humanistic limitations of the scientific approach, and the need for complementing it by means of philosophical reflection. Finally we shall conclude by outlining the basic features of the new epistemological synthesis.

(a) Experience vs. Reasoning: The Epistemological Root of the Current Cultural Split

As is well known, our current Western culture is deeply split. Not without reason, people often speak of the so-called two-culture phenomenon. The expression, however imprecise, conveys a widely felt sense of humanistic dissatisfaction motivated especially by a lack of common ground between science and philosophy. We must now examine the roots of this situation, in particular the motives that account for the lack of communication between scientists and philosophers.

(aa) The Philosophical Interest of Scientists

One major conclusion can easily be drawn from our entire preceding discussion. It regards the philosophical interest of scientists. The widespread cliché claims that scientists are not interested in philosophy and are even opposed to it. The actual situation is exactly the opposite. Scientists are so much interested in philosophy that a general rule seems to apply: the greater the scientific originality of a researcher, the profounder his involvement in philosophy. Let us consider briefly some documentation of this statement.

The spirit of science, as is universally acknowledged, is embodied especially in Galileo, the father of science itself. But Galileo would never have accepted a sharp distinction—much less a separation—between science and philosophy. For the union of the two appeared to him obvious both because of the Renaissance mentality of his times and the keen awareness of his own philosophical gifts. This is shown, in particular, by his almost pedantic insistence on obtaining the right title at the court of the Grand Duke of Tuscany. When negotiating to obtain the most coveted position of his life, he stressed that he wanted to be called not only the Mathematician but also the Philosopher of the Grand Duke. [93] Actually, Galileo gave a good deal of attention to philosophy throughout his life. He went so far as to challenge the contemporary Aristotelians in their own ground and used to claim that he was a much better interpreter of their master's thought than they themselves. Galileo's great respect for philosophy is epitomized in a moving if defiant manner in the Dedication of his *Dialogue*, the book that marks the watershed between the prescientific and the scientific eras. In it he asserts that the differences between man and man can be enormous. He finds the ultimate explanation of such differences in the widely varying philosophical attitudes of individuals. To him, the philosopher seems to be the only authentic man.

Such differences depend upon diverse mental abilities, and I reduce them to the difference between being or not being a philosopher; for philosophy, as the proper nutriment of those who can feed upon it, does in fact distinguish that single man from the common herd in a greater or less degree of merit according as his diet varies. [94]

Galileo's conviction about the importance of philosophy for science is widely shared by original scientific thinkers. An episode from Heisenberg's memoirs is indicative. The young man, who had already found an oustanding scientific teacher in the person of Sommerfeld, dates the beginning of his scientific career from an afternoon walk he took with Bohr, the first time he met him. What was the reason for Bohr's great influence? Heisenberg has but one term for it: Bohr was a genuine philosopher. In his own words, when reminiscing about the event many years later:

And I felt that here there was a real philosopher who tried just to get concepts by which you can handle things.... So this new way of theoretical physics did actually occur to me for the first time in just this very conversation with Bohr. [95]

---

[93] Letter to Belisario Vinta; May 7, 1610. In Galileo, *Opere* (note 12), pp. 892f.

[94] G. Galilei, *Dialogue concerning the Two Chief World Systems, Ptolemaic and Copernican*, trans. S. Drake (Berkeley: University of California Press, 1962), p. 3.

[95] In Kuhn *et al.* (note 73); interview November 30, 1962; p. 14. For details see Heisenberg, *Der Teil* (note 49), pp. 59-64.

Actually, we know from history that Bohr had a lively interest in philosophy proper. He attended Høffding's lectures at the University of Copenhagen, read his writings as well as those of other philosophers, and kept in continual touch with Høffding himself. [96] In the light of these facts, one should not be surprised that Bohr was such an inspiring scientific leader—not only for Heisenberg, but for the many who flocked to his Institute to learn from him how to improve their thinking processes in order to become better quantum physicists.

The central reason why reflective scientists are interested in philosophy is pointed out by Heisenberg himself while discussing the attitude of his lifelong friend, the theoretical physicist Wolfgang Pauli. This reason is the *seriousness of science*. Science does nothing less than investigate the very intelligibility of the world. Pauli was an exceptionally sharp-minded person. Despite his youth during the years in which modern quantum theory was being created, he used to be called "the conscience" of the theory, owing to his unremitting criticism of any sloppy thinking. This was the person who became increasingly concerned by philosophical problems, when reflecting on the inadequacy of rational thinking alone. In Heisenberg's words:

You know, physicists really do very serious things; they think about the structure of the world. After all, that's what we do.... It was not for Pauli a kind of funny game. It was certainly not meant as opium; it was the contrary of opium for Pauli. Pauli was so extremely sceptical that he very soon reached that point where he becomes sceptical about sceptics—where it turns around. That is the point which is unavoidable for everybody who wants to be consistent.... It's very important if one is consistent and then, of course, one sees that rational thinking is only a limited approach to the world. [97]

The same conviction about the importance of philosophy is expressed by Einstein in his obituary of Mach. Echoing his classical description of those who deserve to be called genuine scientists (Chapter 4, Section 1a), he insists that a burning interest in epistemology must characterize such scientists. In his own words:

For when I turn to science, not for some superficial reasons such as money-making or ambition, also not (or at least not exclusively) for the pleasure of the sport, the delights of brain-athletics, then the following questions must burningly interest me as a disciple of this science: What goal will and can be reached by the science to which I am dedicating

---

[96] For details see M. Jammer, *The Conceptual Development of Quantum Mechanics* (New York: McGraw-Hill, 1966), pp. 172-179.

[97] In Kuhn *et al.* (note 73); interview February 27, 1963; p. 20.

myself? To what extent are its general results "true"? What is essential and what is based only on accidents of development? [98]

Theoretical scientists, such as the ones cited, are more likely to acknowledge the importance of philosophical reflection on science. But also thoughtful experimentalists agree on this point. An impressive case is that of Claude Bernard. This great experimentalist expresses the conviction that experiment alone is not sufficient to persuade the mind adequately about the existence of truth. The reason of this deficiency, in his view, is the purely "objective" nature of experiment. The mind needs also "subjective truths." In Bernard's own words:

The experimental method is concerned only with the search for objective truths, not with any search for subjective truths.... Subjective truths are those flowing from principles of which the mind is conscious, and which bring it the sensation of absolute and necessary evidence. [99]

(ab) Scientists' Rejection of Aprioristic Systematism

Given the widespread interest of scientists in philosophy, it may appear surprising that there is currently such a lack of communication between scientists and philosophers—up to the point that people feel justified to speak of the so-called two cultures. To detect the root of the situation, let us examine rapidly the reasons why scientists feel dissatisfied, as a rule, with the contemporary philosophy of science.

To avoid polemical attitudes as far as possible, we start from a concrete case where, originally, no polemic was intended. Reichenbach, one of the leading contemporary philosophers of science, reports that he once asked Einstein how the latter had come to discover his theory of relativity. [100] Einstein's answer, in Reichenbach's words, was that "he found it because he was so strongly convinced of the harmony of the universe." After this brief mention, Reichenbach goes on to dismiss Einstein's declaration as a "creed" of purely psychological nature, with no philosophical significance whatever. He refuses to consider scientific discovery as a subject of philosophical importance. According to his view, philosophy should consist only in the logical analysis of a scientific theory, once this has been completed. In Reichenbach's own words:

The philosopher of science is not much interested in the thought processes which lead to scientific discoveries; he looks for a logical analysis of the completed theory, including the relationships establishing its

---

[98] Quoted in G. Holton, *American Journal of Physics* **29** (1961), 806.
[99] Bernard (note 8), pp. 28f.
[100] H. Reichenbach, "Philosophical Significance of Relativity," in Schilpp (note 7), pp. 289-311; quotations in text are from pp. 292f.

validity. That is, he is not interested in the context of discovery, but in the context of justification.... It [philosophy of physics] incorporates the physicist's beliefs into the psychology of discovery; it endeavors to clarify the meanings of physical theories, independently of the interpretation by their authors, and is concerned with logical relationships alone.

Reichenbach's methodological standpoint, widely shared by his colleagues, may or may not be justified by philosophical theory. However, if it is, one consequence is unavoidable: it becomes hard for the scientist to see what a relevant role philosophy itself could still play relative to science. For the crucial philosophical issue that presents itself to the scientist is precisely that of explaining why a given discovery, as expressed in a theory, should be considered true. In other words, according to the scientist, there is an intrinsic connection between the two so-called contexts of discovery and justification. Hence the philosopher cannot concentrate on the latter to the exclusion of the former without arousing the suspicion that he wants to elude the main problem he is supposed to solve. This criticism was made, for instance, by Leonard Nash, a chemist and historian of science, while discussing Reichenbach's position. In his own words:

If he accepts the Reichenbachian dichotomy, the philosopher of science no longer concerns himself with the science practiced by scientists.... That logical analysis of completed scientific theories offers a complete philosophy of science is a delusion.... Indeed, if we confine ourselves to logical analysis of finished structures, our understanding must forever rest incomplete simply because we cannot then understand why those structures are what they are. [101]

In the light of the preceding introduction, we can understand at once the general attitude of scientists toward philosophers. They accuse the latter of *irrelevancy* in their endeavors. The point is made frequently, in different ways. For instance, Pauli is reported to have remarked:

It belongs to the confession of faith of the positivists that one should accept facts blindly, as it were.... Possibly this is an attitude which is logically self-contained. Only, if it is, I do no longer know what it means to understand nature. [102]

Another example of the same attitude is that of Bohr. He knew from experience the philosophical difficulties entailed by scientific discovery, especially with regard to quantum theory. Thus he felt upset that people manifested no surprise at the existence of this theory. "For—as he put

---

[101] L. K. Nash, *The Nature of the Natural Sciences* (Boston: Little, Brown, 1963), pp. 295f.

[102] Translated from Heisenberg, *Der Teil* (note 49), p. 280.

16

it—when one does not begin by being amazed by the quantum theory, one cannot possibly have understood it." [103]

In detail, the criticisms leveled by scientists against philosophers are mainly two. The first can be called *systematism*. Scientists accuse philosophers of being too concerned with systematization and thus of disregarding the epistemological complexity of concrete science. Sometimes they are very bitter on this score. Bernard's attack against Auguste Comte and his positivism is a classic example of this attitude. He writes in his private notes:

What is a positivistic philosopher? ... He is a man, as Comte himself says, who makes generalities his speciality. Now there are no worse spirits than these as far as science is concerned, no matter how brilliant and sublime they may be. This kind of man was born particularly during the scholasticism of the Middle Ages and they are its residues. These are the people who intend to argue about everything in general and about nothing in particular, because they know nothing in detail. [104]

The deep-seated reason why scientists distrust philosophers is manifested clearly in a passage by Einstein. Replying to two friendly epistemologists who had commented on his work, he starts out by acknowledging the mutual interdependence of science and epistemology. In his own words:

They are dependent upon each other. Epistemology without contact with science becomes an empty scheme. Science without epistemology is—insofar as it is thinkable at all—primitive and muddled. [105]

After this declaration of principle, however, Einstein goes on immediately to dissociate the epistemological attitude of the scientist from that of the philosopher. The parting ground, in his mind, is the systematism of philosophers. As he puts it:

However, no sooner has the epistemologist, who is seeking a clear system, fought his way through to such a system, than he is inclined to interpret the thought-content of science in the sense of his system and to reject whatever does not fit into his system. The scientist, however, cannot afford to carry his striving for epistemological systematic that far. He accepts gratefully the epistemological conceptual analysis; but the external conditions, which are set for him by the facts of experience, do not permit him to let himself be too much restricted in the construction of his conceptual world by the adherence to an epistemological system.

---

[103] *Ibid.*

[104] Translated from C. Bernard, *Philosophie: Manuscrit Inédit* (Paris: Boivin, 1937), p. 35.

[105] In Schilpp (note 7), pp. 683f.

The second major criticism uttered by scientists against philosophers is that of *apriorism*. They accuse philosophers of reducing science to a too rigid rationalistic structure of axiomatic-deductive type. In particular, they attack Kant for having systematized epistemology by giving too much importance to the logical structure of classical mechanics, as though this were the ideal of knowledge. This interpretation was actually detrimental to the scientific understanding of the world, as Heisenberg, for instance, points out:

Even Kant's philosophy, intended as a critique of premature dogmatization in scientific concepts, could not prevent the torpescence of the scientific concept of the universe—it may even be said that it encouraged it. For, once the main reasoning of classical physics had been accepted as the *a priori* of physical investigations, the belief arose, through an obvious though false extrapolation, that it was absolute, i.e., valid for all time, and could never be modified as a result of new experiences. [106]

To sum up, we can begin to see what is the profound reason that opposes scientists to philosophers. It is not, properly speaking, that the former are uninterested in the reflective study of knowledge while the latter dedicate all their efforts to it. It is rather a different conception of the nature of knowledge entertained by the two classes of thinkers. Or, at the very least, it is a matter of different emphasis. Philosophers ordinarily give the impression to scientists that they conceive knowledge in such a rationalistic and logical key that their conclusions are largely irrelevant if not misleading as far as scientific knowledge itself is involved. We must now hasten to formulate the conception of knowledge characteristic of scientists themselves.

(ac) The Epistemological Root of the Current Cultural Split

What is the conception of knowledge commonly entertained by scientists? We can synthesize it in one word by speaking of *experience*. To realize the import of the term, let us examine a few typical statements on this subject.

Experience, in the scientists' conviction, stands for everything that is original and tangible in science itself. Hence the emphasis that they lay on it, frequently with polemical undertones. As an example of this mentality, the following statement by Bernard on induction is illuminating:

In a word, induction must have been the primitive, general form of reasoning; and the ideas which philosophers and men of science constantly take for *a priori* ideas are at bottom really *a posteriori* ideas. [107]

---

[106] Heisenberg (note 37), p. 22.
[107] Bernard (note 8), p. 46.

In a similar vein, Born extols the virtues of what he calls "the empirical standpoint." He defines it as follows:

This standpoint denies the existence of *a priori* principles in the shape of laws of pure reason and pure intuition; and it declares that the validity of every statement of science (including geometry as applied to nature) is based on experience. [108]

Going on to explain his mind, Born recognizes that his position may be labeled empiricism. He does not like the word because of its ambiguity, since classical empiricism is hardly an adequate interpretation of science. Nonetheless he is willing to put up with such a designation owing to the necessity of stressing the experiential character of scientific knowledge. In his own words:

I do not think that there is any objection to this form of empiricism. It has the virtue of being free from the petrifying tendency which systems of *a priori* philosophy have. It gives the necessary freedom to research, and as a matter of fact modern physics has made ample use of this freedom.

In a synthetical passage on the epistemological significance of modern physics Heisenberg stresses the liberating power of experience. Experience alone, he points out, was able to free the mind from the rigidity of basic concepts. These concepts had proved to be unsatisfactory, but man was not able to replace them with better ones until he was illuminated by new experimental research and theoretical reflection on it.

Coming back to the contribution of modern physics, one may say that the most important change brought about by its results consists in the dissolution of this rigid frame of concepts of the nineteenth century. Of course many attempts had been made before to get away from this rigid frame which seemed obviously too narrow for an understanding of the essential parts of reality. But it had not been possible to see what could be wrong with the fundamental concepts like matter, space, time and causality that had been so extremely successful in the history of science. Only experimental research itself, carried out with all the refined equipment that technical science could offer, and its mathematical interpretation, provided the basis for a critical analysis—or, one may say, enforced the critical analysis—of these concepts, and finally resulted in the dissolution of the rigid frame. [109]

As a consequence of the preceding, we can determine with precision the reason that opposes scientists to philosophers on epistemological grounds. It is the *experiential character of knowledge*. Knowledge, as

---

[108] Born (note 76), p. 39.
[109] Heisenberg (note 11), p. 198.

perceived by scientists in their daily work, demands that man make himself constantly attentive to new and surprising manifestations of nature—that he be continually open to experience. But philosophers as a rule seem to be interested only in clarity of definitions and rigor of logical deductions. At the very least they appear unwilling to study, with all the patient care that is needed, the experiential complexity of knowledge. As a result, scientists—even the most philosophically minded ones—feel inclined to give up the effort for reaching an understanding with philosophers. For to them such an effort appears doomed to failure. The disconsolateness of the situation is summarized aptly in an utterance by Bohr. After a lifelong interest in philosophy, he pronounced in the end a sweeping condemnation of philosophers. With bitter resignation he indicated that it was impossible to find an understanding with them. In his own words:

I felt ... that philosophers were very odd people who really were lost, because they have not the instinct that it is important to learn something and that we must be prepared really to learn something of very great importance.... First of all I would say ... that it is hopeless to have any kind of understanding between scientists and philosophers directly. [110]

To summarize, we detect a twofold root of the current cultural split: one objective and epistemological, the other subjective and psychological. Objectively one cannot say that scientists as such are against philosophy, but rather that they entertain epistemological views that differ from those ordinarily defended by professional philosophers. In other words, objectively the split is an affirmation of philosophy, not a rejection of it. Subjectively or psychologically speaking, however, it is true that scientists frequently reject philosophy and overemphasize science. The ground of this attitude is easy to see. Scientists, at least the reflective ones, cannot avoid realizing the vital importance of philosophical issues. Yet only too often they feel confronted with philosophers who appear not to take seriously enough the philosophical significance of science. As a consequence, scientists frequently react by overshooting the mark. Instead of criticizing philosophers, they reject philosophy itself. This reaction discloses the bitterness of a betrayed expectation. Specifically, as has been seen, scientists castigate philosophy as commonly practiced because of its lack of interest in experience. As Bernard put it synthetically, "Philosophy does not learn anything nor can it learn anything new because by itself it neither experiments nor observes." [111] Needless to add, however, rejection of philosophy in the name of science is far from being a necessary consequence of science itself. It is rather a philosophical position that tries

[110] In Kuhn *et al.* (note 73); interview November 17, 1962, p. 4.
[111] Translated from C. Bernard (note 104), p. 37.

to gain acceptance by passing itself off as science. We must now examine this widespread philosophical position to assess more adequately the significance of science itself as a cognitive undertaking.

(b) The Insufficiency of Public Knowledge

What epistemological ground normally underlies the overemphasis of science to the exclusion of philosophy? Mainly, it is the enthusiasm aroused by science as public experience. Science has manifested itself as a form of knowing that is experientially public, and as such quite fruitful (see Chapter 5, Section 2). Hence the inference: why should one bother about any other form of knowledge? Actually, in moments of unguarded enthusiasm, it is only too natural to reject philosophy as rationalistic and misleading while stressing science as experiential and certain. For instance, Claude Bernard writes: "Reason and reasoning alone are the source of all our errors. Feeling is a safer guide." [112] We grant the importance of the public-experiential character of science, but must hasten to ask a crucial question. Is public knowledge sufficient as a source of information for man? In particular, does the publicity of knowledge justify the existence and validity of science itself? We begin our discussion by studying what science, as public knowledge, is silent about.

(ba) The Silence of Science

At this point of our epistemological research it is clear that, in a true sense, we can speak of a silence of science. In the first place, *science is silent about itself*. This is but an immediate conclusion from our preceding investigations about objectivity and progressivity of scientific knowledge (Chapter 5, Sections 3 and 4). For we have seen that, in the last analysis, these two properties of science, although quite real, remain problematic and open to doubt unless studied by a typically philosophical approach. But, if this is the case, should we perhaps not infer that science is silent in an even more serious sense, affecting man as such? To analyze this question we need but follow the considerations of an outstanding contemporary theoretical physicist. Schrödinger has given much thought to the matter.

The point of departure is a reflection on the public-experiential nature of the scientific approach. To do science, the scientist must, necessarily and exclusively, be object-oriented. That is to say, the scientist must restrict his investigation to those properties of reality that can be observed on an interpersonal basis. This is the *objectifying* aspect of the scientific endeavor. It consists in the resolute methodological step of leaving

---

[112] *Ibid.*, p. 19.

entirely out of consideration any subjective or nonobservable entities. In Schrödinger's formulation:

The scientist subconsciously, almost inadvertently, simplifies his problem of understanding nature by disregarding or cutting out of the picture to be constructed, himself, his own personality, the subject of cognizance. [113]

The objectifying attitude of the scientific approach is not only permissible, but amply justified by its fruitfulness. Nevertheless, a disturbing situation arises. Man, the very creator of science, is cut off from the reality that science itself studies and is made into a sort of anonymous and disembodied spectator. As Schrödinger points out, "the scientific picture of the real word around me is very deficient." The deficiency, as he goes on to explain, arises from the reductionism that necessarily follows the objectifying approach of science. Reality is reduced to what can be observed of it. Man is acknowledged as being present only insofar as he can be observed, namely as a body. Schrödinger summarizes:

So in brief, we do not belong to this material world that science constructs for us. We are not in it, we are outside. We are only spectators. The reason why we believe that we are in it, that we belong to the picture, is that our bodies are in the picture. Our bodies belong to it.

The reaction to the preceding consideration may well be an impatient "so what?" Science is justified in studying reality by reducing it to its observable aspects, isn't it? Doubtless it is, but the question at issue is another. It is to know whether science suffices to give man adequate information about reality as a whole, including himself. Let us suppose for a moment that science does suffice. What happens then? Schrödinger stresses that, in this hypothesis, man as a person evaporates from the picture. Reality, as far as observable, becomes a self-contained structure, complete in itself, where man as such is superfluous, simply ceases to exist. In other words, the scientific representation of reality—if taken as exhaustive—explains everything, but only at the condition of explaining man away. As Schrödinger puts it:

But then comes the impasse, this very embarrassing discovery of science, that I am not needed as an author. Within the scientific world-picture all these happenings [the movements of my body] take care of themselves, they are amply accounted for by direct energetic interplay.... The scientific world-picture vouchsafes a very complete understanding of all that happens—it makes it just a little too understandable. It allows you to imagine the total display as that of a mechanical clock-work, which

---

[113] E. Schrödinger, *What is Life? and Other Scientific Essays* (New York: Doubleday Anchor Books, 1956). Texts quoted belong to the essay "Nature and the Greeks," and are to be found on pages 105-108.

for all that science knows could go on just the same as it does, without there being consciousness, will, endeavor, pain and delight and responsibility connected with it—though they actually are.

The obvious consequence of the preceding is that science is, in a sense, quite informative. It says very much of importance about everything that can be observed. But in another very true sense, *science is completely silent*. In fact, being intrinsically dependent on the objectifying approach, it is unable to tell man anything about those questions he is most interested in. In other words, if science were the only source of information about reality, there would be no aesthetics, no ethics, no religion—nothing that affects the human personality as such. In brief, reality would contain nothing genuinely significant for man as such. In Schrödinger's words:

... the scientific world-view contains of itself no ethical values, no aesthetical values, not a word about our own ultimate scope or destination, and no God, if you please. Whence came I, whither go I? Science cannot tell us a word about why music delights us, of why and how an old song can move us to tears.

To sum up, the objectifying feature of science, typical of its nature as public knowledge, confronts man with a humanistic choice. He must decide whether or not to admit the existence of another form of knowledge, not object-oriented but subject-oriented. In other words, he must decide whether or not to accept philosophy as a valid form of cognition in its own right. If one accepts philosophy, the objectifying feature of science need not give rise to any humanistic concern. In fact science, as seen, must cut off man as a person from the reality it studies. But, if philosophy is genuine knowledge, it can itself explore the personal aspects of man that science leaves out. Thus a genuine humanism, in the traditional sense of the term, remains possible. If, on the contrary, one rejects the cognitive validity of philosophy, one must be ready to face the humanistic consequences of such a decision. That is, one must acknowledge that humanism is no longer possible in the traditional sense of the term. Science, in fact, if it is considered as the exclusively unique form of knowledge, must deny the existence of everything that cannot be observed—in particular, the existence of man as a person. This is certainly a heavy price to pay. And yet, many seem willing to pay it. For science presents itself to them as so convincing and philosophy as so discouraging that they opt for the former at the expense of the latter. However, if the option is adopted, a further, far-reaching question presents itself: what are the consequences of such an option as far as science itself is involved? We must consider briefly the import of this issue.

(bb) The Destructiveness of Scientism

The humanistic controversies of the mid-19th century have produced a special term to designate the option under examination here. The term is *scientism*. It should not be misconstrued as science. It is rather a philosophical tenet—the view that science is the universal and uniquely acceptable form of knowledge. To attain our present aim, it is enough for us to inquire about the consequences of scientism for science.

The immediate consequence of scientism is *reductionism*. The connection between the two positions is obvious. The sole criterion according to which scientism judges the reality of something is its observability. Indeed, according to scientism, nothing is real but that which can be scientifically observed and only insofar as it can be observed. It follows that scientism reduces reality to what can be observed of it; that is, scientism necessarily entails reductionism. The reductionist interpretation of science is widely accepted as unobjectionable.

What are the implications of reductionism for science? We may limit our considerations to one example: biology. Reductionism is currently rampant in biological circles especially because of the successes of molecular biology. Many biologists tend to reduce the living being to the physicochemical interactions of its components, as studied precisely by molecular biology. How do thoughtful scientists react to such a view? The scornful terms with which they designate it are illuminating. They speak of "central dogma" and "monomania." Is such a hostile reaction justified? It is, at least insofar as reductionism threatens the healthy development of biology itself. Already Poincaré scoffed at certain myopic naturalists of his time by saying:

Should a naturalist who had never studied the elephant except by means of the microscope think himself sufficiently acquainted with that animal? [114]

Currently the reductionist threat to biology has grown much worse. Simpson, for instance, speaks of a "crisis in biology" and indicates its source in reductionism.

It is ridiculous to base a philosophy of science or a concept of scientific explanation wholly on the nonbiological levels of the hierarchy [atom, molecule, etc.] and then to attempt to apply it to the biological levels without modification. [115]

To have a complete picture of the situation, however, one should not think that reductionism limits its threat to the theoretical aspect of

---

[114] Poincaré (note 62), p. 21.
[115] G. G. Simpson, *Biology and Man* (New York: Harcourt, 1969), p. 8. See the whole essay, "The Crisis of Biology," pp. 3-18.

biology. Barry Commoner, the great ecologist, stresses that the reductionist conviction entails a threat to life itself. He caustically speaks of the statement that denies any intrinsic distinction between life and nonlife as of a self-fulfilling prophecy. In his words:

One of the conceits pressed upon us by the illusory successes of molecular biology is the idea that life is, after all, nothing but a mixture of chemical reactions and that 'the boundary between life and nonlife has all but disappeared.' If we fail to appreciate the profound connection between ignorance and death and persist in our unwitting efforts to disseminate into the environment substances known chiefly for their power to kill, this statement may, after all, turn out to be true. [116]

Although reductionism is the main consequence, scientism also manifests its ominous influence in numerous other forms. One of these can be called *actualism*. It is the view that identifies reality with what can be observed here and now. It implies the conviction that nothing else should be deemed worthy of consideration by man. The connection between scientism and actualism is stressed for instance by Bridgman, whose operationalism is but a stringent form of scientism (see Chapter 5, Section 3ca). He writes:

The common feature in all these things to which we ascribe "physical reality" is that they may be determined in terms of instrumental operations made at the point and the instant in question. [117]

Another view that follows from reductionist scientism is a sort of *dogmatic subjectivism*. Reality is reduced to what can be observed—not by scientists in general, but by the individual scientist. Hence the individual becomes not only the center but also the arbiter of reality. He feels entitled to decide what should be recognized the right to exist and what should be denied it. We must speak of dogmatism here because there is dogma, by definition, wherever man makes himself into the peremptory arbiter of what is true and what is false. Dogmatic subjectivism is particularly disturbing from the humanistic point of view. For it brutally compels man to declare nonexistent one of two realms, both of which present themselves to his intimate experience as very real. The two realms are external reality, on the one hand, and man's own interior self, on the other. Schrödinger has emphasized this distressing consequence of scientism:

One of the two thus seems irrevocably doomed to a ghostlike existence, either the objective external world of the scientist, or the self of con-

---

[116] B. Commoner, *Science and Survival* (New York: Viking Press, 1967), p. 46. See the whole chapter, "Greater than the Sum of Its Parts," pp. 30-46.

[117] P. W. Bridgman, *The Nature of Thermodynamics* (New York: Harper Torchbooks, 1961), pp. 216 f.

sciousness which by thinking constructs the former, withdrawing from it in the process. [118]

Scientism engenders two general attitudes concerning the overall significance of science. The first is *pragmatism*. Scientism pragmatizes science in that, as we have seen, it robs it of any humanistic meaning according to the traditional acceptation of the term. Schrödinger has emphasized the negative result of this attitude for science itself. Speaking out of experience and addressing scientists, he points to the fact that pragmatically interpreted science cannot keep itself going, because it can no longer be pursued as an end in itself. He writes, while referring to the economic pragmatism of Mach and Kirchhoff (see Chapter 5, Section 3ca):

Call to mind that sense of misgiving, that cold clutch of dreary emptiness which comes over everybody, I expect, when they first encounter the description given by Kirchhoff and Mach of the task of physics (or of science generally): "a description of the facts, with the maximum of completeness and the maximum economy of thought".... In actual fact (let us examine ourselves honestly and faithfully), to have *only* this goal before one's eyes would not suffice to keep the work of research going forward in any field whatsoever. [119]

The second general attitude fostered by scientism concerning the overall significance of science is *technicalism*. We have already discussed this phenomenon at some length (see Chapter 4, Section 2c). But now we can see the roots of the domineering technicalist ideology as increasingly prevailing in our own days. Technology takes on the character of overriding ideal, of which pure science must be the servant. [120] The reason is precisely the influence of scientism. Scientistically interpreted science has no meaning left for man but the technical products it can bring about.

In the light of the preceding, it is clear that scientism is far from being an adequate interpretation of science. It is rather a perversion of the scientific attitude and tends to suppress genuine science as such. [121] This being the case, one should ask why scientism exerts such a powerful fascination on contemporary minds. The answer can be found—paradoxically enough—in the writings of some of the most enthusiastic promoters of the scientistic mentality. Like any other form of outstanding human success, science tends to originate its own *mythology*. The term is employed by

---

[118] Schrödinger, (note 113), p. 214.

[119] E. Schrödinger, *My View of the World,* trans. C. Hastings (London: Cambridge University Press, 1964), pp. 3f.

[120] See especially J. Ellul, *The Technological Society*, trans. J. Wilkinson (New York: Vintage Books, 1967).

[121] For the negative effects of scientism on science, especially on physics, see, for instance, "The Fate of Physics in Scientism," in S. L. Jaki, *The Relevance of Physics* (Chicago: University of Chicago Press, 1966), pp. 461-500.

Ernst Mach. While discussing the worldview of the French encyclopedists, he scoffs at their naïve reliance on their science as though it would have been absolute knowledge. What those philosophers called science was, in Mach's opinion, no more than a myth—just as despicable as those entertained by the old animistic religions. In his own formulation:

... the world-conception of the encyclopaedists appears to us a *mechanical mythology* in contrast to the *animistic* of the old religions. Both views contain undue and fantastical exaggerations of an incomplete perception. [122]

Another warning against transforming science into a myth was voiced by T. H. Huxley. Speaking of the rising public enthusiasm for science among his contemporaries, he excoriated it in very strong terms. He spoke of "stupidity" and "superstition." Thus, for instance, responding to a friend who was complaining that science was still meeting opposition on the part of public opinion, he said:

You may depend on it, victory is on your side—we or our sons shall live to see all the stupidity *in favour* of science. [123]

Returning to the same point in one of his public addresses, Huxley pointed to history as a warning. He wrote:

History warns us that it is the customary fate of new truths to begin as heresies and to end as superstitions.... Against any such a consummation let us devoutly pray; for the scientific spirit is of more value than its products, and irrationally held truths may be more harmful than reasoned errors. [124]

To sum up, not much remains to be said to stress the insufficiency of that public form of knowledge which is called science. If there were no other knowledge besides the public one, not only would man as a person be eliminated from the world explained by science, but science would eventually destroy itself as well. This conclusion, of course, should not be interpreted as an attack against science. We have no intention to deny here what we have affirmed all along, namely that science is genuine knowledge and should be recognized as such. Our only contention is that no one has the right to claim that science is the uniquely reliable form of knowledge, much less that its cognitive reliability is self-evident. For those who advance such a claim do a signal disservice to science itself. They transform it into a myth that clamors for its debunking.

---

[122] Mach (note 51), p. 559.
[123] In C. Bibby, ed., *The Essence of T. H. Huxley: Selections from His Writings* (New York: St. Martin's, 1967), p. 234.
[124] *Ibid.*, p. 14.

(c) *The Personal Character of Knowledge*

We are trying to outline the comprehensive epistemological perspective of scientific humanism. To this end, obviously, two conditions have to be fulfilled. One is to acknowledge the importance and limitations of public knowledge. The other is to realize the personal dimension of knowledge and the consequences it entails. So far we have dealt extensively with the first theme. We must now deal briefly with the second theme, beginning with the personal character of science itself.

(ca) *The Personal Dimension of Science*

In the light of our entire preceding investigation, the personal dimension of science emerges just as clearly as its public dimension. To be sure, many people—being afraid of philosophy—will continue to call science an impersonal endeavor. But theirs is a desperate attitude. Polanyi rightly speaks of the "desperate craving to represent scientific knowledge as impersonal." [125] The desperation of such a stand is obvious in that, as Polanyi continues, the creativity of man would be stamped out from the scientific process. In his own words:

... the scientist would be left uncommitted ... he would say nothing more than a telephone directory ... he would have a machine to speak for him, impersonally.

The basic point we must admit if we really esteem science is its personal dimension. Indeed, if science is genuine knowledge, it must lead to truth. But truth can be achieved only through the personal judgment of man. In fact, cognition implies certainty. But certainty can only be had through a personal commitment of myself to a given proposition that I, after careful consideration, judge to be true. Polanyi puts it beautifully:

Intellectual commitment is a personal decision, in submission to the compelling claims of what in good conscience I conceive to be true. [126]

Recognition of the personal dimension of science is not really novel nor unusual. We meet with it again and again in the writings of reflective scientists, including some who would gladly find a reason for refusing to acknowledge it. For instance, the physicist Campbell writes:

The judgment that there is absolute truth in any proposition must ultimately depend on a judgment which has only relative truth. For, after all, it is I who judge that other people agree with me; or, if it is held

---

[125] M. Polanyi, *Personal Knowledge: Towards a Post-Critical Philosophy* (New York: Harper Torchbooks, 1964), p. 169.

[126] *Ibid.*, p. 65.

that there is universal agreement, it is still I who judge that there is this universal agreement. [127]

This view is so unavoidable that even such a thoroughgoing empiricist as Bridgman cannot help pointing it out—albeit with a deeply felt discomfiture, as is noticeable from the context. He says:

In any event the "correct" or ultimately accepted "scientific" point of view takes its origin with some individual.... Consensus does not enable the human race to get away from itself, and for that reason consensus appears of less significance than is usually esteemed. [128]

In brief, it is clear why the overemphasis laid on the public or "impersonal" dimension of science can be so misleading as to amount to a misunderstanding of science itself. The impersonalist interpretation reduces science to a mere shadow of itself. As Campbell put it, those who defend this interpretation "exhibit to the outside world only the dry bones of science, from which the spirit has departed." [129] The challenge consists in accepting both dimensions of science and exploring what significance emerges from them for man living in the scientific age.

(cb) The Role of Philosophical Reflection

Polanyi, the indefatigable defender of science as personal knowledge, has rightly pointed out that science acquires genuine meaning for man only if he learns to pause and starts to reflect on it philosophically. As long as it remains purely on its public level, scientific experience fails to be appreciated in its richness and profundity. Even worse, it can become a sort of screen that impedes man from coming into direct contact with the very things science discloses to him. In his own words:

As observers or manipulators of experience we are guided *by* experience and pass *through* experience without experiencing it *in* itself. The conceptual framework by which we observe and manipulate things being present as a screen between ourselves and these things, their sights and sound and the smell and touch of them transpire but tenuously through this screen, which keeps us aloof from them. [130]

Philosophical reflection—which Polanyi does not blush to call "contemplation"—reverses the dehumanizing trend of impersonalism. It makes man enter into direct contact with the concrete objects that science studies. Thus, in fact, Polanyi goes on to say in the context cited:

---

[127] N. R. Campbell, *Foundations of Physics: The Philosophy of Theory and Experiment* (formerly *Physics, The Elements*) (New York: Dover, 1957), p. 264.

[128] P. W. Bridgman, *The Way Things Are* (New York: Viking Press, 1961), p. 129.

[129] N. Campbell, *What Is Science?* (New York: Dover, 1952), p. 98.

[130] Polanyi (note 125), p. 197.

Contemplation dissolves the screen, stops our movement through experience and pours us straight into experience; we cease to handle things and become immersed in them. Contemplation has no ulterior intention or ulterior meaning; in it we cease to deal with things and become absorbed in the inherent quality of our experience, for its own sake.

Polanyi insists in his writings that knowledge, as effectively practiced by man, is eminently tacit: "tacit knowing is in fact the dominant principle of all knowledge." [131] But tacitness is merely a manifestation of the personal-experiential character of knowledge itself. For to know—experientially and personally—means, concretely, to encounter things, to dwell in them, to participate somehow in their existence. But by so doing, of course, man knows without realizing explicitly what knowledge itself is.

... tacit knowing is more fundamental than explicit knowing: we can know more than we can tell and we can tell nothing without relying on our awareness of things we may not be able to tell. Things which we can tell, we know by observing them; those that we cannot tell, we know by dwelling in them. All understanding is based on our dwelling in the particulars of that which we comprehend. Such indwelling is a participation of ours in the existence of that which we comprehend.... [132]

Given the inexplicitness of knowledge as a lived experience, the basic contribution of philosophy should lie in making man explicitly aware of what knowledge actually is. This philosophy ought to do by means of a systematic reflection conducted by the subject on his own knowing. If philosophy is interpreted in such a way, one can easily see the importance of philosophy for the genuinely human understanding of science, as well as of knowledge in general. It is not a question of concentrating on the logical structure of knowledge. For a philosophy that is mainly interested in logic takes knowledge for granted, so to speak. It assumes knowledge as a fact and proceeds to dissect and organize its contents according to rigorous schemes. Philosophy as advocated here, on the contrary, starts out by being surprised by the very existence of knowledge. Then it proceeds to examine not just the contents of knowledge, but the whole experience that makes up the cognitive phenomenon. Philosophy thus makes man reflectively aware of his knowing by enabling him to realize the structure, validity, and limitations of his own cognitive experience.

(cc) The Cognitive Contribution of Philosophy

If philosophy is taken in the sense just defined, its importance for man living in the scientific age is obvious. Philosophy humanizes science

---

[131] Polanyi, *The Study of Man* (Chicago: University of Chicago Press, 1963), p. 13.
[132] Polanyi (note 125), p. x.

by making man realize the meaning of science itself as an experience of the whole person. And yet, a possible ambiguity can still persist on this subject. It regards the cognitive import of philosophy itself. How should it be interpreted: as a mere subjective awareness or as a source of objective information? In other terms, has philosophy to be taken simply as a psychological explication or as authentic cognition existing in its own right? Many philosophers interested in science opt for the first alternative. To them philosophy is essentially a subjective analysis whose only goal is to make explicit one's own subjective convictions or beliefs. Polanyi, for instance, says:

I believe that the function of philosophic reflection consists in bringing to light, and affirming as my own, the beliefs implied in such of my thoughts and practices as I believe to be valid; that I must aim at discovering what I truly believe in and at formulating the convictions which I find myself holding; that I must conquer my self-doubt, so as to retain a firm hold on this programme of self-identification. [133]

The subjectivist-psychological interpretation of philosophy is widespread. What are its consequences when this view is applied to the philosophical study of science? Since man cannot avoid being consistent in his thinking, the outcome is inevitable. If philosophy is no more than an explication of a subjective belief, philosophy can only manifest science as a subjectively entertained conviction. Science then becomes an ideology whose objectivity is unproved and unprovable. The only justification of science, therefore, is the subjective belief of its adherents. Polanyi hints that much in the context cited:

Science exists only to the extent to which there lives a passion for its beauty, a beauty believed to be universal and eternal. Yet we know also that our own sense of this beauty is uncertain, its full appreciation being limited to a handful of adepts, and its transmission to posterity insecure. Beliefs held by so few and so precariously are not indubitable in any empirical sense. Our basic beliefs are indubitable only in the sense that we believe them to be so.

The conviction animating the research of this book is the one expressed in the other alternative mentioned. Philosophy is taken here as a source of objective information. In other words, I maintain that philosophy must be acknowledged an authentic cognitive ability. The reason why I make this claim is my persuasion that science itself must be ascribed a genuine cognitive ability. Indeed, as we saw at length, the cognitive ability of science is brought into jeopardy if philosophy itself is denied a genuine ability to know. For, then, the justification of science has to

---

[133] *Ibid.*, p. 267.

be sought in several criteria, which are themselves covertly philosophical but at the same time prove deleterious to the objectivity of science itself. Such criteria, as we have seen, are instinctivism (Chapter 5, Section 3ca and 3cb), consensualism (Chapter 5, Section 4bb), and scientism (Chapter 5, Section 5bb).

As a result of our investigation, therefore, our stand must be clear. Logical consistency compels us to accept *both* science and philosophy as genuinely cognitive undertakings in their own right. In other words, from this point on we shall assume that there is an effective *complementarity* between science and philosophy as far as cognition is involved.

(d) The Humanizing Complementarity of Science and Epistemology

To conclude our introduction into the epistemological perspective of scientific humanism, a few considerations must be added about the new conception of knowledge befitting the scientific age and the means for ensuring its existence. The new conception of knowledge arises from ascribing genuine cognitive ability to both science and philosophy as autonomous yet mutually complementary forms of knowing. Accordingly, the new conception of knowledge will be one in which experience and reflection do not go their independent ways but are mutually integrated into a harmonious whole. But since we have not yet such a harmoniously integrated form of knowing, how should people proceed in order to bring about its existence? The answer is, by developing consistently the humanizing dynamism contained in the complementarity of science and epistemology. As an illustration, the following are two main traits of a better knowledge that can be expected to arise from such a complementarity.

The first cognitive trait we can look forward to in the new humanism is a truly *critical attitude* in man's intellectual outlook as opposed to dogmatism and prejudice as well as naïveté. To be sure, a critical attitude on the part of man as a thinker is an age-old requirement, which has constantly been stressed by succeeding civilizations. Yet man has only too often failed to meet this requirement, especially by overemphasizing the truth content of what he actually knew. Whitehead, for instance, has repeatedly stressed the attraction of what he calls "the fallacy of misplaced concreteness." [134] Indeed, it is a common tendency for man to exaggerate the truth of what he knows, to take for granted that his own idea can be identified with the entire intelligibility of the object known. In particular, as far as science and philosophy are concerned, history shows that—when they operate separately—they tend to increase rather than diminish the dogmatism and prejudices of man. Philosophers, not too rarely, have behaved in

---

[134] Whitehead (note 78), p. 52 and *passim*.

such a way that their discipline has come to be seen as the embodiment of apriorism and unwillingness to face concrete human issues. But also scientists are often guilty of cognitive arrogance. Science can easily give the impression, because of their behavior, of being a contemporary version of mythology and superstition, to speak with Mach and T. H. Huxley. Against such a dehumanizing tendency fostered by the separation of science and philosophy, we can readily see the humanistic advantage that arises from the cooperation of the two in the epistemological area. Man is warned against the danger of falling into extremes and, at the same time, is stimulated to become better aware of the inexhaustible intelligibility of reality. Hence the first sense arises in which science and epistemology deserve to be seen as humanistically complementary. This sense has to do with the fact that man is made more critically conscious of the complexity of truth and, by the same token, is enabled to pursue truth itself more adequately. In brief, man is made more human by being made into a better knower.

A second sense in which science and epistemology deserve to be considered humanistically complementary is their fostering another major trait of genuinely human knowing, *openness to total reality*. This feature, to be sure, is but the positive side of the foregoing one, for a critical attitude of the mind should lead to cognitive openness. Nonetheless, we may spend a few words to mention it explicitly because of the important consequences it entails.

The way epistemological reflection on science leads man to complete openness of mind can be seen by considering the example of Niels Bohr. This great philosopher-scientist was fond of quoting Schiller's couplet: "Fullness alone leads to clarity, and in the abyss does truth dwell." (*Nur die Fülle führt zur Klarheit, und im Abgrund wohnt die Wahrheit.*) [135] Humanists, as a rule, rightfully stress the openness to totality that is inspired by poetry. What is remarkable here is that the scientist—by way of reflection—rediscovers the conviction of the poet, by realizing perhaps even better its content of truth. For, after all, the poet is indeed open to totality, but only in a very vague and subjective sense. The scientist, on the contrary, has an experience of the objective intelligibility of reality as well as of its inexhaustibility. In any case, whatever the relationships between science and poetry may be, the remarkable fact remains: genuinely creative and reflective man—either scientist or poet—realizes and stresses the need for making oneself open to totality in order to be fully human.

The importance of the epistemological openness to totality fostered by the cooperation of science and philosophy is clear even in the realm

---

[135] Translated from W. Heisenberg (note 49), p. 284.

of science proper. For, as has been seen, science tends to produce its own philosophical mythology and thus it runs the risk of making itself insensitive to basic aspects of reality, which it should take into consideration. An example is contained in Bohr's criticism of the positivistic tenet —which claims to speak in the name of science—dictating that man should employ only clear concepts and remain silent about everything else. Bohr's retort: "But this prohibition would prevent us even from understanding the quantum theory." [136] Even more vital to the scientific endeavor is the openness of mind fostered by the cooperation of science and philosophy concerning the so-called human or social sciences. For here more than ever the danger is great of misunderstanding the very object of inquiry—man—by arbitrarily ruling out as meaningless some essential aspects of his nature. The tendency is very widespread among social researchers to take for granted, for instance, that man's moral aspects, his struggling toward values and ideals, should be left entirely out of consideration from an objectively scientific investigation of man himself. Needless to say, however, as the famous sociologist Gunnar Myrdal remarks in this connection, such a methodological position is far from justified from the scientific point of view. Rather, the opposite is the case. For, as he points out, "this is a bias and a blindness, dangerous to the possibility of enabling scientific study to arrive at true knowledge." [137]

Above all, the humanistic impact of the open-mindedness generated by the cooperation between science and epistemology comes into prominence when religious questions are involved. Religion has always been an essential part of humanism in the traditional acceptation of the term. Even currently, for many people, the religious meaning of reality constitutes the core of their humanistic evaluation of man and universe. And yet, only too frequently, those who are enthusiastic about the achievements of science, simply take for granted that religion is irrelevant. They deny any objective foundation to religious convictions, as though religion and science would be utterly incompatible. The serious threat that arises from this situation as regards the attainment of a genuinely comprehensive humanism in the age of science is plain to see. For if people are divided on the religious issue, they will never cooperate to build a truly encompassing humanistic doctrine. They will rather fight against each other in the name of contrasting humanisms—and man, as a whole, will be the

---

[136] *Ibid.,* p. 283.

[137] G. Myrdal, *Value in Social Theory: A Selection of Essays on Methodology,* ed. P. Streeten (New York: Harper & Row, 1958), p. 61 and *passim*; see especially pp. 150f. For similar views concerning the vital importance of value considerations for a proper development of economics see O. H. Taylor, *A History of Economic Thought: Social Ideals and Economic Theories from Quesnay to Keynes* (New York: McGraw-Hill, 1960), pp. xii, 144, 331, and *passim*.

loser. Happily, however, there is no justification for a war between science and religion if man only allows himself to be illuminated by the complementary attitudes of science and philosophy concerning knowledge. Bohr, for one, used to insist that the realm underlying religion deserved to be recognized just as much reality as the one underlying science. In his conviction, the complexity and vagueness of religious formulations were not to be taken as evidence of a self-contradiction of religion itself. On the contrary, he held that the reality of the religious realm was the greater, the more difficult it was for man to speak of it in human terms. As he put it, synthesizing his thought:

If in the language of religions of all times there is use of images, parables and paradoxes—this can hardly mean anything else but that there is no other possibility of grasping the reality to which they refer. But that does not mean that it is not a genuine reality. [138]

Bohr also displayed the same open-minded conviction with regard to the reality of the entities investigated by philosophy in general and metaphysics in particular. He acknowledged that these entities could be expressed only by means of figurative language. Nevertheless he insisted that such a figurative approach was necessary so that man could come closer to that fullness that constitutes reality in its totality. To summarize with him:

Through these images we can surely come in some way closer to the real factual situation (*wirklichen Sachverhalt*). The factual situation itself we should not deny. "In the abyss does truth dwell." [139]

To sum up, the humanizing complementarity of science and epistemology ought to be considered a fundamental datum. The two are complementary in the strict sense of the term because only by taking both of them simultaneously into account can man develop a knowledge that is adequately human for the scientific age. As a consequence, thoughtful man should now strive with greater determination and confidence toward a new epistemological synthesis. For only thus can he become more genuinely human—by becoming a more alert and consistent knower.

---

[138] Translated from Heisenberg (note 49), p. 123.
[139] *Ibid.*, p. 285.

CHAPTER 6

SUPEREMINENT INTELLIGIBILITY:
THE ONTOLOGICAL PERSPECTIVE OF SCIENTIFIC HUMANISM

We have attempted a comprehensive survey of the new epistemological perspective disclosed by science. We must now proceed in our philosophical investigation. To find our bearings at this point of our research, let us recall again the overall aim of our study. We want to realize the significance of science for man as such. But man is principally characterized by the ability to know, and knowledge presents several basic aspects. The first of these aspects is the one we have just examined: knowledge as an activity of the subject. The second aspect is knowledge as a source of information about the object. To continue our research we must now reflect on this second aspect of knowledge as manifested by science. Our aim here is to become reflectively aware of what science has to say about objective reality and the implications that arise from it for man. To speak in philosophical terminology, our present task is ontological. As defined in the Preliminaries (Section 3c), ontology is the reflective study of the object. The philosopher tries to become consciously aware of what makes the object intelligible by reflecting on knowledge as a source of information about the object itself. We intend to study here the ontological significance of science.

To obtain in practice the goal enunciated, we need but follow the general guidelines adopted in this book. Our basic conviction is that science is an autonomous and important form of knowledge. As a consequence, we begin by analyzing the ontological insights brought to light by science. Then we examine the principal ontological problems raised by the success of science. We conclude by outlining the main traits of the new ontological synthesis that should be typical of scientific humanism.

1. THE CONNECTION BETWEEN SCIENCE AND ONTOLOGY: ONTOLOGICAL INSIGHTS OF SCIENCE

The first step in our present investigation is an obvious one. Both science and ontology are interested in the objective intelligibility of reality, but from different angles. Science is interested in intelligibility as observable, ontology is interested in intelligibility as explorable through personal reflection. It follows that the ontologist who takes science seriously must begin his study by exploring the connection between science and ontology. In other words, he must try to realize the typical ontological insights that are brought to light by the very success of science. This is therefore our program in this first section.

(a) The Ontological Source of Science: The Principle of Intelligibility of Nature

The opinion is frequently uttered in philosophical circles that science has no humanistic significance because it has no concern for ontology. But a moment of reflection suffices to show that the actual situation is exactly the opposite. Science cannot even start to exist without a very strong ontological conviction. The reason is obvious. Science is but an effort to understand nature. Nature, however, at first presents itself to man as hopelessly complicated. Thus, how could man start to do science without adopting a strong ontological conviction, namely that nature—despite all appearances—can be intrinsically understood? The philosophical justification of such a conviction may not always be clear in the mind of the scientist. But about the conviction itself there can be no doubt. As Heisenberg put it boldly, to the scientist it is obvious that nature "must" make sense. In his own words:

Nature must make sense. Somehow nature must make sense because otherwise we couldn't understand anything. History shows that we can understand nature somehow and so finally in history always nature has made sense. [1]

The ontological source of science may be given a name: the *principle of intelligibility of nature*. With Schrödinger this may be defined briefly as "the hypothesis that the display of nature can be understood." [2] But, if the intelligibility principle is so essential for science, we should strive to understand it a little more so as to realize better the humanistic relevance of science itself. To begin with, it should be noted that the principle has an essentially philosophical character. It is a great contribution of perennial philosophy to the modern mind. For the principle can be traced all the way back to the Ionian philosophers who first conceived the possibility that man could grasp the structure of nature with his in-

---

[1] W. Heisenberg, in T. S. Kuhn, J. L. Heilbron, P. Forman, L. Allen, eds. *Sources for History of Quantum Physics* (copyright American Philosophical Society); interview February 25, 1963, p. 19.

[2] E. Schrödinger, *What is Life? and Other Scientific Essays* (New York: Doubleday Anchor Books, 1956), p. 103.

tellect. Schrödinger has rightly stressed the immense importance of this philosophical discovery. In his own summary:

For the first time the idea turns up that it should be possible to trace back the manifold of phenomena to a few simple fundamental principles—what later times called laws of nature; the idea that in nature everything comes to pass in a natural way; the hope that once the true principles have been found and the laws and regularities that follow from them clearly understood, the helpless amazement and fear in the face of nature could be overcome, and the uncertainty of one's expectations would be greatly reduced. This was an immense anticipation. It *was* the basic idea of natural science. [3]

To understand the intelligibility principle in depth, we must examine a frequently employed concept, strictly related to it: the concept of *simplicity*. The term often sounds confusing to the nonscientist. He feels baffled by its paradoxical usage. Certainly, the scientific understanding of nature is far from simple, and tends to become less and less so. Yet scientists insist on using the term. Clearly, what they mean cannot be that nature is easy to understand. Rather, what they intend to stress is that, despite appearances, nature is intimately intelligible. An example of this usage can be found in the reminiscences of Heisenberg. He writes:

In regard to nature, I believed strongly that its interconnections were, in the end, simple. Nature is so made up—this was my conviction—that it could be understood. Or, perhaps, I should put it more precisely the converse way: our thinking ability (*Denkvermögen*) is so constituted, that it can understand nature. [4]

The antecedent conviction about the intrinsic intelligibility of nature plays an essential role in the making of science. This is illustrated for instance by a passage where Galileo describes his study of the uniformly accelerated motion of falling bodies. [5] When he published his *Discourses* in 1638 he could look back to well over 30 years of research on the subject. Obviously, there was nothing particularly simple—in the ordinary sense of the word—about his discovery that all bodies fall to the ground with the same speed and their speed increases uniformly in proportion with the time of fall. And yet, when Galileo gave an account of his work, he kept speaking in terms of simplicity. To him, the motion he had discovered was "naturally" accelerated. Simplicity had been his guide

---

[3] *Ibid.*, p. 186.

[4] Translated from W. Heisenberg, *Der Teil und das Ganze: Gespräche im Umkreis der Atomphysik* (Munich: Piper, 1969), p. 142.

[5] Translated from Galileo Galilei, *Discorsi e Dimostrazioni Matematiche intorno a Due Nuove Scienze,* A. Carugo and L. Geymonat, eds. (Turin: Boringhieri, 1958), pp. 170-180; cf. pp. 186f. See the excellent historical-critical note by the editors on pp. 768-774.

because, so to speak, he had just let nature guide him by the hand.   In his own terms:

... as regards the study of naturally accelerated motion, an observation has led us almost by the hand. It is the observation of the custom and rule of nature itself in all its remaining works. This consists in employing the means that are the most immediate, the most simple, the most easy.

To clarify the meaning of his words, Galileo gives some examples.   He points to the swimming of fish and the flying of birds.   In his mind the behavior of such animals is the most simple and most easy.   Then he applies this heuristic conception to the study of falling bodies.

Therefore, if I observe that a stone—when falling from a height and a state of rest—comes to acquire always new increments of speed, why should I not think that these additions take place according to the most simple and most obvious rate (*simplicissima atque magis obvia ratione*)? But, if we reflect attentively, no addition, no increment is more simple than that which increases always in the same way.

As a consequence, Galileo formulates his law of uniformly accelerated motion:

... we conceive with the mind that that motion is uniformly and also continually accelerated which in any equal periods of time acquires equal increments of speed.

The decisive importance of the ontological conviction animating the searching scientist can be seen perhaps even more dramatically in the case of the discovery of the energy quantum. [6]   Galileo, at least, had his senses to guide him in the study of the falling bodies.   Planck could start from nothing more than sets of numbers.   For years experimentalists had been studying the intensity curve of the so-called blackbody radiation, and theoreticians had tried to find a mathematical formula to describe the experimental results.   Then, finally, Planck hit upon the mark.   He found a formula that perfectly fitted the observational data.   At this point it might well have appeared that scientific research into the matter was ended. Indeed, what more could man do about that radiation than observe and describe it?   And yet, Planck considered his remarkable attainment the beginning, not the end of his search.   The reason, as he himself put it later, was that this attainment was just "a luckily guessed mathematical generalization" (*eine glücklich erratene Interpolationsformel*).   So Planck

---

[6] For details and sources see E. Cantore, *Atomic Order: An Introduction to the Philosophy of Microphysics* (Cambridge, Mass.: The MIT Press, 1969), pp. 67-71.   Planck's statements are quoted from M. Jammer, *The Conceptual Development of Quantum Mechanics* (New York: McGraw-Hill, 1966), pp. 19-22.

searched further. To what end? So that, as Planck said himself, his mathematical statement could be transformed into "a statement of real physical significance."

Planck, obviously, was spurred on by the unshakable conviction that nature was intrinsically intelligible. That is, it was not enough for him to know how energy behaved, but it was also necessary to know why it behaved as it did. Continuing his investigation, Planck came to the conclusion that energy had to have a discrete structure, rather than a continuous one, as hitherto universally assumed. This was the great discovery of the quantum, which caused such a renewal in the spirit of physics. Planck himself, having been brought up in the tradition of classical mechanics, was far from enthusiastic about his result, which seemed to contradict the entire tradition of his science. Why, then, did he accept it? His, as he himself was to explain later, was "an act of desperation." The reason: ". . . a theoretical explanation *had* to be supplied at all cost, whatever the price." No more persuasive proof could be desired about the decisive significance of ontological conviction for the development of science.

To conclude, the connection between science and ontology is obvious in a first sense. Science is ontological because it originates and develops as a consequence of an ontological attitude on the part of the researcher. Without ontology, genuine science would not exist. Or, if science were to exist, it would be no more than a shallow and ultimately meaningless empiricism.

(b) The Ontological Insights of Science

There is a second sense in which science manifests a close connection with ontology. It consists in a series of insights about the intelligibility of nature that the success of science first brought to the consciousness of man. The most important of these insights are described in the following paragraphs.

(ba) Intrinsicness

The basic ontological insight of science regards the genuinely intrinsic character of the intelligibility of nature. Prescientific philosophers, as has been mentioned, had already conceived the notion of intrinsic intelligibility of nature. Nevertheless, a brief comparison between the prescientific and the scientific mentalities suffices to bring out the strikingly new way of conceiving the intelligibility that characterizes science.

Aristotle and the Schoolmen thought that what made nature intelligible was something intrinsic to it. They went so far as to define nature itself in terms of its internal ability to originate motion and rest. But the genuine intrinsicness of nature's intelligibility always eluded their grasp. This is clear, for example, if we consider the Aristotelian conception of local motion. Motion was, for Aristotle and his medieval followers, something totally extrinsic. Bodies moved because they were either pushed or pulled from outside, but not because they possessed any internal source of motion of their own. As a result, the prescientific mind came to explain the whole of nature in a purely extrinsic—and ultimately nonintelligible—way. The most massive manifestation of such a mentality is the Ptolemaic system—that extravagant array of spheres, circles and epicycles. Why the recourse to such a construction, obviously anthropomorphic, if not because the planets and stars were thought unable to move by themselves, and needed some sort of external agency to account for their rotations? But, in the end, what kept the celestial spheres themselves moving? The ultimate explanation was so extrinsic as to be animistic. Aristotle spoke of intelligences moving the spheres; the Schoolmen interpreted them as angels. In short it is clear that prescientific man, despite his best efforts, never understood in any profound way the intrinsic intelligibility of nature.

From the very beginning, the scientific mind fixed its attention on the intelligibility of nature as something truly intrinsic. Its first step consisted in ridding the conception of nature of all anthropomorphisms. Nature was considered intelligible because it was assumed to be thoroughly regular, hence completely reliable in its behavior. This is, for instance, Galileo's conviction:

Nature, deaf to our entreaties, will not alter or change the course of her effects; and those things that we are here trying to investigate have not just occurred once and then vanished, but have always proceeded and will always proceed in the same style.[7]

The second step of science consisted of detecting the internal source of nature's regularity, namely of bringing to the fore the agents that make nature intelligible from within. The attainment of this goal as concerns local motion was the great feat of Newton through his discovery of universal gravitation (Chapter 2, Section 1b). The enthusiasm with which the civilized world greeted Newton's accomplishment underscores the ontological significance of his work. For what moved people so much was the hitherto unsuspected realization that nature was indeed open to the mind of man. Gone forever were the intellectually humiliating anthropomorphic constructions of prescientific times. Man no longer had to content himself with the "saving of appearances" when studying nature (Chapter 5, Section 2b). On the contrary, man could now truly understand nature from

---

[7] In S. Drake, ed. and trans., *Discoveries and Opinions of Galileo* (New York: Doubleday Anchor Books, 1957), p. 136.

within; he could penetrate to the very core of nature itself. But, of course, Newton's achievement was just the beginning of man's understanding of nature by means of science. All other scientific syntheses—for instance, the Darwinian and the quantum-mechanical ones—are events of the same type. They enable the mind of man to realize that intelligibility is really an intrinsic property of nature itself.

(bb) Universality

The second major ontological insight of science regards the universal character of nature's intelligibility. In fact, the development of science has led man to an increasing awareness that nature is intelligible as a whole. Orderliness and harmony have been revealed as typical properties of nature as such (Chapter 2, Sections 1c, 2c, and 3c).

To clarify the significance of this important insight, we can speak of the universal intelligibility disclosed by science as presenting a twofold aspect, extensive and intensive. Extensively, science has revealed that regularity and orderliness are present in exactly all areas of reality accessible to the investigation of the scientist. Intensively, science has revealed that nature's intelligibility is a manifold one, exhibiting ever new and surprising manifestations. Even if we remain within the realm of physics, we can easily see the variety of nature's intelligibility when we contrast the results of classical mechanics with those provided, respectively, by relativity and quantum theory. Classical mechanics proves that nature is intelligible, but purely in the sense that man can grasp the successive localization of bodies in space and time. Relativity adds a finer understanding of space and time. Quantum theory discloses the reasons why material bodies, though comprised of many components, behave as wholes—the components associate according to precise rules, give rise to well-defined properties of the aggregates, regenerate the aggregates themselves if these have been destroyed in the interaction with some other bodies.

Humanistically, the most significant manifestation of the universal intelligibility disclosed by science regards the study of man himself. Man for a long time assumed that he was inaccessible to objectively systematic investigation. Science refuted this assumption. Man, just as everything else that can be observed, is explorable by means of the scientific approach. To be sure, genuine science never claims to be able to know everything that can be known about man. For the proper realm of science is limited to what can be observed on an interpersonal level, while man has also a personal or nonobservable side in his being. This granted, it must be admitted that science has been able to include man himself within the universal intelligibility of nature. He is made understandable in all his observable manifestations, including the cultural products of his personal life as motivated by values and meanings. As two leading contemporary anthropologists remark:

Values and significances are of course intangibles, viewed subjectively; but they find objective expression in observable forms of culture and their relations—or, if one prefers to put it so, in patterned behavior and products of behavior. [8]

To sum up, it is no exaggeration to speak nowadays of universality of intelligibility. Thanks to science, we know now that intelligible orderliness is coterminous with nature or observable reality itself.

(bc) Interconnectedness

A third comprehensive feature of nature's intelligibility brought to light by science deserves a special mention. This feature stresses that intelligibility of nature is not just a universal property but also a unifying one. Everything is interactively joined to everything else so as to form an immense holistic structure. This structure is what we call nature. The central contribution of science to ontology consists therefore in the disclosure that we live in a world that is not fragmented but rather constitutes a whole in the true sense of the term. The basic aspiration of man has always been to understand the whole of observable reality as a unified entity. Prescientific man spoke in this regard of an all-encompassing cosmos, of one single universe. Yet he was never able to see an intrinsic unifying connection among the many areas of nature. His world was divided in many ways. There were the mineral, vegetable, animal, and human realms. There were the sublunar and the heavenly kinds of matter. Science has unified it all. Its comprehensive discovery, which proves more and more enlightening, is—to put it in Heisenberg's terms—the existence of "the great interconnectedness" (*der grosse Zusammenhang*). [9] As a result of science, therefore, scientific man lives now in a thoroughly intelligible world.

To sum up, we begin to see another reason for speaking of science as humanistic. Science makes man aware, in an original way, of his intimate relationship to nature. Man and nature are intimately connected because the fundamental desire of man is to know, and science has shown that nature is knowable throughout.

---

[8] A. L. Kroeber and C. Kluckhohn, *Culture: A Critical Review of Concepts and Definitions* (New York: Vintage Books), p. 342.

[9] This is the central theme of the philosophical dialogues conducted by Heisenberg with the leading creators of contemporary atomic physics: Einstein, Bohr, Planck, and the like. The dialogues themselves are recorded in Heisenberg (note 4).

2. THE PROBLEM OF CHANGE

The conclusion we have reached is that science provides vital insights to clarify the ontological meaning of nature and man. In other words, science makes nature and man understandable to man himself. To pursue our investigation we must now ask the pivotal question: does science suffice alone to make nature and man understandable? Opinions differ sharply on this issue. To explore the ontological perspective of scientific humanism we must now consider the issue systematically. Three major problems are usually acknowledged in this area. They regard the understanding of change, of man, and of the absolute or ultimate source of nature's intelligibility. To avoid prejudging the answers, we shall examine each problem separately, following the general guidelines adopted in this study. We start by analyzing the understanding provided by science. Then we shall consider the results that follow if one admits that science suffices alone in providing understanding. Finally we shall outline a plan of research for the ontological philosopher interested in these problems.

The first topic that offers itself to our consideration is that of change, or nature's perennial dynamism. What, if any, are the respective contributions of science and philosophy to its understanding?

(a) The Scientific Understanding of Change

Changeability is perhaps the most obvious aspect of nature. Nothing remains fixed and immutable; everything varies and is altered. History of philosophy shows, however, that changeability is one of the most difficult issues ever tackled by the human mind. Prescientific man was already fascinated by it, but he never arrived at a real understanding. By turns, he either denied the reality of change or reduced all reality to change. As is well known, Parmenides was the standard-bearer of the first interpretation, and Heraclitos of the second. Other prescientific philosophers suggested a variety of additional interpretations. Some spoke of change as a perennial cyclical return. Some reduced change to the rearrangement of minute particles, actually unchangeable. A famous proponent of this view was Democritos, with his atoms. Still other thinkers—notably Aristotle—tried to explain change by taking simultaneously into account the main views of their predecessors. Although much insight was gained, it is fair to say that prescientific man never really succeeded in understanding change. He was baffled by his inability to perceive an intrinsic intelligibility of nature underlying the continual changeability of nature itself.

Where prescientific man failed, scientific man succeeded, and to a remarkable degree. He discovered that, although everything varies in nature, something remains immutable. This is regularity of change itself. The affirmation of such a regularity constitutes that statement called scientific law. Hence scientific man can take pride in having achieved a genuine understanding of change. In a sense, this is what science is all about. Born, for instance, expressed this conviction in a remarkable passage:

He [the scientist] finds nothing at rest, nothing enduring, in the universe. Not everything is knowable, still less is predictable. But the mind of man is capable of grasping and understanding at least a part of creation; amid the flight of phenomena stands the immutable pole of law. [10]

(b) The Ambiguity of the Scientific Understanding of Change

The ability of science to make change understandable needs no stressing. We have seen at some length that science truly succeeds in explaining both the external and the internal kinds of change. External change is the same as motion. Science explains it by tracing its features back to some basic property of nature (see Chapter 2, Section 1). Internal change is growth and evolution. Science explains it by detecting the interactions that make life grow and evolve (see Chapter 2, Section2). But, since science is so successful in making change understandable, should one say that it suffices alone to this end? A widespread view holds that it does. To test the significance of the understanding provided by science, let us suppose that this is well founded and examine two of its most prominent formulations, the mechanistic and the contingentistic opinions.

(ba) The Mechanistic Opinion

The most systematic doctrine that claims the self-sufficiency of science in making change understandable goes ordinarily under the name of *mechanism*. Generally, people assume that what it stands for is strictly a matter of science. Historical investigation shows that the situation is far more complex. The origin of mechanism is to be traced back to the philosophical perspective embodied in the Cartesian system.

Descartes' thought, as has been hinted (Chapter 5, Section 2a), was essentialistic. Being impressed with the certainty and rigor of geometry as embodied in Euclid's treatise, he devised a "physics" or systematic view of the world patterned on the Euclidian model. The starting point was two ideas or intuitions of essences. One was the idea of spirit, which he conceived as pure thinking (*res cogitans*); the other was the idea of matter, which he conceived as pure extension (*res extensa*). From these two ideas, Descartes claimed to be able to explain, by means of

---

[10] M. Born, *The Restless Universe*, trans. W. M. Deans (New York: Dover, 1951), p. 278.

rigorous logical deduction, all the observable properties of the world. In particular, he interpreted change in a very straightforward manner. Since matter was just space extension, change could be nothing but a rearrangement of matter in space. To this end, Descartes conceived space as a *plenum*; matter and space were coextensive. Change resulted simply from the interactions among the various particles that made up matter. These interactions were conceived in terms of mutual impact. The particles of matter, however, were thought to be intrinsically unchangeable. In brief, therefore, in Descartes' view change was but another name for the outcome of mechanical impact.

The connection between mechanism and science began early in the public mind. Descartes himself hardly contributed anything of importance to science proper. But his mental scheme about the universe as understandable in an axiomatic-deductive way was widely influential among scientists, at least as a heuristic guide. To his influence we owe, in particular, the mechanical laws describing the impact of rigid bodies, and also the laws stating the conservation of both kinetic energy and momentum. But mechanism, as we know it now, emerged out of a paradoxical symbiosis between the Cartesian and the Newtonian views about the structure of matter. Newton himself had been a very bitter foe of Cartesianism (see Chapter 5, Section 2a). Yet his *Principia* was formulated in the axiomatic-deductive pattern that Descartes had postulated for the perfect knowledge of nature. Hence, the mechanicians—especially the French—who developed mathematically the work of Newton did not find it objectionable to fuse the Cartesian and Newtonian views into a single system. They postulated that the essence of matter was indeed extension, as hypothesized by Descartes, but not extension alone. To it they attached the property of gravitation, as discovered by Newton. Furthermore, they changed entirely the conception of cosmic space. Since it seemed that gravitational attraction was acting at a distance, they conceived space as a *vacuum,* instead of a *plenum.* With this basic restructuring, mechanism was fully developed. It came to be widely accepted by scientists and laymen alike as a necessary consequence of scientific discoveries. In particular, it was hailed as the doctrine that made natural change completely understandable.[11]

The hold exerted by the mechanistic view on the human mind is best exemplified in the so-called *Laplacian ideal.* Laplace was a gifted mathematician who brought the Newtonian synthesis to the highest degree of internal rigor and systematicness. As a consequence of his work,

---

[11] For a brief summary of the mechanistic mentality prevailing in the 19th century see J. B. Stallo, "First Principles of the Mechanical Theory of the Universe" in his *The Concepts and Theories of Modern Physics* (1881). Reprinted by Harvard University Press in 1960.

Laplace adopted a strictly deterministic interpretation of natural events. His position is expressed most clearly in the opening statement of his famous *Philosophical Essay on Probabilities*:

All events, even those which on account of their insignificance do not seem to follow the great laws of nature, are a result of it just as necessarily as the revolutions of the sun. [12]

In the light of his determinism, we can easily see why Laplace came to give such a prominence to mechanism. The key was his idea of science, thoroughly identified with classical mechanics. Laplace thought that every event in nature was the necessary and totally determined outcome of preceding natural events. He also thought that classical mechanics was able, at least in principle, to describe with perfect accuracy the state of nature at any given instant of time and to calculate with complete precision what other states of nature were going to be. The Laplacian ideal was but the idealized expression of such overall convictions.

Laplace admitted that, in practice, it was not possible to know everything by means of mechanics. But he blamed this limitation on the accidental situation that man needed extensive time to perform all the measurements and calculations required. Hence he conceived a superman capable of operating at lightning speed. Such a being, he contended, would know absolutely everything that takes place in reality—present, past, and future. As he put it, in the context cited:

We ought then to regard the present state of the universe as the effect of its anterior state and as the cause of the one which is to follow. Given for one instant an intelligence which could comprehend all the forces by which nature is animated and the respective situation of the beings who compose it—an intelligence sufficiently vast to submit these data to analysis—it would embrace in the same formula the movements of the greatest bodies of the universe and those of the lightest atom; for it, nothing would be uncertain and the future, as the past, would be present to its eyes. The human mind offers, in the perfection which it has been able to give to astronomy, a feeble idea of this intelligence.

There is no denying the tremendous attraction exerted by the mechanistic conception, especially as focalized in the Laplacian ideal. However, in the first place, is this worldview compatible with science as it effectively exists? In the second place, does it suffice to make nature —especially its most prominent feature, change—truly understandable? The answer to the first question is definitely negative. We have seen, in fact, that one of the major epistemological insights of science is the dis-

---

[12] P. M. Laplace, *A Philosophical Essay on Probabilities*, trans. F. W. Truscott and F. L. Emory (New York: Dover, 1951). Texts cited are on pp. 3f.

covery of the dynamism of human knowledge (Chapter 5, Section 2c). But if knowledge is dynamical—that is, if it advances by detecting totally new and unexpected features of nature—it will never be possible to encompass the entire intelligibility of reality in a single axiomatic-deductive system, as postulated by Laplace.

But, if mechanism is not science, it is philosophy. Is it at least a good philosophy—namely a reflective doctrine capable of making man grasp the intelligibility of nature, especially of change? For an answer to this question, we need but summarize the considerations of Emile Meyerson, a chemist and renowed mechanistic philosopher. His views are propounded in his classic book, whose title is a program—*Identity and Reality*.

Meyerson is a mechanist because, as he explains himself, he wants to make reality understandable. But a being cannot be understood, he contends, unless it is somehow permanent in time; that is, it remains identical to itself. Hence he adopts mechanism as the doctrine that makes nature and change understandable. In his own words:

The external world, nature, appears to us as infinitely changing, becoming incessantly modified in time ... we must needs understand, and yet we cannot do so except by supposing identity in time.... How can I conceive as identical that which I perceive to be different? ... I can suppose that the elements of things have remained the same, but that their arrangement has altered; consequently with the same elements I can create different manifolds just as with the aid of the same letters one can compose a tragedy or a comedy.... The possibility of this conciliation depends evidently upon the particular nature of our concept of displacement. Displacement is and is not a change.... Displacement, then, appears to me as the only intelligible change. [13]

Why does Meyerson speak of space-time rearrangement or displacement as "the only intelligible change"? In other words, why does he say that displacement "is and is not a change"? The ultimate reason is his conception of science, which echoes that of Laplace. Science to him is essentially a matter of equations, where the term is taken in all its etymological strictness, namely as "the expression of complete equality, of identity between the antecedent and the consequent...." [14] The comparison he gives is that of a chemical reaction. Before and after the reaction the same elements are present, but arranged in different ways. On this basis, Meyerson defines science as knowledge that demands complete identity. In his own words:

Poinsot says: "In perfect knowledge we know but one law—that of constancy and uniformity. To this simple idea we try to reduce all others,

---

[13] E. Meyerson, *Identity and Reality*, trans. K. Loewenberg (New York: Dover, 1962), pp. 92f.

[14] *Ibid.,* p. 226.

and it is only in this reduction that we believe science to consist." No truer or more penetrating word; the purely empirical law seems external at once to things and to our mind, impenetrable, opaque. Only the laws which affirm identity, which flow from it or lead to it, appear to us as adequate either to the essence of things and to our understanding; these, alone, we know "in perfect knowledge." [15]

Once Meyerson has made clear that he accepts mechanism as the only adequate interpretation of nature, he goes on to examine the consequences of his own position. These consequences are staggering—all the more so because they are embraced by a person who claims to do justice to science. Meyerson stresses that a scientific equation implies identification between antecedent and consequent. Thus he argues that, according to the scientific view of nature, "everything has been preserved, everything has remained as it was—that is, time has exercised no influence." [16] In other words, Meyerson maintains, the only result of science is to show that everything continues to be as it was before. That is, the contribution of science consists in making change disappear. Meyerson insists on this —that science, by explaining change, actually explains it away. As he puts it, in the context cited:

On the whole, as far as our explanation reaches, *nothing has happened.* And since phenomenon is only change, it is clear that according as we have explained it we have made it disappear. Every explained part of a phenomenon is a part denied; to explain is to explain away.

If science denies the reality of change, what about the meaning of time? In Meyerson's view, time becomes an empty word, too, with no objective validity. For the abolition of change implies the abolition of time and the elimination of progress:

And time itself, whose course no longer implies change, is indiscernible, unimaginable, non-existent. It is the confusion of past, present, and future—a universe eternally immutable. The progress of the world is stopped. [17]

In his relentless consistency, finally, Meyerson notes that the conclusion just reached is no paradox, but simply the unavoidable result of mechanistically interpreted science. The world must be reduced to immobility, exactly as Parmenides had contended:

In appearance it is a paradoxical result. At a glance, however, we can take in the ground covered, and verify the fact that we have not lost ourselves on the way, that the terminus has really been demanded by

---

[15] *Ibid.*
[16] *Ibid.,* p. 227.
[17] *Ibid.,* p. 230.

the starting point. We have searched for the causes of phenomena and we have searched for them with the help of a principle which, we know, is only the principle of identity applied to the existence of objects in time. The ultimate source of all causes can thus only be identical to itself. It is the universe immutable in space and time, the sphere of Parmenides, imperishable and without change. [18]

To sum up, we have obtained some valuable insights into the character of mechanism and natural change. Many people assume that mechanism is either science or a necessary outgrowth of it. In particular, they are convinced that science, through mechanism, suffices alone in making change understandable. Our analysis has shown that the situation is far different. Mechanism is not science but a philosophical opinion that played a considerable historical role in the development of classical mechanics. [19] In the end, however, mechanism proves to be incompatible with science itself. This is the case especially with regard to the understanding of change. For mechanism, if taken consistently, is far from able to explain change. It rather denies its existence since—literally—it explains it away. As a consequence, it is false to say that science suffices alone to explain natural change by means of the mechanistic approach.

From our discussion, we begin to realize the reason for speaking of a philosophical problem of change. Indeed, if one assumes that science is the only form of knowledge able to make change understandable, it may be quite difficult not to end up with mechanism and the denial of change itself. Prominent thinkers, as has been seen, have repeatedly proved through their example that this is most likely to happen. Of course, however, another alternative is also possible. One can hypothesize that science is exclusively competent in understanding change and, by so doing, adhere to contingentism. We must now explore this alternative interpretation.

(bb) The Contingentistic Opinion

The starting point of contingentism is attractively simple. Since philosophical discussions are acknowledgedly open to dispute, why not be satisfied with considering everything that comes to pass (*contingit*) as something that is self-evident and does not require further investigation? In particular, since science provides a genuine understanding of change, why not be satisfied with the understanding provided by science itself? To continue our study, we must now briefly explore this doctrine.

---

[18] *Ibid.*, pp. 230f.

[19] The idea of the universe as a machine was a powerful heuristic guide for the development of science, notably astronomy. For a historical overview see E. J. Dijksterhuis, *The Mechanization of the World Picture,* trans. C. Dishoorn (Oxford: Clarendon Press, 1961).

Despite a widespread conviction to the contrary, it goes without saying that contingentism is not science but merely a philosophical opinion. This is clear from the very definition given. For the contingentistic view does not refer necessarily to science as such, but applies just as well to pre-scientific knowledge. In point of fact we know that, historically, it was mostly mechanism and not contingentism that was assumed as typical of the scientific mentality. The question we must examine is the following: how does contingentism fare in connection with science, particularly as regards the understanding of change? For the sake of clarity, let us examine this question under a twofold hypothesis, one antecedent and one subsequent to the existence of science.

If science did not yet exist, what could contingentism contribute to the existence of science? If a person were consistent with his contingentist view, he could only oppose the rise of science. In fact, as has been seen, science is eminently a product of man's own creativity (Chapter 1, Section 1). That is to say, to originate science man must strive to go beyond the appearances; he must conceive theoretical interpretations of what he observes, then devise experiments to test the validity of his own interpretations. But, of course, all of this runs diametrically against the contingentistic view as defined. Consequently, a consistent contingentist should oppose science as a needless and hopeless waste of time. In particular, contingentism would make it impossible for science to provide any theoretical understanding of natural change, thus depriving it of one of its greatest attainments (see Chapter 6, Section 2a).

Contingentism, as currently adhered to, however, takes science for granted. What does, then, contingentism amount to? Nothing but the thesis that science is uniquely competent as a form of knowledge—in particular, that science alone suffices in making change understandable. Is such a thesis compatible with contingentism itself? On the surface, no doubt. But, then, what is science reduced to? The contingentist must be consistent. If his conviction is that change is self-evident and should not be explored any further, this conviction must apply to the changes of science, too. Science, accordingly, must be considered valid only because it is instinctive to do so or because of the consensual criterion. In any event, there is no possibility left to make the changes of science understandable as authentic progress instead of endless revolution. And this, in turn, entails the destruction of science, as has been seen in our discussions of instinctivism, consensualism, and scientism (see Chapter 5, Sections 3cb, 4ba, and 5bb). In the end, therefore, contingentism far from vindicates the unique importance of science, as it claims to do. It rather does its best to empty science itself of any validity.

From our humanistic angle, the most convincing reason for rejecting contingentism is its disastrous implications about the significance of natural change. The glory of science is the discovery of an intrinsic intelligibility in natural change. Contingentism effectively denies such a discovery. Indeed, if change is the most prominent feature of nature, and change must be taken at its face value as purely endless mutation, it follows that nothing can be actually understood by man, that nothing makes sense any more. To be sure, one can still continue to speak of natural order, but what is the implication of the term now? The anthropologist Loren Eiseley, meditating on the continual changeability of nature, suggests the possibility of interpreting natural orderliness as an illusion. In his own words:

From the oscillating universe, beating like a gigantic heart, to the puzzling existence of antimatter, order, in a human sense, is at least partially an illusion. Ours, in reality, is the order of a time, and of an insignificant fraction of the cosmos, seen by the limited senses of a finite creature. [20]

But, if natural order is an illusion, what are the consequences for the very meaning of man? Man is the product of the continual changeability of nature. He, in other words, is inherently interconnected with the flow of natural events. Accordingly contigentism, in the consistency of its logic, must call into question the meaning of man himself. That is to say, man too runs the risk of becoming an illusory reality to himself. To cite Eiseley again: "Viewed in the light of limitless time, we were optical illusions whose very identity was difficult to fix." [21]

To close, our analysis has shown that one cannot deny with impunity a philosophical dimension to the understanding of change. Science, to be sure, provides understanding of change. But this understanding, though genuine, is by itself one-sided and inadequate, ultimately ambiguous. The ambiguity is obvious in that the understanding provided by science can be easily exaggerated to the point of denying either the reality of change or the meaningfulness of change itself. As a result, science would be harmed, and man would feel lost. We must now hasten to examine how philosophy should contribute to make change understandable and meaningful.

(c) The Ontological Problem of Change

What is the reason for asserting that philosophy should give a decisive contribution to the understanding of change? Following the general guidelines of this book, the answer is obvious. Human knowledge is both object-oriented and subject-oriented, public and personal. Hence, when dealing with change, it is not enough to try to understand it on the object-oriented or public level proper to science. To acquire satisfactory under-

---

[20] L. Eiseley, *The Unexpected Universe* (New York: Harcourt, 1969), p. 46.
[21] *Ibid.*, p. 36.

standing, the subject-oriented or personal reflection of philosophical type is also needed. How should philosophy proceed to make change reflectively understandable? Once again, following our general methodology, the answer is clear. Philosophy should aim to bring out—explicitly, critically, and systematically—the reasons for the reality of change as well as its structure and implications. This is the ontological approach to the issue of change.

Traditionally, philosophers have spoken in this connection not so much of change as of *becoming*. The difference in terminology can be useful to clarify the difference in perspective. The philosopher wants to explore, properly speaking, not change as such, but change as intelligible. But, what makes something intelligible is its existence, its being. Accordingly, what makes change intelligible from the philosophical angle is its special kind of existence. What does this amount to? Obviously it amounts to becoming or coming to be, since change is nothing else but the process through which something comes to be. For the sake of our investigation we shall limit ourselves here to outlining the main aspects of the problem of change or becoming that philosophy should analyze reflectively.

For a guiding thread, we can start out from a basic distinction. Although change is an all-encompassing manifestation of nature, not all changes are of the same kind. Some, at least on the surface of the phenomenon, can be called change only in an outward sense, whereas some others amount to intrinsic change. The first kind embraces all those changes whose existence can be perceived only in relation to a reference system external to them. The second kind embraces all those changes whose reality can be perceived directly, without need of referring them to anything external. Examples of the first kind are motion, time, and space. Examples of the second kind are origin or birth, growth, destruction or death.

To begin with the extrinsic kind of change, let us fix our attention on *motion, time,* and *space*. Everybody, scientist and nonscientist alike, takes implicitly for granted that these concepts stand for something objective, something that is real and not merely a figment of the mind. Human activity as a whole is based on the reality of such entities. Science, for its part, is mainly an exploration of these entities and of their consequences. And yet the reality of these entities is very difficult to understand. When one attempts to pin down what they stand for, the first impression is that their reality seems to dissipate. The task of ontology is to explore this reality systematically. Its starting point in the scientific age, of course, must be the discoveries of science. The philosopher should try to realize explicitly how much and how far scientific statements on motion, time, and space are based on objectively existing reality. In other words, he should try to make reflectively clear what man actually knows about these entities—how far his conceptions of them are objective and how far they are subjective, and the like. Needless to add, through this research science becomes more significant because the import of its basic concepts is clarified. But man himself benefits from it because he is made able to realize more adequately his proper place within the overall context of nature, whose whole structure is dominated by these three entities— motion, time, and space.

When taking into consideration the intrinsic kind of change, the attention of the philosopher should be concentrated at first on the peculiar reality of *change* itself. Change is a reality different from all others: it does not subsist in itself, its entity consists in continually passing away. And yet, the reality of change affects the reality of all other beings. Indeed, things—insofar as they are submitted to change—do not exist, properly speaking, but rather cease to exist according to one way of being and start to be according to another. As a consequence of change, therefore, the question arises about the reality of the changing things themselves. In what sense can man speak of such things as objectively real and intelligible, as opposed to being merely subjective and illusory appearances? A major task of the ontologist in the scientific age is to bring out the reasons for speaking of an objective intelligibility of nature, despite its all-pervasive changeability. He must make clear what is changeable and what is permanent in nature. In particular, he must try to make understandable the very structure of change itself.

To explore intrinsic change in detail, the philosopher must distinguish between two types of such change, namely *nonessential* and *essential* alterations. Both types affect objectively the individual involved, but in a different manner. For, when a being undergoes an essential change, it ceases altogether to be what it was—whereas this does not take place in the case of nonessential change. Thus, to clarify this issue, the philosopher should explore critically what is meant by the basic manifestations of nature's continual changeability. Everybody speaks, in particular, of *origin* or birth, of *growth*, of *destruction* or death as realities. But, what do these terms actually stand for? It is up to the philosopher, in association with the scientist, to give a reflectively satisfactory answer.

The ontological investigation of change or becoming should reach far and deep. One fundamental topic is that of *structure*. Things can change intrinsically because they have a structure; that is, they are made up of many interconnected components. But, what is the relationship between structure and components? Is the structure a whole—a unified and organic entity that deserves to be called an individual—or is it merely the sum total of its parts? This question is especially vital with regard to the understanding of *life*. There are certainly good grounds to assume that life emerged from nonlife in the course of time. But, if this is what happened, what actually is life? Is it true to say that life is nothing but nonlife, as one often hears? For, if this were the case, what would still be the meaning of evolution? It seems hardly justified to interpret evolution as simply an increase in quantitative complexity. But, if evolution has brought about qualitative changes in the world, it is necessary to make clear what these changes amount to.

A major ontological issue is raised by the concept of *matter*. Everybody takes rightly for granted that matter exists objectively. However, what do we mean by this term, especially if we take into account the results of contemporary physics? For the mechanist, there was no problem. Matter was exactly what could be imagined—passive, inert, pure extension moving around in space and time through push and pull. But relativity and quantum theory have shown that matter is an extremely complex reality. It is subject to total alteration—from ponderable matter into nonponderable energy. It is itself the autonomous seat of activity and orderliness. For matter, even the nonliving one, is able to bring about well-defined structures that are characterized by specificity and repeatability as well as individuality. Atoms, molecules and the like are such kind of structures. Therefore, it is clear that, to have a satisfactory conception of matter in the scientific age, one must strive to attain it by means of systematic and critical reflection of the ontological type.

A comprehensive area of ontological research on change regards the reality and the meaning of *interaction* and *development* as well as *creativity* in the world. In prescientific times these topics were discussed under the headings of causality and finality or teleology. Science has disproved many of the traditional assumptions in this regard. However, no one can say that the issues involved are unimportant, much less self-evident. Indeed, possibly even more than his prescientific counterpart, scientific man needs to know what it means to act, to create, to produce something new. In particular, he needs to know the overall meaning of the unceasing dynamism that permeates nature and man himself. For otherwise man would end up by living in a world that makes exactly no sense to him, a truly inhuman world.

To sum up, it is clear what vital contribution the ontological study of change should bring to authentic humanism for man living in the scientific age. This study should make man aware of the genuine import of the discoveries of science concerning the intelligibility of nature. In particular, it should enable man to find his proper place in the overall flow of nature's dynamism.

3. The Problem of Man

The second major humanistic area in which there is frequently tension between science and ontology is man's understanding of himself. In brief, the subject at issue can be called *humanness*. What constitutes man as such? In particular, what form of cognition—scientific or philosophic—is competent to decide the issue? Traditional humanists frequently deny any competence to science as regards the understanding of man, whereas scientific thinkers often assume that science suffices completely to explain man. Following our general methodological principles, our discussion will develop as follows. First, we shall examine the reasons that justify speaking of science as a real understanding of man. Second, we shall scrutinize the claim that science suffices by itself, in an exclusive way, to understand man in his fullness. Third, we shall outline the structure of the ontological problem of man as such.

(a) *The Scientific Understanding of Man*

As has been hinted (Chapter 6, Section 1 bb), the humanistic investigator should gladly acknowledge the universal intelligibility that science has been able to detect in observable reality. In particular, he should welcome the ability of science to understand man as an observable being.

Unquestionably, the attainments of science in the study of man are complex and present different degrees of certainty. The reasons for this state of affairs are multiple, beginning with the fact that the so-called human or sociopsychological sciences are comparatively recent. Nonetheless, it is clear that these sciences, when taken together, are able to provide an understanding of man, at least in principle.

Scientific understanding, as is well known, refers exclusively to reality as it is observable, and takes place following two main steps. The first step is the detection of observable regularities; the second is the discovery that the regularities observed are themselves the consequences of some basic observable property of the object under investigation. In the light of this definition, there should be no objection to accepting science as capable, at least in principle, of understanding man himself. For man, in the first place, is a living being with a specific physical structure. But science is able, through the theory of evolution, to trace back man's own structure to the structures of other animal species and account for its peculiarities. In the second place, man is a living being characterized by a number of natural tendencies or spontaneous urges. But science is increasingly able to find orderliness in the manifestations of such tendencies, and it is also able, more and more, to trace them back to their observable sources. Ethology, for instance, studies how man's spontaneous behavior originated in the mutual interactions of the animal forms of life that preceded his appearance. Psychology classifies the various spontaneous manifestations of man and traces them back to their biohereditary and environmental sources. One remarkable proof of science's success in this area is the psychogenetical explanation of concepts, especially by Jean Piaget and his school. [22] But if the very development of knowledge, which implies the creativity of the person, can be understood scientifically, who could deny science's ability to understand all of man's other observable manifestations? In the third place, man is a social being that presents many characteristic forms of familial, political, and social behavior. Social science, in its various branches, is more and more able to find order in such manifestations and explain them by tracing them to their historical and psychological roots. Finally, man is a being with a distinctive cultural tendency. Here again, science succeeds increasingly in providing genuine understanding. For anthropology is able to discover basic patterns of constancy and change in what at first sight appears to be just a disconnected multitude of cultural expressions.

To sum up, science is far from being fully developed and entirely satisfactory in regard to its investigation of man. Nonetheless, it must be admitted that it is able, at least in principle, to make man understandable insofar as he is observable. Science not only detects many observable regularities in man, but traces them back to some basic observable causes. This is clearly a great achievement from the humanistic point of view. For humanism is fundamentally a matter of understanding; above all, it implies the understanding of man by himself.

(b) The Ambiguity of Scientific Understanding of Man: The Deterministic Opinion

Having accepted the scientific understanding of man and its humanistic significance, we must hasten to inquire whether such an understanding is unique and exclusive. In other words, we must ask whether science suffices alone in explaining man to man. The affirmative answer is widely taken for granted. It constitutes the so-called scientific view of man, which is entertained by many psychologists and social scientists. To test the validity of such a view, let us examine the positions of some of its most prominent supporters.

The famous psychologist B. F. Skinner is quite frank in disclosing the ontological foundation of the view under consideration. He takes it as self-evident that the scientific approach—when applied to the study of

---

[22] For an introduction to Piaget's thought see his *The Psychology of Intelligence,* trans. M. Piercy and D. E. Berlyne (Totowa, N. J.: Littlefield Adams, 1966).

man—presupposes a complete determinism of man himself. In his own words:

If we are to use the methods of science in the field of human affairs, we must assume that behavior is lawful and determined. We must expect to discover that what a man does is the result of specifiable conditions and that once these conditions have been discovered, we can anticipate and to some extent determine his actions. [23]

As for the natural objection that the principle enunciated is not scientific but rather philosophic, Skinner acknowledges its validity. But he replies by saying that science as such demands a deterministic interpretation of reality. Thus, in fact, he goes on to comment in the text cited:

Science is more than the mere description of events as they occur. It is an attempt to discover order, to show that certain events stand in lawful relation to other events.... But order is not only a possible end product; it is a working assumption which must be adopted at the very start.

In the light of the preceding, it is clear what kind of conception of man results from the so-called scientific view of man himself. It is a thoroughgoing deterministic conception. Humanness—taken in the ordinary sense of autonomous personality—is utterly and inexorably interpreted out of existence. In particular, Skinner takes pains to spell out that science, as understood by him, demands the total rejection of human freedom as something objectively real.

The free inner man who is held responsible for the behavior of the external biological organism is only a prescientific substitute for the kinds of causes which are discovered in the course of a scientific analysis. All these alternative causes lie *outside* the individual.... These are the things which make the individual behave as he does. [24]

Concerning the widespread hesitation against accepting such a sweeping denial of humanness, Skinner reacts by insisting on the austere objectivity of science. In his conviction, it is the task of science to destroy commonly entertained humanistic views. He sees in this destruction a source of consolation rather than of despondency for realistic man.

... it has always been the unfortunate task of science to dispossess cherished beliefs regarding the place of man in the universe.... We may console ourselves with the reflection that science is, after all, a cumulative progress in knowledge which is due to man alone, and that

---

[23] B. F. Skinner, *Science and Human Behavior* (New York: Free Press, 1965), p. 6.
[24] *Ibid.*, pp. 447f.

the highest human dignity may be to accept the facts of human behavior regardless of their momentary implications. [25]

It cannot be denied that, for all its starkness, the deterministic interpretation of man strongly appeals to current public opinion. The reason is the ideal character it ascribes to science. Many feel enthusiastically insistent that the whole of reality, including man himself, be thoroughly understandable. As Bernal put it:

For the first time in human history we can hope to trace precisely the whole field of knowledge from nebulae to politics. . . . We find a system of box within box of units, aggregating at a certain stage to form larger units which can then aggregate in turn. [26]

The deterministic view of man is widely accepted as a necessary consequence of science. But many of its supporters are far from optimistic in assessing it. Leading scientists give expression to very great concern in its regard. As an example, Jean Rostand, the biologist-philosopher, reverts time and again to the issue with obvious pain. He seems to find a bitter pleasure in noting that the result of science is that of exposing, as complete deception, what man ordinarily considers beautiful and inspiring. Thus, for instance, he muses that human initiative, creativeness and responsibility, are really nothing but an illusion. Everything should be ascribed to the action of the genetic endowment and the environmental influxes that are at work in a person.

Whatever an individual is—good or bad—has no causes other than the molecular make-up he received from his parents, and the external influences that have worked on him. Our thanks or blame must fall to chemistry and luck. [27]

In his almost brutal honesty, Rostand stops at nothing. Love itself—the highest of human manifestations—becomes nothing but biochemical necessity. In his own words:

Whether one likes or not, and whatever the idealism one may subscribe to, the whole edifice of human love . . . is constructed upon the minimal molecular differences among a few derivatives of phenanthrene (estrogen and testosterone). [28]

Struggling to synthesize his view of man, Rostand falls back on a dualistic interpretation. Man is indeed a completely determined machine.

---

[25] *Ibid.*, p. 449.

[26] J. D. Bernal, *Science in History* (New York: Hawthorn, 1965), p. xviii.

[27] J. Rostand, *The Substance of Man*, trans. I. Brandeis (New York: Doubleday, 1962), p. 15.

[28] *Ibid.*, p. 36.

Yet he is a machine that is able to observe its own behavior—and labors under the illusion of guiding it.

What are we? What is our position in nature? What is the sense of our existence, the value of our activity? ... it would even seem that the only capacity thought possesses is that of observing the action of the machine it has the illusion of commanding. The so-called act of the will appears reducible to a pattern of coordinate reflexes ... no less under compulsion than the caterpillar creeping toward the light...[29]

What is, in the last analysis, the role of science with regard to man according to the deterministic interpretation under discussion? Logic compels even the most convinced adherents of this view to agree on the answer. What science does, in the study of man, consists eventually in "dissolving" him, as the celebrated anthropologist Lévi-Strauss put it. That is to say, deterministic science must ultimately be reductionistic—it must do away with the human phenomenon as such. In his own words:

I believe the ultimate goal of the human sciences to be not to constitute, but to dissolve man.... [It is] incumbent on the exact natural sciences: the reintegration of culture in nature and finally of life within the whole of its physico-chemical conditions.[30]

The statements of the proponents of the so-called scientific view of man are illuminating for the sake of our investigation. We have started with the intention of finding out whether science alone suffices in making man understandable to himself. The statements of those who adhere to such a hypothesis betray a conception of man that is far from satisfactory, as we know by reflecting on our own personal experience of human beings. Hence, the inference we must draw is clear. Science is far from able to explain man by itself. In fact, if science is supposed to explain man entirely, it merely succeeds in explaining him away. But this would amount to the very destruction of science. For science would then become reductionism, which, in turn, is nothing but dogmatic scientism (see Chapter 5, Section 5bb).

Rostand, consistently with his scientism, concludes in desperation: "Man is a miracle of no interest."[31] We reject such a conclusion because it is both unjustified and harmful. Our own conclusion is different. We hold fast to the proven thesis that science supplies a genuine understanding of man (see Chapter 6, Section 3a). But, on the other hand, we avoid giving in to the instinctive dogmatism that takes for granted that science

---

[29] *Ibid.*, pp. 57f.

[30] C. Lévi-Strauss, *The Savage Mind* (Chicago: University of Chicago Press, 1966), p. 247.

[31] Rostand (note 27), p. 63.

is the only reliable form of knowledge—in particular, that science alone suffices in making man understandable to himself. After all, such a dogmatism is not consistent with the spirit of science itself. Thus, as a result, we assume the right and duty of philosophy to contribute its share to the understanding of man.

(c) The Ontological Problem of Man

Ontology, by definition, is the study through which man tries to become consciously aware of the intelligibility of reality that has been brought to his notice by experience. In the light of the preceding, we see at once what is the core of the ontological problem of man living in the scientific age. It consists in finding out—by taking into account both science and philosophy—what man *as* man actually is. In other words, to put it more directly: I, as a man who esteems science, feel called upon to investigate reflectively what constitutes the specificity of my nature in the light of all the information that science gives to me about myself.

The root of the problematicity of man is not to be sought in science as such. Man is problematic by nature because of the *duality as well as unity* of his own being. Man is dual—we are both observable and non-observable, both public or impersonal and private or personal. Man is one—we have an individual identity that manifests itself through all our public as well as private expressions. Science has sharpened our problematicity in that it has given special emphasis to our observability. For science has shown not only that we are observable but also that we are largely understandable in our personal life by means of the observational approach. Hence it is most natural to feel bewildered. One has the impression that there is no dividing line between man as a person and man as an observable entity. This accounts for the tendency to ascribe such an exclusive importance to science as a means of understanding man. However, the very existence of science is, if anything, convincing evidence of the inner personality of man. For science, as has been seen at length (Chapter 1, Sections 1 and 3b; Chapter 4, Section 1) is exactly one of the best manifestations of man's own typical creativity, dignity, love, and dedication to ideal. This, then, justifies the insistence on philosophical reflection as an indispensable means for understanding man in general, and especially man living in the scientific age.

How is man to proceed in order to understand himself while trying to take into account both science and philosophy? The general layout of the investigation as well as its main steps are suggested by the principles just enunciated. Man is both observable and nonobservable; man is one. The issues to be explored, therefore, are marked out: What constitutes the typical observability of man? What characterizes his nonobservable personality? What, above all, is implied by the assertion that man is a unitary being?

Beginning with the observability of man, the basic attitude to be adopted is, of course, to welcome all the illumination that science can give about man himself, especially through the theory of evolution. In this regard, one point is firm. Evolutionary biology provides a genuine understanding of man as an animal. To synthesize with Simpson, the leading contemporary zoologist:

The general position of man within the animal kingdom, within the vertebrate subphylum, and within the mammalian class is absolutely established and beyond any doubt. [32]

However, once one has accepted that science is able to explain man as an animal, one should hasten to explore reflectively the *specific animality* of man. This is the first goal to be achieved if one wants to understand man adequately. For, as Simpson goes on to remark, rightly: "It is a fact that man is an animal, but it is not a fact that he is nothing but an animal."

The question at issue here underscores at once the vital role ontological reflection has to play so that man may achieve a satisfactory understanding of himself. Indeed, if one were to rest content with the data of science, it is hardly possible not to fall into the deceptive interpretation of man that, with Simpson, can be called the "nothing-but fallacy." We have just seen how widespread this fallacy is among those who claim that science alone is competent in making man understandable. Man is reduced to nothing but animality, even nothing but chemical-physical reactions.

As Simpson rightly notes in the context cited, it is misleading in general to speak of any living species—for instance, dog or oyster—as nothing but animals. For such a formulation conveys the impression that there is no specific difference between the two kinds of animal—when, of course, the difference is profound. But, if this consideration applies to all species of animal, it applies above all to man since his animality is clearly of a very special kind. Hence the nothing-but interpretation is not only misleading in the case of man but also positively dangerous. The reason is that, if man adopts such an interpretation, he is impeded from arriving at a satisfactory understanding of himself. In Simpson's own words:

Such [nothing-but] statements are not only untrue but also vicious for they deliberately lead astray enquiry as to what man really is and so distort our whole comprehension of ourselves and of our proper values.

---

[32] G. G. Simpson, *The Meaning of Evolution: A Study of the History of Life and of Its Significance for Man* (New Haven: Yale University Press, 1967). See the whole chapter, "Man's Place in Nature," especially pp. 282-284.

What is then the first goal of the philosopher who intends to contribute his share to the understanding of man living in the scientific age? It consists in clarifying what is implied by the obvious fact that man is an animal but of a very special kind—namely, the fact that man's animality is actually *humanness*. In other words, the philosopher should aim at making systematically explicit what constitutes the specific or unique animality of man. This is what scientists themselves aspire to know, as is stressed by Simpson in the conclusion of the passage mentioned:

... man is an entirely new kind of animal in ways altogether fundamental for understanding of his nature. It is important to realize that man is an animal, but it is even more important to realize that the essence of his unique nature lies precisely in those characteristics that are not shared with any other animal. His place in nature and its supreme significance to man are not defined by his animality but by his humanity.

The second goal of the philosopher, concerning the understanding of man living in the scientific age, consists in clarifying more precisely what is meant by the *personality of man* himself. Traditionally, personality —namely the ability to think and act creatively—has always been recognized as the typical feature of man. However, prescientific thinkers had very little factual information in this area. They hardly knew about the structure of the human psyche, how man learns and thinks, how man is dependent on his physicosocial environment and yet also creatively original in its regard. Science has changed the cognitive situation entirely concerning all these subjects, both directly and indirectly. Directly, science has shed light on the personality of man by means of its psychological and sociological findings. Indirectly, science has illustrated the import of the personality of man through the very creativity that is implied by the achievements of science itself. The philosopher should take all of these data into systematic account and formulate a more satisfactory doctrine of the human personality.

The third goal of the genuinely scientific and humanistic philosophy of man arises from the synthesis of the two previous ones. It is concerned with arriving at a better awareness of man as a *holistic entity*. To be sure, man has always thought of himself as one entity, not many. Yet prescientific thinkers were hardly able to formulate a satisfactory doctrine in this regard. Ordinary man, at least in the West, still speaks of himself as though he were made up of two quasi-autonomous entities, one observable and one nonobservable: body and soul, matter and spirit. Accordingly, many thinkers have tended to overemphasize either one or the other side of man. The philosopher who is illumined by science can expect to arrive at a better understanding of man as one single being. This he can do because, through science, he is made aware of the specificity of man in a more comprehensive way than was accessible to his nonscientific counterpart.

To close, there is such an objective issue as the ontological problem of man living in the age of science. The problem arises not because one overrates science nor because one underrates it. Rather, the problem is but a manifestation of the complex richness of man—who is one, yet twofold, observable and nonobservable. The problem has to be solved by philosophy in close cooperation with science because either of these forms of knowing is competent by itself to explore only one of the two sides of man. Hence, if man really intends to understand himself as one being, he needs to study himself both scientifically and philosophically. Such a problem, as outlined, is obviously a difficult one. However, one should not get discouraged by the difficulties. For, if man has proved his intellectual ability by developing both science and philosophy, he has no reason for doubting his ability to make the two converge and throw satisfactory light on himself.

4. THE PROBLEM OF THE ABSOLUTE

In this ontological chapter our overall intention is to realize reflectively the new light thrown by science on the intelligible structure of reality as a whole. To carry out our program, we have first examined the general ontological insights of science. Then we have explored the new perspectives under which scientific man is led to perceive the intelligibility of nature and of man himself. At this point we must still enlarge and deepen the scope of our research. The new issue we must consider is that bearing on the ultimate intelligibility of reality as such. For short, we may call it the problem of the absolute.

Before addressing ourselves directly to the issue, some preliminary remarks are in order. To begin with, are we entitled—within the cultural context of science—to discuss the question announced? The objection is worth considering. For science, as a specific form of knowledge, must necessarily limit its competence to what can be observed. But the absolute, whatever it may be, is ordinarily thought to be unobservable. Hence, is one on justifiable methodological grounds when speaking of the absolute in relation to science? Would it not be advisable to avoid a discussion that is quite difficult and inescapably controversial? Indeed, the problem of the absolute touches unavoidably on the delicate question of the relationships between science and religion. But, as history shows, this is a very contentious matter and, to all appearances, an all but hopeless one.

One should acknowledge the difficulty of exploring the ultimate intelligibility of reality by taking science into systematic account. Nonethe-

less, the difficulty should not deter but rather spur the consistently thoughtful person living in the scientific age. Indeed, what characterizes science
is a persistent search for truth, but only at the public or observational
level. Thus, the consistently thoughtful person should feel urged to search
for truth as far as the human mind can go—that is, by continuing to
search at the personally reflective level, without stopping at the observational one. For, after all, this is exactly what intellectual consistency
amounts to. But this is also what is meant by the problem of the absolute
—namely, the need of finding the ultimate truth or intelligibility of reality
as opposed to any merely relative intelligibility. On the other hand, the
central importance of the problem of the absolute is not merely stressed
by the philosophical inquirer. In point of fact, scientists themselves not
rarely lead the way in discussing the issue. The reason is the experiential
nature of science. Science, as a specialized form of knowing, is competent
to discuss adequately only observable reality. However, science is also an
experience of the whole person. As such, it cannot avoid making man
aware of the absolute (see Chapter 3, Section 3).

To avoid misunderstandings, still a word must be added to define
with precision the topic of our investigation. We are not concerned with
religion as such but only with ontology. Thus, our only intent here will
be to clarify the ultimate meaning of reality as accessible to philosophy
when reflecting on science. With these specifications, it is both possible
and profitable to face the problem of the absolute in the scientific age.
This is the plan we are going to follow. First, we shall study historically
the ontological root of the cultural split that prevails in our contemporary
civilization. Second, we shall discuss the typical ways with which the
scientific mind perceives the absolute. Third we shall explore a certain
latent ambiguity in the scientific fashion of perceiving the absolute. To
close, we shall outline the main traits of the ontological problem of the
absolute.

(a) The Clash of Prescientific and Scientific Ontologies: The Galilean Tragedy

Much has been made, with reason, of the biblical question in the events
that led to the condemnation of Galileo by the Roman Inquisition. However, if one really wants to learn from history, it is not sufficient to stop
at the obvious fact that Galileo and his clerical opponents disagreed in
interpreting the Bible. For, indeed, why did the two parties disagree so
much and what lesson should we draw from the episode for the enlightenment of our own times?

To begin at the beginning, it is essential not to miscast the two
protagonists of that momentous clash: Galileo on the one hand, and Maffeo

Barberini (Urban VIII) with his counselors on the other. Frequently, uninformed or superficial commentators take sides by extolling their hero and casting aspersions on the opposing party. In particular, the idea is widespread that Galileo was great because he rebelled against religion, whereas the pope and his theologians were villainous because they tried to suppress science in the name of religion. History, as opposed to prejudice, shows that the situation was far more complex—and ominous. For the two parties agreed on the essential importance of religion, and both were also open to learning according to their lights. Even more, they had a sincere respect for each other. And yet, they could not avoid clashing against each other with such far-reaching consequences for our entire civilization.

As regards the religious seriousness of Galileo, there can be no doubt about it. He always was and remained a convinced Catholic. All his conduct and writings bear evidence to it. Since we cannot enter into details, it may suffice to quote a telling passage from a letter of his to Francesco Cardinal Barberini, nephew and right hand of the pope who had decided his condemnation. The letter was written from Florence shortly before the trial. In it Galileo tries to defend himself by appealing to the sincerity of his religious motivation in all his work, especially in writing the incriminated *Dialogue*. In his own words:

Everyone will understand that I have been moved to become involved in this task only by zeal for the Holy Church, and to give to its ministers that information which my long studies have brought to me, some of which might perhaps be required by them, being obscure matters and remote from the accustomed doctrines. [33]

The passage quoted, as well as the entire letter in the same vein, is particularly persuasive. In fact, were one to think otherwise, this would amount not only to casting a gratuitous aspersion on Galileo's sincerity, but also to making a fool of him—and a very self-defeating one, indeed. For, how could he hope to persuade those who wanted his condemnation—and accused him of both ignorance and arrogance—by assuring them that he had done all he had done just to illumine them? The reason can only be one. Since Galileo was sure of speaking the truth, he appealed to an argument that, in his own mind, was convincing precisely because it was truthful. But, of course, his adversaries could easily twist it into further evidence of his alleged conceit and arrogance.

Concerning the attitude of Urban VIII and his counselors, a distin-

---

[33] Quoted from L. Geymonat, *Galileo Galilei: A Biography and Inquiry into His Philosophy of Science,* trans. S. Drake (New York: McGraw-Hill, 1965), p. 144. The letter was dated October 13, 1632. The full Italian text can be found in Galileo Galilei, *Opere,* F. Flora, ed. (Milan: Ricciardi, 1953), pp. 1049-1054.

guished contemporary historian—hardly suspect of clericalism—begins his celebrated book on Galileo's trials by denying the widespread assumption that Galileo's adversaries were "bigoted oppressors of science." [34]   Countering such a view, de Santillana takes a firm stand, which he documents abundantly in the course of the book.   In his own words, continuing the text cited:

It would be possibly more accurate to say that they were the first bewildered victims of the scientific age.   They had come into collision with a force of which they had not the faintest notion.

The point at issue in our discussion is now clear.   If Galileo and Urban agreed on the importance of both religion and learning, why did they clash so fatally?   The answer, too little known, is philosophical —specifically, ontological—in nature.   Galileo and Urban clashed because they entertained antithetic views about the intrinsic intelligibility of nature and the ability of man's mind to understand it.   Theirs was literally a clash between ontologies: of the prescientific and the scientific kinds.

History has preserved a precious document that explains Barberini's stand in relation to Galileo's science.   It contains the gist of a meeting that took place between the two personages under very significant circumstances. [35]   Galileo had already been admonished but not condemned by the Roman Inquisition (1616).   Barberini, a powerful cardinal, had been instrumental in averting his condemnation.   Some time afterward, before being elected pope, Barberini himself met privately with Galileo, whom he esteemed highly, and tried to show to him all his understanding as well as misgivings concerning the latter's position.   To start off, he said he was willing to admit the possibility that all of Galileo's views about the structure of the universe were sound and even that they described nature in a perfectly satisfactory way.   But, after this initial admission, he posed the rhetorical question:

Can one opine that God would not have been able to think and act in such a way as to dispose and move the planets and stars in another manner [than the one implied by Galileo's interpretation]?   And yet, this manner would be such that all celestial phenomena, and everything that can be said about motion, order, location, distance, and disposition of stars could be saved [so as to fit Galileo's own description]?

---

[34] G. de Santillana, *The Crime of Galileo* (Chicago: University of Chicago Press, 1955), p. 2.

[35] The document is contained in a book by Agostino Cardinal Oreggi, a theological consultant of Urban VIII, published in Rome in 1629.   The Italian translation of the text is to be found in Galileo, *Opere* (note 33), p. 1052, note 1.   My translation is from the latter book.   A paraphrase of the same text is to be found in de Santillana (note 34), pp. 165f.

To a modern mind, of course, the question seems nonsensical, for its two parts appear to contradict each other. We shall revert shortly to the mentality that it betrays, but for the moment we must continue our exposition. Barberini must have felt quite sure about the position expressed because he went on to throw an unacceptable challenge to Galileo.

If you deny my argumentation, you must prove that there is an intrinsic contradiction to think that things can exist in a way different from what you have hypothesized. God, in fact, in his infinite power, can do whatever does not imply contradiction.

Galileo, naturally, could not reply that God would have been unable to create a world different from the one accounted for by his science. Thus Barberini concluded by rejecting the possibility of science as an intrinsic knowledge of nature. Such a knowledge, in fact, appeared to him an arrogant pretension to limit the divine cognition and power. As he put it:

If God can have been able to think and act in such a way as to dispose things in a manner different from the one you have thought—so that everything would be saved as said above—then we should not limit to this one manner [that implied by Galileo] the divine power and knowledge.

What was the reason that, in Barberini's view, justified the uncompromising rejection of science as an attack on God himself? The reason was a philosophical one, as can easily be seen by recalling the positions of the prescientific mind concerning knowledge. Prescientific thinkers normally shared the essentialist conception of knowledge. Genuine knowledge, they assumed, consisted in grasping the essences of things (see Chapter 5, Section 2a). That is, genuine knowledge was conceived not as a realization of fact but as a realization of principle, with aprioristic characters. Thus, the genuine knower was supposed to be the one who knew not only how things were, but also why they were the way they were, according to an intrinsic or absolute necessity.

It is clear that the essentialist conception of knowledge, as outlined, constituted a practically impossible and theoretically absurd goal for the human mind when applied to the cognition of nature. For the mind of man, when investigating nature, must inescapably depend on a reality that is beyond its control as far as the manifoldness of its intelligibility goes. Hence man can hardly dream of discovering the aprioristic reasons that cause things to be the way they are. As a consequence, prescientific thinkers adopted their typical convictions about occultism and externalism concerning the human knowledge of nature (Chapter 5, Section 2a). In addition, they came to consider any intrinsic knowledge of nature not only as impossible but also as a blasphemous attempt at appropriating a prerogative that belonged to God alone.

In brief, the position of prescientific man relative to the knowledge of nature was one of unrelieved resignation. Nature was thought to be intrinsically unknowable. Hence, the cognition of the observable world was reduced to phenomenism. Man was expected to be interested only in the "saving of the appearances"—that is, in collecting and arranging sense data according to accurate descriptive schemes that remained forever at the surface of phenomena (Chapter 5, Section 2b). The best example of such schemes was the Ptolemaic system. This was clearly a man-made construction, which provided no intrinsic information about the effective structure of the heavens. Hence sophisticated thinkers—Barberini was one of them—never overrated its importance. They were ready to drop it in favor of any other more satisfactory scheme. But the principle in their minds was firm. No interpretation of nature was thought able to give intrinsic knowledge. Thus, in particular, we can understand why Barberini could easily admit, by way of hypothesis, that Galileo might have been entirely right in his description of the world while, at the same time, he rejected the latter's claim to scientific objectivity.

The reason why prescientific man, as visible in the Barberinian position, rejected intrinsic knowledge of nature as an attack on the prerogatives of God himself followed from the aprioristic conception of knowledge described. If man could know nature in an aprioristic way, his cognition would obviously be indistinguishable from that of the very creator of nature. But this hypothesis was held to be inherently blasphemous. The ground was that the knowledge of man is unavoidably a dependent one, while that of the creator is independent by definition. As a conclusion, people like Barberini thought that the acknowledgement of complete ignorance in relation to the intrinsic intelligibility of nature was the only appropriate way for man to acknowledge the transcendence of the creator.

We do not know what Galileo answered, if anything, to the Barberinian declaration. But we know quite well that he felt he had to adopt the diametrically opposite position concerning both the intrinsic cognoscibility of nature and the attitude that man, conscious of his dignity, had to take about it. For he, as a scientist, had experienced that nature was intrinsically accessible to the mind of man (see Chapter 6, Sections 1a and 1ba). Furthermore he, having a superior philosophical mind, could not help seeing a contradiction between the Barberinian position and the immemorial Judeo-Christian conviction concerning the knowledge of nature. This traditional view he formulated briefly elsewhere by saying that "God is no less excellently revealed in nature's actions than in the sacred statements of the Bible." [36]

---

[36] In Drake (note 7), p. 183.

From the whole of Galileo's behavior it is clear that it was especially the theological conviction just cited that prompted him to reject the Barberinian position.   Indeed, if nature was the revelation of God, man not only had the right to consider it as intrinsically intelligible but, in addition, he had the duty to investigate such an intelligibility.   In particular, how could man assume an attitude of complete ignorance with regard to nature and claim to honor the creator thereby?   If God was the creator of man, he was the source of his intelligence, too.   Thus, God could only be honored by man if the latter used his intelligence to know him better through the study of nature.   Especially preposterous appeared to Galileo the view of those theologians who claimed that ignorance about nature was the condition imposed by God himself on man, since—according to their opinion—the Bible sufficed to reveal God.   Indeed, was God to be conceived as a capricious tyrant who, after having given man the incomparable dignity of his intelligence, decided to suppress it by prohibiting its exercise?   As Galileo himself put it, in his biblical controversies:

But I do not feel obliged to believe that that same God who has endowed us with senses, reason, and intellect has intended to forgo their use and by some other means to give us knowledge which we can attain by them. He would not require us to deny sense and reason in physical matters which are set before our eyes and minds by direct experience or necessary demonstrations. [37]

In the light of the preceding we can understand why, when Barberini became pope, Galileo thought that the time had come for him to defend publicly the rights of science against his theological adversaries.   His reason was twofold.   One was the learning and friendliness of the new pope toward himself.   The other reason was the desire to help the Church meet positively the challenge of the new scientific culture.   Accordingly, Galileo wrote his celebrated *Dialogue* (published 1632).   In this work, however, he had to take into account the opinion of the pope himself, but without attacking him directly.   This task—given the irreducibility of their respective positions—was all but impossible.   At any rate, with the wisdom of hindsight, it is clear that Galileo did not acquit himself of the task very ably.   For he simply resorted to irony.   He apparently thought that exposure alone would serve to manifest the absurdity of the position of his adversaries.

Simplicio, the dogmatic Aristotelian, is made the spokesman for the antiscientific mentality.   From time to time he utters grave declarations about the ignorance of man with regard to nature.   In particular, toward the end of the book, he stresses the Barberinian position that Galileo in-

---

[37] *Ibid.,* pp. 183f.

tended to ridicule.   In this connection, a word of clarification is needed. Galileo was convinced—wrongly—that the phenomenon of the tides offered compelling evidence for the motion of the earth.   It is on this subject that he decided to expose his adversary.   Salviati, the spokesman for science, discusses at length the tides and their importance in order to prove the Copernican hypothesis.   To it, Simplicio replies characteristically that he finds Salviati's scientific views "more ingenious than many others."   Nonetheless, he adds in the same breath, he cannot "consider them true and conclusive." [38]   What is the reason for this apparently inconsistent reaction?   Simplicio appeals grandly to "a most solid doctrine that I once heard from a most eminent and learned person, and before which one must fall silent."   The allusion to Barberini and his view is unmistakable.   Simplicio, then, goes on to apply the Barberinian position to the phenomenon of the tides.   He admits that the appearances—the motion of water—lead the mind to infer the motion of the earth, which contains the water.   But, at once, he poses the rhetorical question to his interlocutors.   Would you mean, he asks, that the interpretation given explains what really must take place in nature?   Certainly not, he adds, because this would restrict the divine power, which can produce the phenomenon of the tides in any other ways he thinks fit.   In Galileo's own words, as put in Simplicio's mouth:

I know that if asked whether God in his infinite power and wisdom could have conferred upon the watery element its observed reciprocating motion using some other means than moving its containing vessels, both of you would reply that he could have, and that he would have known how to do this in many ways which are unthinkable to our minds.   From this I forthwith conclude that, this being so, it would be excessive boldness for anyone to limit and restrict the divine power and wisdom to some particular fancy of his own.

To such an unvarnished declaration of principle, which manifests a total impossibility of communication, Salviati can only reply with disconsolate, if subtle, irony.   He praises Simplicio's view as an "admirable and angelic doctrine."   He also extols the latter's intention of emphasizing God's transcendence.   Thus the book ends.

When the *Dialogue* appeared, Urban VIII was understandably less than amused by Galileo's effort to discredit the ontological view he considered so essential.   Thus, being unable to realize Galileo's profound insight about the intrinsic intelligibility of nature, and feeling called upon by his pastoral responsibility to decide an issue that threatened religious peace, he had no other choice but to strike.   Galileo was condemned.   The

---

[38] Galileo Galilei, *Dialogue Concerning the Two Chief World Systems, Ptolemaic and Copernican*, trans. S. Drake (Berkeley: University of California Press, 1962), p. 464.

condemnation was ostensibly based on religious grounds. But, as is clear from our investigation, the real motive was a philosophical—strictly, onto-logical—one. Galileo was condemned by the prescientific mentality, which could not conceive, even in principle, how nature could be intrinsically intelligible to the human mind. [39] In other words, although God figures prominently in the Galilean controversy, the controversy itself refers to him only indirectly. Properly speaking, in fact, there was no religious disagreement between Galileo and his adversaries. Both parties were unanimous in defending the importance of religion. The fact, however, that the Galilean controversy was not strictly religious does not necessarily simplify the situation; on the contrary, it makes it even more ominous. For it shows how people are touched in the inmost recesses of their per-sonality not only by religion but also by science. Religion deals with the ultimate meaning of reality. But science, too, has something to say with regard to this meaning, owing to its ability to penetrate the intrinsic in-telligibility of observable reality. It follows that unless science and the ontological foundations of religion are harmoniously integrated, people living in the scientific age can hardly fail to be split in the middle of their personality. For people cannot live without a sense of ultimate meaning. But, if this sense of meaning is not realized as common to both science and religion, people feel moved to take sides for the one against the other.

How should one designate the historical lesson to be drawn from the Galilean controversy? It is no exaggeration to speak of *cultural tragedy.* To be sure, Galileo's personality was not harmed by his condemnation. Being a superior man, he knew how to make a fecund synthesis between genuine science and genuine religion (see Chapter 3, Sections 3ca and 3cc). [40] But Galileo's condemnation caused untold harm to mankind as a whole. The reason is obvious. The superficial opinion was given credit that science and religion are incompatible in principle. But, since both science and religion touch upon ultimates that are most dear to the heart of man, it is easy to see how this superficial opinion became the source of great cultural damage. Man was pitted against man in the name of ultimates and absolutes. As a consequence, man—the whole man—was the real loser. We may well date from this episode the start of the so-

---

[39] The theoretical motive of Galileo's condemnation was the one discussed. Urban VIII himself referred to it a number of times, both during and after the trial. But, of course, additional nontheoretical reasons led to the condemnation. Prominent among these was the bitter enmity of Galileo's adversaries. For details see de Santillana (note 34).

[40] De Santillana (note 34, p. 326) characterizes Galileo's religious attitude as that of an "anticlerical Catholic." This designation is felicitous as it stands in the passage quoted. But it should be understood rightly. That is to say, Galileo was never an anticlerical as Voltaire and the French philosophers of the 18th century were, but he stood in the genuinely Christian tradition of Dante Alighieri, Michelangelo Buonarroti, and countless others who were both great Christians and great men.

called two-culture tension, with all its attending evils for mankind. Our effort here should be to pave the way for a better understanding of the absolute, an understanding toward which *both* science and ontology have much to contribute for the good of man as such.

(b) The Scientific Perception of the Absolute

We have tried to remove a major historical and psychological obstacle that is blocking the way toward a proper discussion of the ontological problem of the absolute as it concerns scientific man. We must now pursue our investigation by exploring the characteristic ways in which the scientific mind comes to perceive the ultimate intelligibility of reality or the absolute. For, to be frank, at this point of our investigation we cannot as yet know with precision what the characteristics of the absolute are, at least as manifested by scientifically explored nature. We know only that there must be some sort of absolute or ultimate source of intelligibility of nature. This, in fact, is but an immediate inference from the success of science itself. Indeed, science presupposes and detects a universal intelligibility in nature. But such a universal intelligibility demands that there be an ultimate or absolute—as opposed to merely relative—explanation of it. To pursue our investigation, therefore, we must now hasten to see what typical features science reveals about such an ultimate source of explanation or intelligibility of nature.

(ba) Immanency

In the light of the Galilean affair one point is unmistakably clear. The basic ontological insight that differentiates the scientific from the prescientific way of conceiving nature is the immanency or intrinsicness of nature's intelligibility. This is the fundamental datum that must be acknowledged by everyone who accepts science (see Chapter 5, Sections 1a and 1ba). Indeed, science simply stands or falls on this insight of immanency as opposed to any kind of externalism. To quote one among many, Simpson is certainly the spokesman for the scientific mind when he stresses this view.

The rise of science depended first of all on the belief that the universe makes sense, that its phenomena are orderly and can be explained without recourse to the miraculous, the mystical, and the ineffable. [41]

The scientific insight about the intrinsic intelligibility of nature entails an immediate consequence as regards the absolute. This, whatever it may be, must be something that explains nature from within. For, other-

---

[41] G. G. Simpson, *Biology and Man* (New York: Harcourt, 1969), p. 44.

wise, admission of an absolute would be incompatible with the acceptance of science. To quote Simpson again:

Appeal to the unknown or to the scientifically untestable always stultifies the progress of science, because it stops the search for material explanations that *are* scientifically testable—which, as a matter of experience, have generally been forthcoming when the search has been continued. [42]

The scientists who reflect on the issue of nature's immanent intelligibility to its logical end tend to assert rather than deny the existence of an absolute. The reason of this attitude seems to be a matter of intellectual consistency. Science seeks for intelligibility as originating from the interior of nature. But the intelligibility of nature, as accessible to scientific research proper, is not entirely satisfactory to the mind because it remains relative rather than ultimate. In fact, science can only tell man that things are intelligible; it cannot tell him why they are so. Accordingly, the thoughtful scientific person feels moved, both by experience and reflection, to assert the existence of an immanent, ultimate source of intelligibility in nature, although this source is inaccessible to science as such. We have already discussed the experience of the scientist in this area (Chapter 3, Section 3a). We may summarize the reflective conviction of the scientific mind by citing a few of Einstein's penetrating comments on this issue.

Although Einstein frequently uses the word "religion" in this context, it is important to realize the ontological character of his thought. In his view, science is necessarily based on an ontological conviction about the rationality or intelligibility of nature, a conviction similar to that entertained by the religious person.

Certain it is that a conviction, akin to religious feeling, of the rationality or intelligibility of the world lies behind all scientific work of a higher order. [43]

The reason why Einstein so strictly connects science to religion becomes obvious if one thinks of the ontological foundations of religion. Religion is based on the acceptance of an absolute as an ultimate source of meaning in reality. But reflection on science, Einstein contends, leads to recognition, precisely, of an ultimate source of meaning or active intelligence in nature.

His [the scientist's] religious feeling takes the form of a rapturous amazement at the harmony of natural law, which reveals an intelligence of such a superiority that, compared with it, all the systematic thinking and acting of human beings is an utterly insignificant reflection. [44]

---

[42] G. G. Simpson, *This View of Life: The World of an Evolutionist* (New York: Harcourt, 1964), pp. 22f.

[43] A. Einstein, *Ideas and Opinions,* trans. S. Bargmann (New York: Crown, 1954), p. 262.

[44] *Ibid.*, p. 40.

(bb) Developmentalness

The second major aspect under which the reflective scientist perceives the absolute may be called developmentalness. What is meant by this term can be seen from a consideration of Simpson's. He discusses the different historical attitudes of the scientist and the natural theologian concerning the cognition of God defined as the author of the cosmos. Natural theology is the branch of ontology that studies reflectively the absolute as accessible to the mind through the consideration of nature or observable reality. Simpson admits the possibility of natural theology. Yet he takes to task famous representatives of this philosophical approach because of their tendency to adopt an aprioristically rigid and unchanging view of God. In his own words:

If we are willing to define God as the author of the cosmos, with no further a priori qualifications, then there must indeed be something in the Bridgewater premise: the study of nature should truly bear on the attributes of God. But the scientific revelation is gradual and changing. It makes no pretension of absolute knowledge and shows no sign of reaching finality. If the "rational" and "natural" theology of Paley, the Bridgewater authors, and other worthies... had really been rational and natural, it would have had to recognize that revelation is continuously modified and never complete. [45]

Developmentalness is a very important feature of the absolute as perceived by science. It offers a key to realizing the reason for the continual state of tension between theologians and scientists all along history. We have seen that theologians rejected Galileo's discoveries in the name of their theology (Chapter 6, Section 4a). But that was just one case of a series. Practically all great advances of science—it suffices to mention here the names of Darwin, Freud, and Einstein—were greeted by theologians with suspicion and donwright opposition. Why? The reason seems clear. Theologians incline to think of God in a static, nondevelopmental manner. Hence they are upset and angered when science, in its progress, shows that God, as the source of nature, is quite surprisingly unpredictable in his manifestations.

It goes without saying that developmentalness, as defined, constitutes a major advance in the genuine perception of the absolute. Indeed, we have seen that science makes man experience the inexhaustibility of reality (Chapter 3, Section 2b) and its mysteriousness (Chapter 3, Section 2c). But if reality is inexhaustible and mysterious, the ultimate source of reality must necessarily be such that its manifestations cannot be grasped once and for all by the human mind. As a consequence, reflective man should

---

[45] Simpson (note 42), p. 214.

feel spurred to know the absolute increasingly better, without ever giving in to discouragement or self-satisfaction. For this is the basic way of honoring the ultimate source of the intelligibility of reality: to strive to realize in practice its man-transcending cognoscibility.

(bc) Transrationality

A third feature of the absolute as experienced by the reflective scientist can be called transrationality. This feature is not completely different from the two foregoing ones, yet it deserves special consideration because it makes explicit what they contain only implicitly. Indeed, immanency as well as developmentalness stress that the absolute is cognoscible to man, but not in an anthropomorphic or rationalistic way. Immanency, in fact, shows that the absolute cannot be thought of in terms of an agent that acts on nature from outside. Developmentalness shows that the absolute cannot be encompassed by any well-defined, static formulation. The feature of transrationality emphasizes explicitly that the absolute is cognoscible only as an entity whose intelligibility necessarily transcends the comprehension or the reason of man.

Man tends instinctively to assume that the absolute, if it is cognoscible, must be cognoscible in rationalistic or anthropomorphic terms. This can easily be seen by recalling some examples. Why, for instance, have theologians repeatedly rejected scientific discoveries in the name of God? Obviously because they thought that their own conception of God was an adequate representation of God himself. But the tendency to anthropomorphization and rationalization of the absolute is not limited to theologians. Also scientists are inclined to commit the same error by overemphasizing the validity of their science. A typical case is that of Laplace, who conceived the ultimate intelligibility of reality in strictly classical-mechanical terms (see Chapter 6, Section 2ba). In brief, we can say that the tendency of man is to conceive the absolute as a kind of superman. That is, man thinks of the absolute as though it would be a creative mind substantially similar to that of man himself, but only with an immensely magnified insight and power. Transrationality, as disclosed by reflection on science, exposes the inherent falsity of any rationalistic conception of the absolute.

Fundamentally, science manifests the transrationality of the absolute by insisting on the inexhaustible intelligibility of nature. The absolute must be inherently superior to the human mind or reason because even nature, though directly accessible to man, proves to be effectively inexhaustible to the human mind. The general contribution of science in this connection consists in dispelling the persistent temptation of man toward rationalism. Man inclines instinctively toward a total mental comprehension of the intelligibility of reality with rationalistic features. Prescientific man dreamed of intuiting essences. He aspired to form a huge axiomatic-deductive system that would include whatever man could ever be able to know. But scientific man also tends to think along practically the same lines. The expression, in fact, is often heard that science gradually strives to obtain a world formula. This formula should leave nothing undiscovered, at least in principle. But science, through its history, demonstrates precisely the emptiness of such rationalistic expectations. The reason is the inherently experiential-developmental character of human knowing as well as the inexhaustibility of nature. If knowledge is truly dynamical—as shown, for instance, by the history of mechanics (Chapter 5, Section 2c)—it is absurd to expect that man will ever be able to obtain a world formula or any similar rationalistic scheme. For a genuinely *new* discovery—totally unthinkable and unprecedented—should never be ruled out as impossible. But, by the same token, if reflective scientific man cannot think of the intelligibility of nature in any rationalistic way, even less can he think of the absolute in a rationalistic manner.

In particular, the transrationality of the absolute revealed by science comes strikingly to the fore when one examines the position of the theorists of evolution concerning the orderliness or intelligibility of life. The nonscientific philosopher, impressed by the evident orderliness of life, speaks confidently in rationalistic or anthropomorphic terms. To him it seems obvious that where there is orderliness, there is an ordinator. And this ordinator is normally interpreted in manlike manner, as an outside agent that guides living structures toward predetermined goals. This is the doctrine of *teleology*, at least in its popular form. To the scientist who reflects on evolution such an interpretation is unacceptable. The contrast between the nonscientific and scientific interpretations is expressed vividly, for instance, by the biologist de Beer in discussing Darwin's explanation of biological adaptation:

Adaptation is purposive but not teleological ... adaptations are directed to an end without implying that they were designed to serve that end. [46]

What is meant by expressions of the kind cited, which seem to be contradictory? Indeed, how can one speak of adaptation as "purposive but not teleological"? The position of the scientist is clear. He does not deny the obvious, namely the marvelous orderliness of the living world. Accordingly, he accepts to speak of purpose in regard to nature. This is, for instance, the way Simpson describes the fundamental scientific attitude:

Adaptation does exist and so does purpose in nature, if we define "pur-

---

[46] G. de Beer, *Charles Darwin: A Scientific Biography* (New York: Doubleday Anchor Books, 1965), p. 106.

pose" as the opposite of randomness, as a causal and not merely accidental relationship between structure and function.... Denial of this does
violence to the most elementary principle of rational thought. Look again,
with Sir Charles Bell, at your own hand, manipulate the fingers.... How
can one estimate the improbability that such a structure arose by sheer
accident, or by any continued series of accidents short of infinity? [47]

But, if the scientist accepts purposiveness in nature, why does he
reject teleology? The reason is the transrational or nonanthropomorphic
character of nature's orderliness as brought to light by the theory of evolution. A key factor of evolution is randomness or chance. Evolutionary
adaptation arises out of a totally unpredictable and uncontrollable encounter
between living organisms and environmental conditions. Simpson speaks
in this connection of "opportunism of evolution." The expression is meant
to convey the fact that evolution cannot be conceived as a planned and
controlled development. For, observationally, evolution takes place by
seizing any opportunity that happens to present itself. In other terms,
the factor of chance is decisive. To summarize with Simpson himself:

...changes occur as they may and not as would be hypothetically best;
and the course of evolution follows opportunity rather than plan....
There are two aspects of opportunism: to seize such diverse opportunities as occur, and when a single opportunity or need occurs, to meet it
with what is available, even if this is not the best possible. [48]

As a consequence of the preceding, one can see why evolutionary
scientists reject teleology while accepting purposiveness. What they cannot
admit is that the ultimate agency that accounts for evolutionary regularities be conceived in manlike terms, as though it would be a kind of human
ordinator. For such an agency would simply not fit the data, especially
as regards the crucial role played by chance. R. A. Fisher, the outstanding
mathematician of evolution, has formulated this conviction strongly:

That evolutionary changes can be explained by some hypothetical agency
capable of controlling the nature of the mutations which occur, is involving a cause which demonstrably would not work, even if it were
known to exist. [49]

The transrational character of the intelligibility of nature revealed
by science accounts for the widespread critical attitudes of scientists against
natural theologians. An example of such an attitude is the negative judgment passed by Simpson on several famous thinkers, beginning with

---

[47] Simpson (note 42), p. 202.

[48] G. G. Simpson, "Opportunism of Evolution," in his *Meaning* (note 32), pp. 160-
185, especially pp. 161, 167f.

[49] Quoted in de Beer (note 46), p. 191.

William Paley and ending with Pierre Teilhard de Chardin. [50]   However, the fact that the scientist frequently opposes the natural theologian does not necessarily mean that he opposes natural theology as such.   In fact, the opposite is true—at least when the scientist decides to reflect on the issue of the absolute to its logical end.   Simpson himself, for instance, makes this point explicitly:

Does that mean that religion is simply invalid from a scientific point of view, that the conflict is insoluble and one must choose one side or the other?   I do not think so.   Science can and does invalidate *some* views held to be religious.   Whatever else God may be, he is surely consistent with the world of observed phenomena in which we live. [51]

To sum up, transrationality is clearly a major insight brought by science to the consciousness of man, reflecting on the ultimate intelligibility of reality and its source, the absolute.   Genuinely religious man has always thought of God as inherently mysterious.   Science confirms and deepens this conviction by showing concretely how the intelligibility of nature surpasses the rational comprehension of man.

(bd) The Vindication of Man's Dignity

Still a fourth insight about the conception of the absolute emanates from the scientific experience.   Although this insight deals directly with man, it is of great importance because it shows what a genuine conception of the absolute must be in order to be acceptable to the scientific person.   In brief, it asserts that no view of the absolute or God is admissible unless it allows full scope to the dignity of man.   Scientists often express such an insight in a rather polemical terminology.   To detect its significance, let us survey briefly two of its formulations.

The objection to a personal God.   One of the main reasons that make it difficult for scientists and philosophers of religion to understand each other is the doctrine of a personal God.   Einstein speaks for many a scientific colleague when he states it flatly:

The main source of the present-day conflicts between the spheres of religion and of science lies in this concept of a personal God. [52]

It may appear surprising that Einstein, for one, rejects so flatly the view of a personal God.   Indeed, has he not repeatedly insisted on a

---

[50] G. G. Simpson, "Evolutionary Theology: The New Mysticism," in his *View of Life* (note 42), pp. 213-233.

[51] *Ibid.*, p. 232.

[52] Einstein (note 43), pp. 47f.

religionlike attitude as necessary to do science (Chapter 3, Section 3cb)? But is not such an attitude experientially similar to the awareness of a personality (Chapter 3, Section 3a)? Einstein himself has explicitly asserted that reflection on science reveals an "intelligence" in nature (Chapter 6, Section 4ba). Intelligence, however, is the most typical characteristic of personality. And yet Einstein's views on this subject need not be considered contradictory. For his statements reveal a profound perception, which is difficult to express in systematic form.

What does Einstein—as spokesman for the scientific mind—intend to say when he rejects the view of a personal God? We find an answer by collating several of his expressions on this subject in the context cited. He speaks of "the doctrine of a personal God" as though it would mean that God is "interfering with natural events." Further on, he urges religious teachers to give up that doctrine because it is the "source of fear and hope which in the past placed such vast power in the hands of priests."

In other passages, Einstein clarifies the reasons of his opposition to the doctrine in question. The first one is the immanent intelligibility of nature (what he calls "the law of causation"), which he regards as essential to the scientific worldview. In his own words:

The man who is thoroughly convinced of the universal operation of the law of causation cannot for a moment entertain the idea of a being who interferes in the course of events.... [53]

The second reason for rejecting a personal God is expressed by Einstein when he defines what he considers "the religiosity of the naïve man."

For the latter [the naïve man], God is a being from whose care one hopes to benefit and whose punishment one fears; a sublimation of a feeling similar to that of a child for its father, a being to whom one stands, so to speak, in a personal relation, however deeply it may be tinged with awe. [54]

In summary, the insight underlying the scientists' widespread refusal —or, at the very least, repugnance—to speak of God in personal terms can be considered clear. What they reject is not, properly speaking, the common practice of ascribing personality to God. Rather, they refuse to accept that personality be predicated of God in such a way as to make him manlike or anthropomorphic. In particular, they refuse to admit that the personality of God be interpreted as something detrimental to the dignity of man. Needless to add, such an insight is far from inimical to the genuine philosophy of religion. On the contrary, it is a most valuable contribution to it. Hence it should be explicitly acknowledged as such.

---

[53] *Ibid.*, p. 39.
[54] *Ibid.*, p. 40.

20

The refusal of religious instinctivism. Another polemical formulation with which the scientific mind insists on the dignity of man relative to religion is to be found in Freud's little book *The Future of an Illusion.* This book is rightly considered to be antireligious. And yet, beneath all the polemic, we can detect some valuable information about the scientific conception of the absolute.

What is the religion Freud intends to stamp out? Freud takes care to answer this question himself when he defines religious ideas as the manifold expression of dogmatism and social conditioning. In his own terms:

Religious ideas are teachings and assertions about facts and conditions of external (or internal) reality which tell one something one has not discovered for oneself and which lay claim to one's belief. [55]

In particular, Freud is outraged by what can be called religious instinctivism, namely the ignoble surrender of man to wishful thinking or illusion. The following, in fact, is another way in which he defines religious ideas:

These, which are given out as teachings, are not precipitates of experience or end-results of thinking: they are illusions, fulfilments of the oldest, strongest and most urgent wishes of mankind. [56]

In brief, Freud is the irreducible enemy of what he calls "infantilism." He attacks religion because he sees it as a factor hostile to the personal maturation of man. He confesses that this was his deep-seated motivation in writing the book.

But surely infantilism is destined to be surmounted. Men cannot remain children forever; they must in the end go out into "hostile life." We may call this *"education to reality."* Need I confess to you that the sole purpose of my book is to point out the necessity for this forward step? [57]

Obviously, Freud's views just surveyed contain nothing that is unacceptable to the genuine philosophy of religion. Hence his views should be welcomed at least as a challenge and warning to keep religion free of misleading interpretations. Indeed, Freud's contribution is quite valuable when he points to the persistent danger that religion may become an impediment to psychological maturation of man. For the sake of the present discussion, however, we must still ask a question. What was the ultimate motivation of Freud himself in writing the book under con-

---

[55] S. Freud, *The Future of an Illusion,* trans. W. D. Robson-Scott and J. Strachey (New York: Doubleday Anchor Books, 1964), p. 37.

[56] *Ibid.,* p. 47.

[57] *Ibid.,* p. 81.

sideration?   At least as far as his subconscious was involved, the answer
is clear.   Paradoxical as it may sound, Freud was animated by an inspiration that deserves to be called religious.   He intended to defend what he
considered authentic religiosity against its many widespread counterfeits.
Thus, in fact, he writes against those who demean man in the name of
religion:

Critics persist in describing as "deeply religious" anyone who admits to
a sense of man's insignificance or impotence in the face of the universe,
although what constitutes the essence of the religious attitude is not this
feeling but only the next step after it, the reaction to it which seeks
a remedy for it.   The man who goes no further, but humbly acquiesces
in the small part which human beings play in the great world—such a
man is, on the contrary, irreligious in the truest sense of the word. [58]

To sum up, scientists may well be unprecise and too polemical when
vindicating the dignity of man as demanded by science, but their contribution to a better understanding of the absolute is, in this regard, unmistakable.   Already Galileo had insisted that genuine religion favors rather than
suppresses the sense of man's dignity (see Chapter 1, Section 3b).   But
he lived in a cultural environment where religion was dominant.   It is
enlightening to see his conviction confirmed by two scientists—Einstein
and Freud—who lived in a milieu where religion was widely considered
passé.

(c) The Ambiguity of the Scientific Perception of the Absolute

At this point of our investigation we may consider one conclusion established.   The scientific search for the intelligibility of nature leads the human
mind to perceive something that deserves to be called the absolute—namely,
the ultimate ground or source of such an intelligibility.   This conclusion
is all the more convincing in that science makes man realize various characteristics of the absolute that the philosopher would have great difficulty
in detecting by himself.   Yet, in our inquiry about the ontological significance of science, we must now again raise a critical question.   If science
leads man to the cognition of the absolute, does it suffice in providing
an adequate cognition of the absolute itself?   Many people take for granted
that this is the case—that if there is any justification to speak of the
absolute, science alone is competent to know about it.   For the sake of
our investigation, let us suppose that this view is well founded and examine
its implications systematically.

---

[58] *Ibid.*, p. 52.   As concerns Freud's implicitly religious inspiration see comment by
E. Fromm in Chapter 7 of this book (text pertaining to note 10).

(ca) The Pantheistic Opinion

One of the most appealing interpretations of the absolute that presents itself as a consequence of science is pantheism. This, at any rate, is what one can infer from the position adopted by Einstein, who was thoroughly convinced that the intelligibility of nature disclosed by science manifested a transrationally superior mind. However, he interpreted such a mind in pantheistic terms. As he put it himself:

This firm belief, a belief bound up with deep feeling, in a superior mind that reveals itself in the world of experience, represents my conception of God. In common parlance this may be described as "pantheistic" (Spinoza). [59]

Why did Einstein adhere to a pantheistic view of the absolute? The decisive reason seems to be his own rather rationalistic conception of science, at least when expressed in philosophical terms. To be sure, Einstein was the person who asserted, on the experiential level, that science ends up in a sense of wonder and mystery see (Chapter 3, Sections 1c and 2c). However, when he tried to formulate his conception of science philosophically, he sounded thoroughly rationalistic. This can be seen for instance in the following passage:

The supreme task of the physicist is to arrive at those universal elementary laws from which the cosmos can be built up by pure deductions. [60]

In the light of the statement quoted, it seems clear why Einstein defined himself as a pantheist. Pantheism implies a very comprehensive and farreaching position. Everything that exists (*pan*) is considered to be part and parcel of the absolute (*theos;* that is, God). But if one interprets the intelligibility of nature, as studied by science, in a rationalistic sense, the pantheistic view is hardly avoidable. In fact, rationalism implies that the mind of man is capable, at least in principle, of exhausting the intelligibility of reality by means of axiomatic-deductive formulations. But, as a result, man must exclude from the realm of existence everything that does not fit into the overall axiomatic-deductive scheme of reality. The conclusion is the pantheistic view. To put it briefly, it asserts that whatever exists is to be identified with what can be observed and explained by science; that is, it has to be reduced to observable reality or nature. This seems to have been the reason that motivated Einstein. We can notice it, at least implicitly, in the following passage:

The more a man is imbued with the ordered regularity of all events the firmer becomes his conviction that there is no room left by the side of

---

[59] Einstein (note 43), p. 262.
[60] *Ibid.*, p. 226.

this ordered regularity for causes of a different nature. For him neither the rule of human nor the rule of divine will exist as an independent cause of natural events. [61]

An additional reason why Einstein inclined to pantheism was undoubtedly his opposition to the doctrine of a personal God. Such an opposition, as has been seen, implies a quite valuable religious insight (see Chapter 6, Section 4bd). As a consequence, it is not surprising that Einstein adopted the pantheistic view. He clearly felt this doctrine satisfied his profound religious attitude, while at the same time making room for his philosophical interpretation of science. The sincerity of Einstein's religious position as well as his motivation can be perceived in the following passage, where he passionately pleads for a synthesis between science and religion:

But whoever has undergone the intense experience of successful advances made in this domain [the rational unification of the manifold] is moved by profound reverence for the rationality made manifest in existence. By way of the understanding he achieves a far-reaching emancipation from the shackles of personal hopes and desires, and thereby attains that humble attitude of mind toward the grandeur of reason incarnate in existence, and which, in its profoundest depths, is inaccessible to man. This attitude, however, appears to me to be religious, in the highest sense of the word. And so it seems to me that science not only purifies the religious impulse of the dross of its anthropomorphism but also contributes to a religious spiritualization of our understanding of life. [62]

Whatever may have been the reasons leading to Einstein's pantheistic confession, the question must be posed: Does pantheism supply an adequate interpretation of the ultimate intelligibility of reality? In particular, is it compatible with the genuine spirit of science? By surveying briefly Einstein's own words and deeds we can easily see that this is not the case.

The first reason that manifests the inadequacy of pantheism as an ultimate interpretation of reality is its being at odds with man's ethical life. Pantheism demands—as a corollary—the complete denial of human freedom. In keeping with his pantheistic principles, Einstein stressed repeatedly his denial of human freedom. In particular, he rejected the notion of man's personal responsibility. As he put it in his philosophical writings:

I do not at all believe in human freedom in the philosophical sense. Everybody acts not only under external compulsion but also in accordance with inner necessity ... a man's actions are determined by necessity, external and internal, so that in God's eyes he cannot be responsible,

---

[61] *Ibid.*, p. 48.
[62] *Ibid.*, p. 49.

any more than an inanimate object is responsible for the motions it undergoes. [63]

When the circumstances called for it, however, Einstein simply denied through practice what he had asserted in theory. He defended human freedom and preached responsibility. This he did, for instance, by extolling the "almost fanatical love of justice and the desire of personal independence" of his Jewish brethren persecuted by the Nazis. Simultaneously, he publicly castigated "those who are raging today against the ideals of reason and individual liberty." [64]   After the battle of the Warsaw ghetto he even declared:

The Germans as an entire people are responsible for these mass murders and must be punished as a people if there is justice in the world.... [65]

As is well known, Einstein took his own moral responsibility quite seriously, up to the point of trying to influence politics. He was a prime mover behind the initiative of having the United States produce the atom bomb and thus defend democracy against the Nazi danger. After the war was won, Einstein bestirred himself to promote a world government to prevent the occurrence of another war. [66]   In brief, it is clear from Einstein's own moral example that pantheism cannot be accepted as an adequate interpretation of reality. The reason is that ethical man must contradict through practice what pantheism demands in theory.

The second reason that manifests the inadequacy of pantheism as the ultimate interpretation of reality is its deleterious effects on science. Science, to be alive, must be creatively progressive. That is, the scientist must keep himself open to unexpected, completely unforeseeable manifestations of the intelligibility of nature. But pantheism, because of its rationalistic foundations, makes man rule out as absurd all those manifestations of nature whose intelligibility cannot be fitted into an overall axiomatic-deductive scheme. Returning to Einstein, we can probably detect in his scientific conduct the negative effects of pantheism just mentioned. Einstein, as is well known, stubbornly refused throughout his life to admit the statistical results of quantum physics as final. As a consequence, during his mature life he contributed nothing but negative criticism to quantum research. What was, in his eyes, the justification of his attitude? He kept repeating the phrase that fitted so well into his pantheistic-rationalistic view of reality: "God does not play dice!" Creative quantum physicists could only reject such an aprioristic argumentation. Niels Bohr—Einstein's great

---

[63] *Ibid.,* pp. 8 and 39.

[64] *Ibid.,* p. 185.

[65] *Ibid.,* p. 212.

[66] See various documents in Einstein, *op. cit.,* Part II.

philosophical antagonist and foremost creator in the quantum realm—offered an appropriate rebuttal of Einstein's rationalistic aphorism. In his own words: "It certainly cannot be our task to prescribe to God how he should govern the world." [67]   Such a theological exchange between scientific giants provides perhaps the most convincing proof of the inherent incompatibility between pantheism and creative science.

To sum up, pantheism is clearly not a necessary concomitant of the scientific attitude. In point of fact, it can easily be detrimental to science itself. As a consequence, we begin to see the ambiguity of science in leading man to the cognition of the absolute. Science does lead man to realize the existence of the absolute. But if it is not supplemented by an engagedly systematic philosophical reflection, it is in danger of transforming the cognition of the absolute into a source of errors harmful to man and science itself.

(cb) The Materialistic Opinion

Among the major conceptions of the absolute that are frequently defended as necessary implications of the scientific mentality, materialism stands at the opposite end of the spectrum from pantheism. Pantheism stresses the spiritual or intelligible character of reality. Materialism reduces the whole of reality to what can be seen and touched of it.

The fundamental reason that materialism is often seen as the obvious consequence of science is not difficult to find. It is of a psychological kind. Science has given a great prominence to the material or observable world by detecting increasingly its variety, richness, beauty, and all-pervasiveness. The very personality of man is—at least in some way explorable by science—dependent on matter. Furthermore, science has aroused a great enthusiasm for matter by giving rise to an endless stream of technological inventions. To many it seems that matter, through science and technology, is able to fulfil completely all of man's expectations and ideals. As a consequence, it is but natural that numerous people assume it to be obvious that matter is the ultimate or absolute reality. Such an assumption finds a ready confirmation in the apparent sterility of so-called spiritualism. Many of those who claim to uphold spiritual realities and values adopt a predominantly negative attitude toward both matter and science. They seem to be bent on preserving a dogmatic past rather than fostering progress. As a result, materialism is widely seen as the embodiment of whatever is attractive in the scientific age.

If the psychological identification between science and materialism is understandable, very little reflection is needed to realize that such an

---

[67] Translated from quotations in W. Heisenberg (note 4), p. 115.

identification is far from being obviously justified. In fact, if materialism were the necessary ontological consequence of science, materialism would be universally accepted by all those who know science. But this is by no means the case. Many other ontological alternatives are possible, and are actually shared by people who accept science entirely. We have already discussed pantheism, and we shall examine some more of these alternatives shortly. As a consequence, materialism cannot claim to be scientific except on the strength of the philosophical arguments it can muster to defend its claim.

For the purpose of our investigation it is sufficient to point out here a certain basic incompatibility between materialism and the spirit of genuine science. Surprisingly, this point is made with emphasis by T. H. Huxley, the great agnostic and declared enemy of spiritualism. Huxley admits a strict connection between science and materialism, but he does not like it at all. In his view, the more science advances "the more extensively and consistently will all the phenomena of nature be represented by materialistic formulae and symbols." [68] However, he goes on to explain, this is bad because it tends to transform science into a kind of dogmatic rationalism that may seriously harm man and drain away the significance of his existence. In his own words:

But the man of science, who forgetting the limits of philosophical inquiry, slides from these formulae and symbols into what is commonly understood by materialism, seems to me to place himself on a level with the mathematician who should mistake the x's and y's with which he works his problems, for real entities—and with this further disadvantage, as compared with the mathematician, that the blunders of the latter are of no practical consequence, while the errors of systematic materialism may paralyze the energies and destroy the beauty of a life.

But Huxley is not alone in criticizing materialism in the name of science. For example, Rostand finds it just as meaningless as spiritualism. "My impression is that materialism is closed tight and that spiritualism opens on nothing." [69]

To sum up, materialism is another proof of the ambiguity with which science leads the mind of man to the cognition of the absolute. Science, by itself, is unable to give man a satisfactory conception of the absolute. Indeed, if materialism were the last insight provided by science, it is hard to see why science started to exist in the first place, and why it continues to flourish.

For science, as has been seen, is essentially a search for ideal—the

---

[68] In C. Bibby, ed., *The Essence of T. H. Huxley: Selections from His Writings* (New York: St. Martin's, 1967), pp. 58f.

[69] Rostand (note 27), p. 221.

ideal being the mental contemplation of the intrinsic intelligibility of reality (see Chapter 4, Section 1). But materialism makes a mockery of such an interpretation of science.

(cc) The Deistic Opinion

Possibly the most widespread conception of the absolute inspired by one-sided enthusiasm for science, at least in times past, is the one that goes under the name of deism. The term is vague, but intentionally so. Many people are instinctively loath to reject all reference to God as conceived as the main element of traditional culture. But, on the other hand, they feel that science alone is competent in judging the ultimate meaning of reality. Hence deism presents itself as appealing. It can be defined, following the etymology of the word, as the view that there is a divinity or ultimate ground of reality, but this is totally conceivable in scientific and rational terms.

Not surprisingly, the historical evolution of deism follows closely that of the popular interpretation of science. Deism arose in the 17th century, when science was beginning to gain in public prestige. Since the current view was that science was practically the same as classical mechanics, the conception of God adopted at the times was mechanistic. God was conceived as the Supreme Architect or Engineer. [70] He was supposed to have first set in motion the machine of the world at the beginning of time. As for his later activity, opinions differed according to successive interpretations of mechanics. Early mechanicians still admitted some occasional interventions of God in the world. These were reputed necessary in order to keep the world from falling apart. However, when the mathematical treatment of mechanics showed that the world could continue to function indefinitely according to mechanical laws without intervention from outside, God was denied any action in the world. The French, with their noted sense of sarcasm, spoke then of the "do-nothing god" (le dieu fainéant). A telling example of the mentality involved is contained in a celebrated reply by Laplace to Napoleon. When he was asked why he had not mentioned the Creator anywhere in his great treatise Mécanique Céleste, Laplace answered: "Sire, I was able to dispense with such a hypothesis." [71] Deism, however, did not disappear from the scene once science had seemingly made God superfluous. Owing to its vagueness and psychological motivation, the deistic doctrine is still abroad among contemporary people. It has simply adapted itself to further developments of science. One example of such an adaptation is the often-quoted phrase by James

---

[70] For a summary see Dijksterhuis (note 19). See also A. Koyré, *From the Closed World to the Infinite Universe* (New York: Harper Torchbooks, 1958), pp. 273-276.

[71] Quoted in Dijksterhuis (note 19), p. 491.

Jeans, a contemporary theoretical astronomer. According to him: "The Great Architect of the universe now begins to appear as a pure mathematician." [72]

Clearly, deism offers further evidence for the inherent ambiguity of science with regard to the cognition of the absolute. It shows that scientific man is prone to adopt the most disgusting compromises when faced with the issue of the ultimate intelligibility of reality. In order to save both religion and science, he tends to undermine both. The point has been made with special emphasis by Rostand. He insists that deism implies a disgraceful conception of God.

We purify God, we simplify him, we pick him bare, we accept his silence and his idleness. We accept the fact that everything goes on down here as though he did not exist. We ask him merely to keep his name. [73]

But, if deism destroys religion—Rostand remarks elsewhere—it treats science no more kindly. Indeed, deism is but another word for rationalism, since the deistic view of the absolute is clearly a self-projection of man. But rationalism tends to close the mind of man to the surprising richness of reality. Hence science tends to be stifled. Rostand finds deism leading to the same meaninglessness of reality that is entailed by materialism.

Seldom does anyone feel that stupefaction in the presence of nature's works which they deserve. Materialism, like deism, clouds our minds to the fabulousness of the real. [74]

(cd) The Atheistic Opinion

To complete our survey, we must still consider another conception of the absolute that is increasingly taken as a necessary consequence of the acceptance of science. The doctrine in question is atheism. Atheism is a complex subject, which arouses much controversy in our own times. For the sake of our investigation, we are going to limit ourselves to essentials. Two questions must be answered: What is meant by atheism when the term is applied to the issue of the ultimate intelligibility of reality? What is the connection between science and atheism—in particular, what consequences does atheism entail for science?

The discussion of atheism is difficult because of the negative structure of the term involved. Etymologically, the term amounts to a denial. It stands for the doctrine that denies the existence of God. But, of course, when people agree on denying something, their reasons for doing it may

---

[72] *Ibid.*, p. 500.
[73] Rostand (note 27), p. 76.
[74] *Ibid.*, p. 296.

be quite different and even contradictory. Accordingly, there are many possible forms of atheism. In view of our goal, we shall concentrate our attention solely on the negative sense of the term. However, even with this limitation, a distinction must be noted in its current usage. The term is employed in a twofold acceptation: one hard or literal, the other soft or derivative. To avoid misunderstandings, we shall use two different words. Atheism, in its hard or literal meaning, is an assertive term. It asserts the certainty that there is no God at all. Accordingly, we shall designate this acceptation by using for it the term "atheism" itself. *Atheism*, therefore, will imply the positive denial of any absolute whatsoever as the ultimate source of the intrinsic intelligibility detectable in nature. The soft or derivative sense of atheism, as currently employed, stresses exclusively a negative attitude. It declares that man cannot know anything certain about the absolute. To designate this view we shall use the word *agnosticism*.

Starting with atheism proper, what is its connection with science? Psychologically speaking, of course, it is obvious. Since science is so successful as a form of knowledge, and since philosophy—notably natural theology—appears to be so inconclusive, atheism is the spontaneous fruit of science. Man inclines to take for granted that there is no absolute because the success of science seems to prove that the study of the absolute is totally dispensable, even obnoxious, when doing science. However, if one leaves psychology aside and faces the issue critically, what is the justification for speaking of atheism as scientific? Discounting equivocations—such as the confusion between atheism and materialism—only one answer is to be found in the literature supporting the doctrine in question. Atheism is predicated as scientific on the presupposition that science is the exclusively reliable form of knowledge. Obviously, the observational approach of science has nothing to say about the existence of the absolute, whose properties are nonobservable. Accordingly, the nonexistence of the absolute is declared evident in the name of science.

At this point of our investigation we need not insist much on the incompatibility of the so-called scientific atheism with the genuine spirit of science. Indeed atheism is not, properly speaking, scientific but scientistic. For, as has just been said, it is based on the dogmatic view that science is a self-evident and exclusively reliable form of knowledge. But scientism is a caricature of science, destructive of its genuineness as well as its meaningfulness (see Chapter 4, Section 5bb). Clearly, therefore, atheism is not to be seen as the authentic result of science but rather as its self-defeating travesty. The conclusion is harsh, but unavoidable. A surprising confirmation of it is found, for example, in the total rejection of atheism by T. H. Huxley. The father of agnosticism found only one category of persons even worse than the theologians he abhorred. These

were the atheists. The reason is obvious. The theologians claimed to
know everything about God—but at least they were not scientific. But,
how could scientific persons presume to prove that there is no God? In
his own terms:

> Of all the senseless babble I have ever had occasion to read, the demon-
> strations of these philosophers who undertake to tell us all about the
> nature of God would be the worst, if they were not surpassed by the
> still greater absurdities of the philosophers who try to prove that there
> is no God. [75]

As regards the soft form of atheism, or agnosticism, we have already
seen its complex psychological connection with science (Chapter 4, Section
3d). What remains to be done here is to explore its consequences for
science itself. If nothing can be known with certainty about anything
objectively absolute, what will become of science? A telling answer, albeit
unintentional, is supplied by Freud.

Freud denies the very ontological foundations of religion, namely the
possibility of an objective doctrine of the absolute. This he does in the
name of science.

> The scientific spirit brings about a particular attitude towards worldly
> matters; before religious matters it pauses for a little, hesitates, and finally
> there too crosses the threshold. In this process there is no stopping;
> the greater the number of men to whom the treasures of knowledge
> become accessible, the more widespread is the falling-away from religious
> belief—at first only from its obsolete and objectionable trappings, but
> later from its fundamental postulates as well. [76]

If the absolute and religion have to be rejected, what remains undoubtedly
certain? Freud's answer is unequivocal: science, and science alone.

> But scientific work is the only road which can lead us to a knowledge
> of reality outside ourselves. It is once again merely an illusion to expect
> anything from intuition and introspection.... [77]

Freud is convinced to have proved that religion, as a doctrine of the
absolute, is an illusion. But then, as a critically minded person, he cannot
refrain asking whether the same judgment might perhaps not be uttered of
science itself.

> Having recognized religious doctrines as illusions... we shall not shrink
> from asking too whether our conviction that we can learn something

---

[75] In Bibby, (note 68), p. 71.
[76] Freud (note 55), p. 63.
[77] *Ibid.*, p. 50.

about external reality through the use of observation and reasoning in scientific work—whether this conviction has any better foundation. [78]

At first, Freud tries to dismiss the whole issue by pleading personal inability to discuss the philosophical questions involved. He wants to focus only on religion.

But the author does not dispose of the means for undertaking so comprehensive a task; he needs must confine his work to following out one only of these illusions—that, namely, of religion. [79]

Dismissal of the objection, however, does not satisfy Freud's critical sense. He soon makes an imaginary opponent accuse him of being "an enthusiast who allows himself to be carried away by illusions." [80] To such an accusation Freud finally concedes that it may have some foundation. But he tries to defend himself by saying that, if he is under an illusion in stressing science, this is not so bad as the religious illusion itself.

I know how difficult it is to avoid illusions; perhaps the hopes I have confessed to are of an illusory nature, too. But I hold fast to one distinction. Apart from the fact that no penalty is imposed for not sharing them, my illusions are not, like religious ones, incapable of correction. They have not the character of a delusion. [81]

Even this declaration, however, leaves Freud dissatisfied in his almost frantic defense of science at all costs. Shortly afterwards, he returns to the issue. He confesses openly that he adheres to science on a purely fideistic-pragmatic basis. Speaking against his imaginary opponent he says:

We believe that it is possible to gain some knowledge about the reality of the world, by means of which we can increase our power and in accordance with which we can arrange our life. If this belief is an illusion, then we are in the same position as you. But science has given us evidence by its numerous and important successes that it is no illusion. [82]

Freud's continual retreat is nothing short of pathetic. Having set out with the conviction that religion is illusory and science absolutely certain, he ends up by testily proclaiming the validity of science in purely fideistic terms. The very last sentence of the book still tries to exclude the possibility that science itself be considered an illusion.

---

[78] *Ibid.*, p. 55.
[79] *Ibid.*, p. 56.
[80] *Ibid.*, p. 83.
[81] *Ibid.*, p. 86.
[82] *Ibid.*, p. 90.

No, our science is no illusion. But an illusion it would be to suppose that what science cannot give us we can get elsewhere. [83]

To sum up, there is apparently no escape from logic. If the conviction about the existence of the absolute—the ontological foundation of religion—is denied or explained away as psychological illusion, science itself comes to grief. Agnosticism entails the ruin of genuine science; either science becomes an illusion or is upheld on purely dogmatic grounds. Even worse, when trying to displace religion, science is made into a religion of its own. As Viktor Frankl, the founder of the so-called third Viennese school of psychotherapy remarks: "As soon as we have interpreted religion as being merely a product of psychodynamics . . . the psychology *of* religion often becomes psychology *as* religion." [84]

As a general conclusion, the ambiguity of science in leading man to the cognition of the absolute needs no further comment. There is no doubt that science, as an experience of the whole person, makes man perceive the absolute in a profound and original way. But there is just as little doubt that science is unable alone to make man attain a critically satisfactory view of the absolute. Consequently, if one really cares for science and man, one must face up to the ontological problem of the absolute with a strict philosophical approach—an approach, of course, that takes the contributions of science into due account. For otherwise, as has just been seen, man would be in the end frustrated of what he wants most when dedicating himself to science. Man strives for understanding, but understanding cannot be had unless the mind attains to an ultimate or absolute explanation of reality. Thus, without an ontological investigation of the problem of the absolute, consistently thinking scientific man would be condemned, in the end, to skepticism or—even worse—dogmatism.

(d) The Ontological Problem of the Absolute

To continue our introductory investigation of science as significant to man, we must now examine directly the ontological problem of the absolute that emerges from the success of science. But, of course, when approaching such a problem, we must move with great caution. For, as is well known, the question involved here constitutes one of the most difficult and contentious issues confronting the human mind. The issue is difficult because it inquires about the very ultimate source of meaning of reality as a whole. The issue is contentious because it is at the center of the violent storms

---

[83] *Ibid.*, p. 92.
[84] V. E. Frankl, *Man's Search for Meaning: An Introduction to Logotherapy*, trans. I. Lasch (New York: Washington Square Press, 1963), p. 210.

and deceptive lulls that have marked the relationships between science and religion ever since modern science arose. Thus the theoretical difficulties can easily be complicated by the emotional reactions that the discussion of religious matters tends to arouse. To avoid possible misunderstandings, these are the guidelines to be followed here. We shall remain strictly within the boundaries of philosophical reflection, without taking into consideration the religious issue as such. We shall face our problem by degrees. First, we shall consider a set of conditions that appear necessary for a satisfactory solution of the problem under examination. Second, we shall formulate an outline of the problem itself.

(da) Conditions for a Satisfactory Solution

Our methodological approach must be guided essentially by respect for science and concern for man. As a consequence, a number of conditions must be fulfilled so that the problem of the absolute may find a solution satisfactory to man living in the scientific age. The following are the conditions that appear necessary and sufficient.

The intrinsic bearing of science on the problem of the absolute. The fundamental requirement for a satisfactory solution of the problem of the absolute in the scientific age is obvious. Science has an inherent ontological significance (Chapter 6, Section 1); it leads the mind of man to perceive the ultimate intelligibility of reality in a characteristic way (Chapter 6, Sections 4a and 4b). Accordingly, man living in the scientific age can face the ontological problem of the absolute satisfactorily only if he acknowledges the obvious, namely the intrinsic bearing of science on the problem itself.

Despite the theoretical obviousness of the requirement outlined, currently the temptation is strong among both scientists and philosophers to ignore it. History clarifies the psychological reasons of such a situation. Science arose with a momentous clash between scientists and philosophers precisely in connection with the interpretation of the absolute (Chapter 6, Section 4a). The development of science was marked by a continual tension between the two sides. Hardly a major scientific advance was left undisputed by the philosophers of the absolute or natural theologians. On the other hand, many scientifically inspired thinkers propounded—as necessary consequences of science—a number of views which were unproved and unacceptable (Chapter 6, Section 4c). It is clear, as a result, that contemporary man feels weary and wary of the endless polemics and their apparent futility. This explains the widespread view, shared by both scientists and philosophers, that science has no significance for man in his search for the absolute. Many thinkers even consider themselves enlightened by adopting such a position. So some scientists claim that science is thereby freed from cumbersome and meaningless metaphysical preoccupations. Some philosophers claim that philosophy is thus freed from the limitations and changeability of the scientific approach. Both sides contend that their position is motivated by genuine respect for truth, man, and the absolute itself.

No matter how sincere the psychological motivation may be for refusing a dialogue between science and philosophy on the question of the absolute, it is clear from our entire discussion that such a position cannot be accepted. For it would call into question the entire significance of both science and philosophy. Indeed, both science and philosophy—in different but complementary ways—are a search for intelligibility. But such a search, of course, cannot be stopped by the consistently thinking person before reaching the absolute, namely the ultimate explanation of the intelligibility of reality. Accordingly, if man living in the scientific age does not want to condemn himself to pure pragmatism (that is, aprioristic denial of the absolute) or dogmatism (that is, uncritical affirmation of the absolute), he must admit the intrinsic bearing of science on the problem of the absolute.

No mysticization, but a fully human development of science. Once the intrinsic bearing of science on the problem of the absolute is recognized, we must hasten to add another condition for the solution of the problem in question. This condition balances the first one and prevents an improper interpretation of it. Negatively speaking, one should avoid "mysticizing" science. That is to say, one should never think that science—because it has a bearing on the problem of the absolute—is now to be regarded as a sort of cryptometaphysics or, even worse, a cryptoreligion. Quite definitely, science must remain science. That is, people must continue to consider and handle it as an autonomous and specific form of knowing, characterized by the principle of observability. However, since the discoveries of science have a direct bearing on the problem of the absolute, it should clearly be the concern of everyone who respects science that its insights also be brought to full development in this regard. This is all the more needed because the problem of the absolute is central for a genuinely human understanding of reality. Hence every effort should be made to integrate the ontological contributions of science into a comprehensive philosophical view of reality, which, in particular, must include a theory about the absolute itself. To restate the foregoing, the methodological guideline advocated here is that science be considered an autonomous but not an exclusive nor a separatist undertaking of man. The reason for this stand is the essential unity of man. Hence science cannot gain genuinely human significance—cannot become truly humanistic—unless and until it is integrated into a genuinely human understanding of reality as a whole.

The reflective verifiability of the solution. A third condition must be satisfied in order to obtain a proper solution to the problem of the absolute. This condition requires that the problem be recognized strictly as a human problem—that is, one that can be solved by thinking man as such, and whose solution can be verified by means of reflection. This condition rules out many possible ways of facing the problem of the absolute that are uncontrollable by man and thus inaccessible to reflective critical examination. Such ways are, for instance, mental intuition, mystical illumination, and the like. It is not our concern here to pass a judgment upon the reliability, if any, of such forms of knowing the absolute. Certainly, however, the problem of the absolute as envisaged here cannot be solved by any such means. For the problem itself, in the perspective we have adopted, is really a human problem—specifically, an ontological one. Hence it must be solved by human philosophical means, and no others. But philosophy, as a specific form of knowledge, has its own characteristic approach for solving problems and testing the validity of the solution proposed. This approach, as has been repeatedly seen, is personal reflection. It follows that the philosophical problem of the absolute can be solved satisfactorily only by means of personal reflection conducted on one's own experience of reality. To be sure, the solution of the problem can and should be enhanced greatly by discussing it with other people. For, since the problem is objective, its solution also has to be objective; that is, it must be capable of being examined on an interpersonal basis. However, in the end, the one who decides whether or not the solution proposed is truly satisfactory must be the individual who is personally concerned with the problem in question. In other words, the solution of our problem cannot be taught by experts. Much less can it be imposed by any sources of external persuasion, such as cultural or social pressures. Rather, each individual must, in the last analysis, be his own expert. Therefore, if a solution is proposed by experts, each one should strive, by reflection, to verify its validity. For the problem of the absolute is truly the problem of each man who seeks the ultimate meaning of reality.

Indicative rather than demonstrative character of the solution. Still one major condition must be fulfilled in order to have a truly satisfactory solution of the problem of the absolute. This condition arises from the twofoldness of man's experience relative to the absolute. Man can perceive with unshakable certainty that the absolute exists (see Chapter 3, Section 3a). Hence one is justified to say that the absolute is cognoscible to man—at least in the basic sense that man can be certain about its existence. On the other hand, when man experiences the absolute, he experiences it as inherently mysterious—transcending his own power of conceptual representation and demonstration (see Chapter 3, Section 2c). Hence one is justified in saying that the absolute is incognoscible to man, at least in the sense that observable things can be known. The condition that arises from this situation for a satisfactory solution of the problem of the absolute is obvious. The solution itself simultaneously must take into account both the cognoscibility and the incognoscibility of the absolute. One can speak therefore of a solution that must be indicative rather than demonstrative: *indicative* because its statements must be based on experience and point to experience; *nondemonstrative* because its statements are not to be seen as logical consequences of obvious or axiomatic premises.

The condition in question is essential for fostering a fruitful dialogue between scientists and philosophers concerning the absolute. Misunderstanding and suspicion, in fact, are fostered by the failure to realize clearly the peculiar cognoscibility of the absolute. Hence each side tends to blame the other for being rationalistic in this regard. Philosophers tend to see scientists as opposed to admitting the existence of the absolute. Scientists tend to see philosophers and theologians as unduly dogmatic in their statements about the absolute. The attitude to be adopted is the one that follows from the experiential as well as mysterious cognoscibility of the absolute itself. If man can experience but not comprehend the absolute, he should indeed strive to know it more and more. But, at the same time, he should never arrogantly claim to possess an adequate cognition of it. This is the illuminating insight reached by some of the most penetrating minds, acquainted with science and philosophy, of our time. Thus, for instance, Simpson closes his discussion of the relations between natural theology and biological evolution by pointing to a sense of mystery. He then questions the ability of theologians to penetrate the mystery itself. In his own words:

There lie the ultimate mysteries, the ones that science will never solve. It does not lessen our religious awe of them if we question whether theology, too, is not powerless to pierce that ultimate veil. [85]

In the same vein, Rostand writes succinctly: "We must not suppress the incomprehensible, but neither must we use it as an explanation." [86]

(db) The Theoretical Structure of the Problem

In the light of the conditions discussed, we can now proceed to outline the theoretical structure of the problem of the absolute. The

---

[85] Simpson (note 42), p. 233.
[86] Rostand (note 27), p. 96.

reason the problem arises is obvious. Both science and philosophy seek for intelligibility in different yet complementary ways. The ontological problem of the absolute arises from the need of pressing the search for intelligibility to its consistently ultimate end. But the issue is complex because of the peculiar cognoscibility of the absolute—accessible to, yet not comprehensible by, the human mind. Accordingly, we can still distinguish two phases in the theoretical discussion of the problem itself.

Analogical use of language. The first question to face when examining philosophically the problem of the absolute is to clarify the validity as well as inherent limitation of man's language with relation to the absolute itself. The problematic character of this situation is clear. No matter how much we try, we can never speak of the absolute in terms other than human. But such terms are inherently inadequate to express the man-surpassing intelligibility of the absolute. And yet such inherent inadequacy should not stop us from speaking about the absolute itself. For otherwise we would make the absolute totally incognoscible. That is, our quest for intelligibility would end up in complete agnosticism.

To realize the necessity of a philosophical study of language when accepting the data of science, we can perhaps find no better guide than Heisenberg. In his discussions with Bohr and Pauli he reverts repeatedly to the subject. To begin with, he stresses the paradoxical experience of scientific man confronted with the intrinsic intelligibility of nature. The paradox results from two apparently conflicting perceptions. On the one hand, the necessity seems inescapable of ascribing to nature some features that are eminently typical of man as a person. On the other hand, one feels uneasy at the idea of ascribing personal features to an entity that is obviously not human. In Heisenberg's own terms:

Is it entirely meaningless to infer the existence of a "consciousness" (*Bewusstsein*) behind the orderly structures of the observable world— these structures being the "intention" (*Absicht*) of such a consciousness? Of course, this very question is an anthropomorphization of the problem. For the word "consciousness" is clearly derived from human experiences. So, properly speaking, one should not employ this concept outside the human area. However, were one to restrict the usage so much, it would be prohibited, for instance, to speak of the consciousness of an animal. Still, one has the feeling that an expression of the kind makes a certain sense. One perceives that the sense of the concept "consciousness" becomes simultaneously broader and vaguer when we try to employ it outside the human area. [87]

---

[87] Translated from Heisenberg, *Teil* (note 4), p. 290.

In the light of the position described, it is clear why Heisenberg roundly rejects the positivistic objection against the cognoscibility of the absolute. The objection seems to be justified enough. If human terms cannot be satisfactory when referred to the absolute, then why not quit speaking about the absolute at all? Heisenberg finds such a view totally nonsensical. As he put it, in the same context:

For the positivists there is then a simple solution: divide the world into what can be said clearly and that about which one should keep silent. So here one should just keep silent. But there is clearly no more non-sensical philosophy than this. For one can say almost nothing clearly. When one has rejected everything unclear, probably what remains are only tautologies totally deprived of any interest.

In summary, according to Heisenberg, the fundamental reasons that should stimulate a philosophical study of language in connection with science are two. One is experiential. Science leads the human mind to perceive a central orderliness in nature, the presence of something like a plan.

In science the central orderliness (*zentrale Ordnung*) is to be recognized in the fact that one can use such metaphors as "nature is made according to this plan." [88]

The second reason for the study of language is the necessity for clarifying the difference between science and nonscience. This difference can be reduced to a simple alternative, symbolized by Ptolemy and Newton. Non-science sees the cognition of nature as merely something subjective, the ability to predict events. Science, on the contrary, insists on the intrinsic intelligibility of nature. In Heisenberg's words:

If we adopt as a criterion of truth the ability to predict phenomena, then Ptolemaic astronomy is no worse than the later Newtonian astron-omy. But if we today compare Newton and Ptolemy, we have the im-pression that Newton, through his dynamical equations, has expressed the trajectories of the stars in a more comprehensive and truer fashion. That is, we have the impression that he, as it were, has described the intention (*Absicht*) according to which nature is constructed. [89]

The foregoing considerations suffice to outline the content of the philosophical problem raised by the analogical use of language. *Analogy* can be defined in general as similarity. The problem arises because of the fact that we must unavoidably speak of the ultimate or absolute intel-ligibility of reality in terms that are similar to those we employ to describe human intelligibility. It shall be the task of the philosopher to determine

---

[88] *Ibid.*, p. 292.
[89] *Ibid.*, p. 288.

—explicitly and critically—the objective justification for using such terms and what content of truth they can communicate to man. In particular, the philosopher must clarify how far such terms can be applied to the absolute and how far they are restricted to the human sphere as such.

Before we close the survey of this topic, still one important consideration should be added to situate the import of the analogical question properly. We have been speaking of the necessity for clarifying the semantic use of language. However, the inference should not be drawn therefrom that we have to deal here only with a theoretically abstract issue. In fact, what is at issue is something that is very concretely human; it affects nothing less, ultimately, than man as a genuinely personal being. Man, to be truly personal, must understand himself. But man, to understand himself, must necessarily realize the truth content disclosed by science because only by so doing can he become aware of his proper station within the overall structure of reality. This is therefore the reason why, for instance, Heisenberg criticizes positivism so strongly. His objection is that positivism makes man unable to perceive the great interconnectedness of reality. In his own words:

But positivism, in its contemporary interpretation, commits the error that it does not want to see the great interconnectedness (*den grossen Zusammenhang*). It consciously wants to keep the latter—perhaps I exaggerate now in my criticism—in a mist. At the very least, it does not encourage anyone to reflect on it. [90]

Clearly, if man fails to understand himself within the overall context of reality, this is much more than an intellectual inadequacy. It amounts to total humanistic failure. For man feels utterly lost in the sea of reality, just as a sailor is lost when he has no compass to guide him.

For if man is no longer allowed to speak and reflect about the great interconnections (*grossen Zusammenhänge*), then also the compass would be lost, with which we may orient ourselves. [91]

The seriousness of the humanistic stake involved should encourage man to make himself open without any prejudice to the totality of reality. In particular, Heisenberg emphasizes, man should explore the reality that is embodied in the terminology and formulations of religion. The reason religion speaks in terms so different from the scientific ones, Heisenberg suggests, apparently lies in the different angle from which religion looks at reality itself. The information it gathers can only be expressed in analogical or metaphorical language. But this information is no less important

---

[90] *Ibid.*, p. 294.
[91] *Ibid.*, p. 295.

than that provided by science. Hence, Heisenberg infers, man should simply rethink religion in the light of science, if needed, but not rule it out.

We know that in religion it is necessary to use a language of images and parables which can never present exactly what is meant.... Positivists may be right in saying that it is nowadays difficult to give a sense to such parables. But then we are confronted with the task of understanding the sense in question, for obviously it refers to a decisive part of our reality. Or, perhaps, we should express it in a new language, if it cannot be formulated in the old one any longer. [92]

Reflective understanding of nature's intelligibility. Once it is accepted that man can know the absolute, even though only in analogical terms, it is not difficult to see what constitutes the core of the ontological problem under examination. The goal is to obtain a comprehensive understanding of the intelligibility of nature by means of personal reflection. In other words, one should strive to attain a coherent doctrine of the how and the why nature makes sense—and this one should strive to do in typically philosophical fashion: explicitly, critically, systematically.

The starting point has to be the intelligibility of nature disclosed by science. That nature effectively makes sense constitutes one of the greatest achievements of the scientific endeavor. Scientists are rightfully proud about it, and they themselves frequently connect such an achievement with the religious view of reality. The connection they stress is not unjustified because science really reveals in nature an intelligible feature, which traditionally has always been thought to constitute the ontological basis of religion. In brief, nature makes sense according to science in a twofold way. First, it proves to be thoroughly permeated by regularity. Second, it presents an overall interconnectedness that gives meaning to the whole. These two aspects justify seeing nature as a manifestation of the absolute, when the term is taken in the traditional ontological-religious sense. As an illustration, we can read the statements of two leading scientist-philosophers who gave much thought to the matter.

Henri Poincaré extols the marvel of nature's regularity as embodied in scientific laws. He scores the blindness of those people who expect miracles as a manifestation of God. In his view, the world is indeed divine, but because of the immanent orderliness that science discovers in it. In his own words:

Law is one of the most recent conquests of the human mind; there still are people who live in the presence of a perpetual miracle and are not astonished at it. On the contrary, we it is who should be astonished at nature's regularity. Men demand of their gods to prove their existence

---

[92] *Ibid.*, p. 288.

by miracles; but the eternal marvel is that there are not miracles without cease. The world is divine because it is a harmony. If it were ruled by caprice, what could prove to us it was not ruled by chance? [93]

Niels Bohr, for his part, is more greatly impressed by the fact that the whole of nature is accessible to the human mind. He perceives this "interconnectedness of meaning" as something truly objective, not merely amounting to a subjective impression. Hence he, too, sees a reason for speaking of God in connection with scientifically investigated nature. In his own formulation:

When, in the end, one speaks of the intervention of God, this is certainly not to be referred to the intrinsic necessity of events as studied by science. What is in question is the interconnectedness of meaning (*Sinnzusammenhang*) which unites an event with another or with the thinking of men. This interconnectedness of meaning, too, belongs to reality, just as the intrinsic necessity studied by science. It would certainly be a much too gross simplification, were one to confine such an interconnectedness to the purely subjective aspects of reality. [94]

But, if scientists are justified in speaking of scientifically investigated nature as a manifestation of the absolute, what role should the philosopher still play in this connection? As has been intimated above, his role is twofold. Starting out from the information that science provides about the intelligibility of nature, he should study—by means of a strictly philosophical approach—how and why nature is intelligible and what inferences must be drawn from such a state of affairs.

In the first place, there is need to study philosophically the *how* of the intelligibility of nature disclosed by science. In other words, there is need to see *in what effective way nature makes sense* through its properties that science brings to light. The reason for this assertion becomes obvious if one stops just for a moment to consider the issue closely. For, if it is true that the properties of nature disclosed by science are marvelous, it is also true that their significance is not self-evident, but rather presents questions of its own. Indeed, for instance, there are good grounds for speaking of regularity in nature. But, how can this be reconciled with the equally well attested datum that nature is thoroughly permeated by chance and randomness? Regularity and chance seem to exclude each other. And yet—science reveals—they are just two faces of the same reality that we call nature. Clearly, once one starts examining topics of this kind, there is not much need for stressing any further that philosophical reflection is indispensable. For, otherwise, the sense that science discovers

---

[93] H. Poincaré, *The Value of Science,* trans. G. B. Halsted (New York: Dover, 1958), p. 13.

[94] Tranlasted from quotation in Heisenberg (note 4), p. 128.

in the world becomes easily clouded, up to the point of disappearing entirely.   Actually, many scientists—being unable or unwilling to apply the philosophical reflective approach to these topics—end up by asserting that nature makes no sense at all.   In their view, the investigation of nature leads to nothing but agnosticism and meaninglessness (see Chapter 4, Section 3d).   Thus, if one respects science and is concerned for man, one cannot avoid admitting the necessity for exploring philosophically how nature really makes sense as a result of scientific research.

In the second place, the philosopher should consider the *why* of the intelligibility of nature brought to light by science.   In other words, he should explore the *ultimate source or reason of nature's cognoscibility*. Once again, there is not much need for insisting on the legitimacy as well as necessity for a philosophical investigation in this area.   This is obvious because, when we speak of nature's cognoscibility, we imply necessarily that nature itself presents a character that makes it closely related to the human mind.   Indeed, cognition is what specifies the mind as such.   And yet nature—though cognoscible—is quite remote from being an essentially mindlike reality.   For it does certainly present mindlike features, such as orderliness and predictability.   But it also presents features that contradict them, such as unpredictability and randomness.   Thus, if nature behaves in a mindlike manner, this is more as though it were programmed to do so than as though it would be itself the source of its own cognoscibility.   The philosopher should therefore have the courage to face this issue in all its complexity—without allowing himself to be frightened by it, but also without attempting a prejudicial solution of it.   For if it is a fact—as science proves—that nature is cognoscible, there must be an explanation of it, and this explanation must be somehow accessible to the human mind.   Otherwise, man would end up with the absurd conclusion that the cognoscibility of nature is totally incognoscible.   But this, of course, would make the scientific endeavor itself an exercise in intellectual futility.   On the other hand, however, the philosopher should not too easily think that he has the answer to the question at issue by simply speaking of God as the author of nature according to the prescientific interpretation of the term.   For this is the whole crux of the present discussion.   Prescientific man saw much too easily and anthropomorphically the action of God in nature.   And, as a consequence, he inclined to entertain a far from satisfactory view of God himself as author of nature.   And yet, one simply cannot rule out as unthinkable the traditional view that saw God as the ultimate source of the meaning of the observable world.   For otherwise it would seem impossible to find any sense at all in the universe.

In brief, thoughtful man living in the scientific age should not shrink back from exploring the problem of the meaning of reality at its very source. This he should do because of his human dignity and as a service to his fellow men, who run the risk of being dehumanized by the overwhelming problems raised by science. But, above all, this he should do because of his inherent vocation as a servant of truth. Such a vocation should keep him inquiring until he has found the genuinely satisfactory answer.

5. REALITY AS SUPEREMINENT INTELLIGIBILITY: THE ONTOLOGICAL PERSPECTIVE OF SCIENTIFIC HUMANISM

We have explored some ontological insights as well as problems brought to light by science. To complete the introductory discussion of our topic we must now outline comprehensively the new ontological perspective that arises therefrom for man living in the scientific age. We shall concentrate our attention on three main points of general interest. First, we shall examine the ontological relevance of science as perceived by scientists themselves. Second, we shall discuss the humanistic traits of the new ontological perspective. Third, we shall comment briefly on the humanistic complementarity of science and ontology.

(a) The Ontological Relevance of Science

At this point of our investigation it is no longer necessary to insist that science, as such, has a great ontological relevance. All of our preceding analysis bears evidence to that. Nonetheless, it may be useful for the overall humanistic purpose of our study to stress this result. For, indeed, ultimately the split between scientists and humanists seems to be reducible to one point. Humanists insist that ontological or metaphysical questions are all-important—and they generally assume that scientists do not care for metaphysics or even oppose it in principle. In the light of the preceding we know that this is a very serious misunderstanding, which calls for explicit clarification.

Up to this point I have purposely avoided using the term metaphysics when discussing intelligibility of reality with relation to science. My reason has been historical-psychological. Unfortunately the term "metaphysics" only too frequently was interpreted rationalistically in the past and this interpretation gave rise to endless as well as fruitless polemics. Thus the very word metaphysics is nowadays likely to stir up emotional reactions that easily becloud the question at issue and make an intelligent exchange impossible. This is why I have consistently spoken of ontology and not of metaphysics. However, at this point of our investigation we have learned to avoid the semantic confusion that surrounds the term metaphysics, and thus we may employ it without any concern. Metaphysics, as taken here, is the same as ontology. Ontology we have defined as the study of the intelligibility of reality that can be attained by reflecting on the data furnished by the observational approach. But metaphysics, etymologically and factually, is just that. It is the study of the intelligibility of reality that the mind can attain by trying to penetrate, through reflection, beyond (*meta*) the information which the observational approach (*physics,* in the original sense of factual investigation of observable reality) can supply. For the sake of our humanistic purpose we must now briefly see what was and is the attitude of the genuine scientific mind toward metaphysics.

Historically speaking, there is no doubt that science first arose and thrived because its great creators were outstandingly gifted metaphysical thinkers. Indeed, science is based on the principle of the intrinsic intelligibility of nature (see Chapter 6, Section 1a, and Chapter 7, Section 4a). But such a principle is nothing if not a metaphysical tenet. In brief, therefore, the transition between nonscience and science is by no means, as is often said, the result of the rejection of metaphysics. On the contrary, the birth of science is the result of a profound metaphysical rethinking— an effort that really took the mind of man *beyond* the sense appearances or what everybody inclined to take as obvious fact. As stated by Koyré:

...the rise and growth of experimental science is not the source but, on the contrary, the result of the new *theoretical,* that is, the new *metaphysical* approach to nature that forms the content of the scientific revolution of the seventeenth century.... [95]

If science was concerned with metaphysics at the beginning, what about subsequent times? Ever since positivism arose in the middle of the past century, the popular view has been that science is positivistic and opposed to metaphysics in principle. Does this view agree with the conviction of creative scientists reflecting on their science? At the height of Comtian positivism in France, the outstanding experimentalist Bernard could not refrain from scolding Comte himself in paradoxical terms. To him it rather seemed that all knowledge, including science itself, was metaphysical. As he put it:

The error of Comte in this matter consists in believing that there is something *positive.* He believes to chase metaphysics by admitting philosophical generalities which he calls positive—nothing at all (*pas du tout*).

---

[95] A. Koyré, *Newtonian Studies* (Chicago: University of Chicago Press, 1968), p. 6. For the metaphysical relevance of science see also Koyré's book significantly entitled *Metaphysics and Measurement: Essays in Scientific Revolution* (London: Chapman & Hall, 1968). Also quite important for the same subject is E. A. Burtt, *The Metaphysical Foundations of Modern Physical Science* (New York: Doubleday Anchor Books). But Burtt commits the error of identifying mechanism with the metaphysics of classical mechanics as such.

All scientific theories are metaphysical abstractions. The very facts are but abstractions. [96]

Bernard was not alone in opposing positivism by insisting on the metaphysical aspects of science. A passage by T. H. Huxley is illuminating in this regard. With his customary brilliance and causticity, the great promoter of science ridicules those who want to do without metaphysics. He compares their situation to that of a naïve enthusiast of scientific progress to whom a microscopist suddenly shows the little animals moving in a drop of the water that he uses to slake his thirst. Such a person simply loses his peace of mind as a result of that experience. But, Huxley insists, the same reaction lies in store for the person who naïvely conceives the results of science as being plain common sense and requiring no philosophical discussion. Huxley goes on to comment:

...the unsuspecting devotee of plain common sense may look for as unexpected a shock when the magnifier of severe logic reveals the germs, if not the full-grown shapes, of lively metaphysical postulates rampant amidst his most positive and matter-of-fact notions... the fish of immortal memory, who threw himself out of the frying-pan into the fire, was not more ill-advised than the man who seeks sanctuary from philosophical persecution within the walls of the observatory or of the laboratory. [97]

Numerous contemporary scientists insist on Huxley's point. They also stress the implication that follows from it: the refusal to acknowledge the legitimacy of metaphysics leads often in practice to the adoption of a dogmatic metaphysical stand. Thus, for instance, the physicist Campbell writes pointedly:

We are all metaphysicians, physicists included. We are all interested in problems which the metaphyisician attempts to solve.... The world is not divided into those who do and those who do not hold metaphysical doctrines, but rather into those who hold them for some reason and those those who hold them for none. [98]

The same conviction is expressed by Schrödinger. In a polemical allusion to Kant, he rejects flatly any antimetaphysical stance as unworkable. He even goes so far as to intimate that the opponents of traditional metaphysics are themselves in danger of replacing past errors with some others that are even worse. As he put it:

---

[96] Translated from C. Bernard, *Philosophie: Manuscrit Inédit* (Paris: Boivin, 1937), p. 32.
[97] In Bibby (note 68), p. 72.
[98] N. R. Campbell, *Foundations of Physics: The Philosophy of Theory and Experiment* (formerly *Physics: The Elements*) (New York: Dover, 1957), p. 12.

The final conclusion of western wisdom—that all transcendence has got to go, once and for all—is not really applicable in the field of knowledge (for which it is actually intended), because we cannot do without metaphysical guidance here: when we think we can, all that is apt to happen is that we replace the grand old metaphysical errors with infinitely more naïve and petty ones. [99]

To sum up, it is obvious that science and ontology—strictly, *metaphysics*—have very much in common. The two are intrinsically relevant to each other. This state of affairs is synthesized best by Bohr when commenting on Schiller's couplet we have already mentioned (Chapter 5, Section 5d). Polemizing against the positivists—especially Philipp Frank's position—Bohr emphasizes that one should try to take all forms of knowledge into account and be very cautious not to explain away the abyss in which truth dwells. In his own words:

"Fulness alone leads to clarity, and in the abyss does truth dwell." Fulness is here not only the fulness of experience, but also the fulness of conceptions, of different kinds, with which we can speak of our problem and of phenomena.... What matters most to me in discussions of this type is that one is not allowed to try to simply explain away the abyss in which truth dwells. In no area [of knowledge] is one allowed to make things too easy for oneself. [100]

(b) Supereminent Intelligibility: The Humanizing Complementarity of Science and Metaphysics

If science and metaphysics are mutually relevant, to conclude we must still touch briefly upon two questions of humanistic significance that arise from this situation. In the first place, what new metaphysical perspective should characterize the scientific age? That is to say, what new conception of reality should man entertain as a result of science? In the second place, how should scientists and metaphysicians relate to each other so as to ensure the development of a genuine scientific humanism?

As regards the first question enunciated, it is not difficult to see that the new metaphysical conception of reality that arises from the successes of science can be briefly defined by speaking in terms of *supereminent intelligibility*. The reasons for this synthetical definition can be summarized as follows.

To begin with, it is clear that *intelligibility* or cognoscibility is a metaphysical outlook that is really novel and typical of the scientific attitude as such. To satisfy oneself that this is the case, it would be enough to

---

[99] E. Schrödinger, *My View of the World*, trans. C. Hastings (London: Cambridge University Press, 1964), p. 7.
[100] Translated from Heisenberg, *Teil* (note 4), pp. 284-286.

contrast with each other the positions of man relative to nature before and after the advent of science. Prescientific man, to be sure, was already speaking of nature as cognoscible. However, his way of knowing was so much hampered by phenomenism and anthropocentrism that he resignedly admitted his inability to penetrate nature effectively with his mind. This position, as has been seen, found its embodiment in the widely accepted tenet that the knowledge of nature could be nothing more than a "saving of the appearances" (see Chapter 5, Section 2b). But the advent of science changed the outlook entirely. For science proved that the mind of man is truly able to go beyond appearances and can understand nature in a genuine way. The understanding supplied by science is both intrinsic and universal. It is intrinsic because, through science, man discovers the reasons why things behave the way they do (see Chapter 2, Sections 1b and 2b). It is universal because, through science, man realizes that intelligibility encompasses the whole of nature (see Chapter 2, Sections 1c and 2c).

There is not much need to insist at this point on the humanistic significance of the new metaphysical outlook brought about by science. For man is characterized by the desire to know—all his being is strained after cognition. As a consequence, it is obvious that the advent of science has genuinely humanized his conception of nature or observable reality. For through science nature comes to life, as it were, by acquiring unexpected meaning. Even the most humble objects have now a profound message to communicate. For instance, the pebble that one kicks along the road speaks to the scientifically illuminated person of the hidden harmonies of the atomic world as well as of the long history of our globe. Likewise, the sparrow that flies past is now no longer just a commonplace bird, but is the carrier and living evidence of that astonishing process called evolution. In brief, because of science there is nothing in nature that is too small or too large, too common, or too hidden for man as a seeker of meaning.

But, if science humanizes nature by disclosing its inherent intelligibility, it is clear that one should strive to characterize such an intelligibility properly. This is therefore the reason why the suggestion is made here to speak of *supereminence* in this context. The term, taken literally, means the quality or state according to which a being is surpassingly elevated. But, as a result of science, we know that such is indeed the intelligibility of nature—namely, a cognoscibility that always exceeds our ability to comprehend it. For science discloses to man that he can truly understand things, and yet he can never expect to succeed in exhausting their understandability (see Chapter 3, Sections 2b and 2c). Speaking of supereminent intelligibility, therefore, may enable man living in the age of science to entertain a more balanced metaphysical outlook, as an onto-logical foundation of a more adequate humanism.

As regards the second question enunciated above, concerning the guide-lines that ought to direct the relations of scientists and metaphysicians, we are now in a position to formulate an answer of principle. If both science and metaphysics deal with the intelligibility of nature and if this intelligibility is of a complex kind owing to its supereminent character, it follows that both extremes—which consist in either separating or con-fusing the two disciplines—must be avoided. By the same token, both the autonomy and the cooperation of the two disciplines must be fostered. A few more words of comment may clarify this issue and thus bring to completion our ontological introduction to scientific humanism.

To begin with, the basic methodological requirement is clear: there must be *no separation or confusion of science and metaphysics*. The reason for insisting on this requirement is, of course, plain at this point of our discussion. Nevertheless, it is important to stress such a require-ment explicitly because of the perennial temptation that lies in wait for man in this area. In fact, no sooner has man reached understanding by either science or metaphysics than he tends to exaggerate the contributions of one discipline at the expense of the other. But, as a consequence, man runs the risk of harming himself seriously by transforming his under-standing into a source of dehumanizing dogmatism.

All searchers for understanding—scientists as well as metaphy-sicians—should take to heart the sobering lessons of history. Indeed, in early times it was metaphysics that prevailed, and thus many mistook it for science itself. We have seen the deleterious effects of such a confusion, as embodied especially in the episode we have called the Galilean tragedy (see Chapter 6, Section 4a). The phenomenon of the so-called two cultures that threatens the self-understanding of modern man finds its original explanation in the dogmatic identification of metaphysics with knowledge as a whole. But in later times it was science itself that became the source of dogmatic exaggeration. In fact many people, infatuated by the stirring achievements of the Newtonian synthesis, simply identified scientific mech-anics with mechanistic metaphysics. As a result, the humanistic damage perpetrated by the original confusion was made even deeper and longer lasting.

The dehumanizing implications of mechanistic metaphysics taken as the exclusively acceptable interpretation of reality are so obvious that we need not dwell on them here. However, as an illustration we can quote a celebrated passage in which the philosopher Burtt synthesized the new metaphysics in all its arrogant self-assurance:

Space was identified with the realm of geometry, time with the conti-nuity of number. The world that people had thought themselves living in—a world rich with color and sound, redolent with fragrance, filled with gladness, love and beauty, speaking everywhere of purposive harmony and creative ideals—was crowded now into minute corners of the brains of scattered organic beings. The really important world outside was a world hard, cold, colorless, silent, and dead; a world of quantity, a world of mathematically computable motions in mechanical regularity. The world of qualities as immediately perceived by man became just a curious and quite minor effect of that infinite machine beyond. [101]

If one takes into consideration the worldview expressed in the foregoing citation, it is obvious why many people came to dislike and even hate science. The root of the situation is to be found in the confusion between science and metaphysics. Indeed, when science is transformed into metaphysics, it can hardly avoid becoming hateful to man as such. For science, in principle, deals only with the reality that is observable on the interpersonal level. Thus, when science is mistaken for metaphysics, it is made to arrogate to itself the right to exclude from the domain of reality everything that cannot be observed according to strict interpersonal criteria. That is, science is made to deny the existence of everything most dear and meaningful to man as a person or inward being. Consequently, how could man not hate such an arrogantly dogmatic and dehumanizing science? In the light of these considerations, it is therefore clear how the second part of the guideline regulating the mutual relationships of scientists and metaphysicians should read. Not only separation and confusion must be banned, but a positive effort must be made to foster the *autonomy as well as cooperation of science and metaphysics*. In fact, only by so doing can man who lives in the scientific age expect to achieve a genuinely humanizing understanding of reality as a whole.

The humanizing complementarity of science and metaphysics is obvious as far as theory goes. However, is it implementable in practice? It is not feasible for us here to examine the question in all its complexity. But this is not really needed. For encouraging signs abound that the metaphysical-scientific synthesis is considered both desirable and attainable —at least in principle—by some of the most penetrating minds among scientists as well as philosophers. To complete our discussion, we shall include a few telling testimonies from contemporary leaders in both fields.

The general compatibility of science and metaphysics is stressed by many. For instance, the theoretical physicist Heisenberg likes to point out an ideal continuity between the efforts of modern physicists and those

---

[101] Burtt (note 95), pp. 238f. As already noted, Burtt wrongly identifies the mechanistic mentality with the ontological conviction of the great founders of classical mechanics. For a criticism of such a mechanistic interpretation of Newton's thought see, for instance W. C. Dampier, *A History of Science and Its Relations with Philosophy and Religion* (London: Cambridge University Press, 1966), pp. 172-176.

of the ancient Greek metaphysicians. Thus, in a recent talk given at CERN —the largest European center for nuclear research—he declared:

It seems to me fascinating to think that there is today a struggle in the most diverse countries of the world and with the most powerful means at the disposal of modern technology to solve together problems posed two and a half millenia ago by the Greek philosophers.... [102]

From the corresponding point of view, the philosopher Hanson used to insist on physics as being essentially a search for intelligibility. For example he wrote:

Fundamental physics is primarily a search for intelligibility—it is a philosophy of matter. Only secondarily is it a search for objects and facts (though the two endeavors are as hand and glove). [103]

Other thinkers go even farther in emphasizing the need for cooperation between science and metaphysics. Thus, for instance, the zoologist Simpson declared that it is not only legitimate but also necessary to go beyond the strictly scientific study of reality. This he did moved by the conviction that, otherwise, science would be demeaned into becoming a mere presupposite of technology. As he put it, concluding his argumentation:

...it is then legitimate to proceed logically from these [scientific] premises to conclusions regarding the nature of man, of life, or of the universe, even if these conclusions go beyond the realm of science in the strictest sense, and that is not only legitimate but also necessary if science is to have value beyond serving as a base for technology. [104]

The supreme importance of a metaphysical synthesis in the scientific age finds convinced supporters among some of the most prominent scientific researchers. Max Born is one of them. Speaking out of experience, he states that reflection on scientific research "leads unavoidably back to those eternal questions which go under the name of metaphysics...." [105] In particular, Born sees metaphysics as indispensable for science. Indeed, he notes, since science is here to stay—for good or ill—the inference is obvious: "we can try to fill it with a true philosophic spirit: the search of truth for its own sake." [106] Such a philosophizing effort is indispensable

---

[102] In W. Heisenberg, M. Born, E. Schrödinger and P. Auger, *On Modern Physics* (New York: Collier, 1962), p. 19.

[103] N. R. Hanson, *Patterns of Discovery: An Inquiry into the Conceptual Foundations of Science* (London: Cambridge University Press, 1958), p. 18.

[104] Simpson (note 42), p. 230.

[105] M. Born, *Physics in My Generation: A Selection of Papers* (New York: Pergamon, 1956), p. 93.

[106] *Ibid.*, p. 54.

because it is what makes science itself truly worthwhile for the scientist as a human being. As he stated:

Occupied by his tedious work of routine measurement and calculation, the physicist remembers that all this is done for a higher task: the foundation of a philosophy of nature. [107]

But Born, naturally, is not isolated in this philosophical conviction. In the same vein, for instance, the geneticist Waddington extols "the natural philosophical aspects of biology." These aspects, he contends, are what makes biology genuinely significant. For they constitute its "contribution toward man's attempt to understand nature and his place in the system of living things." [108]

In sum, it is now clear that the way lies open in front of the man who intends to honor his native call toward self-humanization as an investigator of the intelligibility of reality. To be sure, effort and patience will continue to be needed. But the humanizing attainment proves to be within his reach if he is determined to consistently dedicate himself to the task.

[107] *Ibid.*, p. 37.
[108] C. H. Waddington, *The Nature of Life* (London: G. Allen, 1961), p. 14.

22

CHAPTER 7

PERSONAL CORESPONSIBILITY:
THE ETHICAL PERSPECTIVE OF SCIENTIFIC HUMANISM

Up to this point we have explored two major manifestations of the human-istic significance of science. ⌐Science is knowledge.⌐ But knowledge presents a twofold polarity. ⌐It is subjective as an activity of the knower and objective as a perception of the reality known.⌐ We have tried to cover the humanistic implications of this twofold aspect of scientific knowing in the two preceding chapters. Should we say now that our introductory analysis of the humanistic significance of science is virtually completed? The view is widespread that reduces knowledge to the twofold polarity mentioned. Accordingly, many people think that science has nothing to say beyond the two fields already discussed. In particular, they deny that science may have any relevance concerning the third major branch of phi-losophy, namely ethics.

Our attitude here should be guided by our fundamental decision of respecting science in its human integrity. But on this basis we must ack-nowledge a relationship of science with ethics. In fact, unless one intends to limit knowledge to its intellectual manifestations arbitrarily, it is clear that knowledge itself also presents an ethical aspect. ⌐For knowing, experi-entially, amounts not merely to receiving a piece of information but also to being internally changed by the information received. Indeed, the knower feels moved to act one way or another following the information that has come to his notice. In other words, knowledge is a source of action.⌐ But, as such, it necessarily entails an ethical aspect. To complete our humanistic investigation, therefore, our goal in this chapter is to make explicit the ethical structure of science. To do this, we shall follow the usual guide-lines. That is, we shall begin by clarifying the ethical insights of science. Then we shall examine the central ethical problem that the success of science raises for man. Finally we shall outline the ethical perspective of scientific humanism.

1. Connection Between Science and Value:
Ethical Insights of Science

The widespread view is that science has no ethical relevance. This view stems from the opinion of the pure factuality of science. Science is interested in facts, not in values—the popular saying goes. To realize the humanistic significance of science, we must now face this issue. Can one speak of value in connection with science when the term is taken strictly in its ethical acceptation? *Value* can be defined as a goal worthy to be striven after by man with all his resources, without reservation or ulterior motive. Thus, can one say that the pursuit of science is animated by such a motivation? If we continue to take science as an activity of the whole person, we can easily see that there is room for such a possibility. Accordingly, we begin our investigation by examining the role played by value in the effective creation of science. Then we shall survey the theoretical insights and practical attitudes inspired by science with relation to value.

(a) *Knowledge as an End in Itself*: *The Ethical Foundation of Science*

What role, if any, does value play in the actual making of science? The answer can be given summarily with Simpson:

The very existence of science demands the value judgment and essential ethic that knowledge is good. The additional and still more fundamental ethic of responsibility makes scientists individually responsible for evaluating the knowledge that they acquire, for transmitting it as may be right, and for its ultimate utilization for good. [1]

In the light of the preceding quotation, the fundamental reason for connecting science with ethics is obvious. There is a psychological relationship of cause and effect between the two. For science cannot start or thrive without ethical valuations. Essentially, the prerequirement of science is that knowledge be conceived as an ethical good, namely as a goal to be pursued for itself and not for any ulterior motive. In actual fact, this is what constitutes the basis of what we have called genuine or pure science—that is, the conviction that scientific knowledge has to be pursued as an end in itself (see Chapter 4, Section 1a). But if science is inherently ethical, some important implications derive from this state of affairs which we must examine explicitly.

Given the overall humanistic interest of our investigation, the central point to be made here is not so much the ethical character of science but

---

[1] G. G. Simpson, *The Meaning of Evolution: A Study of the History of Life and of Its Significance for Man* (New Haven: Yale University Press, 1967), p. 313.

the *novelty of the ethical outlook* brought to the consciousness of man by the very existence of science. This outlook regards the acquisition of knowledge about nature as a self-justifying goal. But such an attitude is quite novel in the history of mankind. In fact, prescientific man never conceived the effort of knowing the world he lived in as an ethical endeavor to be pursued with one's total dedication. At least, he never conceived it with a conviction and engagement in any way comparable to those adopted as a matter of course by the scientific person. The reason for such an attitude of prescientific man is clear in the light of our previous analysis. Man has always wanted to know, from time immemorial. However, the lack of scientific attitude persuaded leading thinkers that the world was intrinsically unintelligible or that it was intelligible only by means of intuitionist speculations (see Chapter 5, Sections 2a and 2b). Hence the idea of knowledge of nature as an ethical value is hardly to be found in the writings of philosophers and masters of mores living in prescientific times. On the contrary, one can easily find in those writings exhortations to shun the study of nature as curiosity both vain and dangerous to the dignity of man. For involvement with matter, as opposed to pure contemplation, was believed to be harmful to the spiritual greatness of man. The advent of science reversed the situation entirely. Knowledge of nature came to be accepted so widely as an end in itself that many people do not even suspect the ethical novelty involved in the process.

But if science is inherently ethical, what about the so-called *impersonalism or value-neutrality of science*? The objection is all the more serious if one takes into account the fact that—following especially the lead of Max Weber—the attitude referred to is widely advocated as necessary for the researcher in the social and anthropological sciences. If science, then, has to be considered impersonal or value-neutral even when studying man, what room can still remain for value in science as such? The widespread view that science is nothing but an ethically irrelevant collection of information seems to reemerge more strongly than ever.

To clarify the question at issue, we must recall the special character of scientific objectivity. Science is undoubtedly objective, but not automatically or self-evidently so (see Chapter 5, Section 3). Hence the possibility arises for science of becoming subjectivistic or personalistic under the very guise of scientific objectivity. In other words, the objectivity of scientific statements cannot be taken for granted but should be tested philosophically to make sure that the subjective prejudices of the speakers have not insinuated themselves into their statements. However, as is well known, many scientists tend to assume as unobjectionable that science is the only reliable form of knowledge. It follows that scientists are inclined to take as objective and scientific views that are merely subjective and express their unexamined philosophical prejudices. If this occurs, it is of course detrimental to science. Hence every effort should be made by the genuine scientist to avoid such a pitfall. In the light of this situation, then, can one realize without difficulty the reason for the insistence on the so-called impersonalism in social research.

The social sciences are especially exposed to the danger of subjective distortion. The motive is obvious. When the object of research is man, since the researchers are themselves human beings, it is more likely than ever that results will be tainted by the hidden biases of the researchers. For, after all, it is most natural for everybody to notice what he expects to see and overlook the remainder. But, then, in the social sciences it is impossible for any investigator to start his research without some preliminary idea of man. Hence the investigator is faced with the danger of seeking a confirmation of his own prejudices in the data of his research. It was precisely to combat such a situation, which threatened to destroy the scientific character of sociology, that Max Weber started campaigning for impersonalism or value-neutrality in science. His choice of terms may have been unfortunate because of the ambiguity involved. But his goal was clear and thoroughly compatible with the ethical character of science as has been described. In fact, all he intended to achieve was to have the social sciences attain genuine scientific objectivity. As he put it, while summarizing his central contention:

What is really at issue is the intrinsically simple demand that the investigator and teacher should keep unconditionally separate the establishment of empirical facts (including the "value-oriented" conduct of the empirical individual whom he is investigating) and *his* own practical evaluations, i. e., his evaluation of these facts as satisfactory or unsatisfactory (including among these facts evaluations made by empirical persons who are the objects of investigation). [2]

The same requirement can be put in less ambiguous terms. Thus, for instance, the anthropologist Kluckhohn speaks of the need for "attached detachment." By this he means that the social scientist should identify himself in some sense with the human beings he is studying "as a participant observer." But, on the other hand, he should also try "to balance his identifications with detached objectivity." [3] In sum, we can easily see that the impersonalism of science, if properly understood, is far from opposing the recognition of science as a striving after value. Actually Weber—in whose opinion science was a genuine "vocation" for man—used to vehemently oppose any attempt to interpret science as though it would entail moral indifference. For, as he put it, stressing his own terms: "*An*

[2] M. Weber, *The Methodology of the Social Sciences,* trans. E. A. Shils and H. A. Finch (New York: Free Press, 1949), p. 11.

[3] C. Kluckhohn, "Common Humanity and Diverse Cultures," in D. Lerner, ed. *The Human Meaning of the Social Sciences* (New York: Meridian, 1959), pp. 251f.

attitude of moral indifference has no connection with *scientific* 'objectivity.' " [4]   In the last analysis, the ultimate reason for expunging biased personalism from science is exactly an ethical motivation.   This motivation consists, as has been seen, in the unshakable conviction that knowledge of nature is an end in itself.   The obvious implication of such a conviction is that science should not be vitiated by any subjectivism.

In short, science as such is permeated by ethics.   It is an encouraging sign that both scientists and humanists are now beginning more and more to acknowledge publicly this fundamental datum. [5]

(b) The Insights of Science Concerning Value

Having realized the fundamentally ethical character of science, we may profitably review the main insights about value that arise out of reflection on scientific research.   Under what typical light does the thoughtful scientist come to conceive ethical values?   The following are the principal features that define the scientific mentality in this area.

(ba) Immanency

The fundamental insight of ethical import arising from science as an experience of the whole man may be termed immanency of values.   The geneticist Waddington put it succinctly: "The [ethical] criteria . . . are immanent in nature as we find it, not superposed on it from outside." [6]

What is the reason that motivates the scientific attitude concerning the immanency of values?   It is not difficulty to find if we recall the general attitude of science toward meaning.   The scientist is convinced that reality makes sense because of immanent grounds.   Indeed, science is based on the principle of the intrinsic intelligibility of reality (Chapter 6, Section 1a).   Science also demands that the ultimate source of meaning of reality or the absolute be detectable as immanent to reality itself (Chapter 6, Section

---

[4] Weber (note 2), p. 60.   See the whole essay, "The Meaning of 'Ethical Neutrality' in Sociology and Economics," pp. 1-47.   See also "Science as a Vocation," in H. H. Kerth and C. Wright Mills, eds. and trans., *From Max Weber: Essays in Sociology* (New York: Oxford, 1958), pp. 129-156.

[5] For science in general see, for instance, E. A. Burtt "The Value Presuppositions of Science," in P. C. Obler and H. A. Estrin, eds. *The New Scientist: Essays on the Methods and Values of Modern Science* (New York, Doubleday Anchor Books, 1962), pp. 258-279.   Also C. P. Snow, "The Moral Un-Neutrality of Science," *op. cit.*, pp. 127-140.   For the social sciences in particular see, for instance, G. Myrdal, *Value in Social Theory: A Selection of Essays on Methodology* (New York: Harper & Row, 1958); also A. W. Gouldner "Anti-Minotaur: The Myth of a Value-Free Sociology," *Social Problems* **9** (1961/62), pp. 199-213.

[6] C. H. Waddington, *The Ethical Animal* (Chicago: University of Chicago Press, 1967), p. 30.

4ba). But ethical values are simply manifestations of the meaningfulness of reality as far as man is involved. Accordingly, the insight of the immanency of ethical values amounts to precisely this—the affirmation that the meaningfulness of reality has to be objective also as far as the behavior of man as an ethical agent is in question. In other terms, when the scientist rejects any form of externalist imposition in ethics he does that because he is against dogmatism and subjectivism of whatever kind—social, political, or religious. On the other hand, however, the scientist does not reject in principle the internal connection of ethics with the ultimate source of meaning in reality, namely the absolute. Rather, he appears favorably inclined in principle to admitting such a connection. In Waddington's own words, continuing the text cited:

However, even if one considers that there is some overriding supernatural being from whom our ethical standards are ultimately derived, it is surely blasphemous to suppose that the nature he has created is such as to deceive us as to his true wishes.

The considerations just touched upon have been discussed with unusual profundity by Heisenberg in a remarkable philosophical page, which we are going to analyze rapidly. [7] To begin at the beginning, the following is the way he defines the problem of values:

The question of values—this is the question of what we should do, what we should strive for, how we should behave. The question is therefore posed by man and relative to man. It is the question of the compass according to which we should orient ourselves when we seek our way through life.

Given the strict relationship between value and man, one should not be surprised, Heisenberg goes on to say, that the "compass" mentioned may have received different names in different religions and worldviews. He admits that one should not belittle the discrepancies that occur in the various formulations. Nonetheless he notes that, according to his impression, there is something common in all of these formulations. The common element consists in "the relationships of men to the central orderliness (*zentrale Ordnung*) of the world."

As to the objection that the knowledge of reality has something subjective in it, since knowing depends on the structure of our mind, Heisenberg concedes it. However, he insists, wherever there is question of how the human subject should behave, one would not be justified at all were one to speak of pure subjectivity.

---

[7] W. Heisenberg, *Der Teil und das Ganze: Gespräche im Umkreis der Atomphysik* (Munich: Piper, 1969), p. 291. The translation is mine.

But also there, where the question is posed about the subjective [ethical] realm, it is the central orderliness [of the world] which is operative. This deprives us of the right to consider the patterns of that realm as being the function of chance or arbitrariness.

A particular merit of Heisenberg's position is to account for ethical failure. Such a failure can take place in reference to both individuals and collectivities. This happens, in Heisenberg's view, because man's connection with the central orderliness is not preserved in its entirety, but is fragmented into partial systems. The partial orderlinesses (*Teilordnungen*) are objective, but inadequate as an ethical norm because of the lack of an overall harmony among themselves. Nonetheless, the objectivity of ethical value remains unimpaired because, even then, man is able to distinguish between ethical good and ethical evil. Experientially speaking, in fact, good is what agrees entirely with the central orderliness of reality, whereas evil is everything else. As a synthesis, formulated in Heisenberg's own words:

But, in the last analysis, it is always the central orderliness which asserts itself, the One (*das "Eine"*), to speak in the ancient terminology—and here we come in touch with the language of religion. When, therefore, there is question of values, the requirement seems to be that we should act in accordance with this central orderliness—this precisely to the end of avoiding the confusion that can arise through separate partial orderlinesses. The effectiveness of the One manifests itself in the fact that we perceive that which is orderly as being good, while we perceive the confused and chaotic as evil.

To sum up, the far-reaching significance of the fundamental ethical insight of science should be noted. Science demands that ethics be anchored on the very core of reality—its objectively ultimate intelligibility. In other words, the scientific attitude is to consider ethics as the consequence of the immanent meaning of reality—this in opposition to the widespread view that sees ethics as the result of the subjective initiative of man.

(bb) Responsibility

Another ethical insight of science is a strong sense of personal responsibility. The term has to be taken here according to its strict etymological meaning. In ordinary language, a person is responsible if he is under obligation to respond, that is, to give an account of his conduct. But this is precisely the conviction that surfaces in the reflective scientific mind—namely, the conviction that the objectivity of value cannot leave man unaffected in his inmost being. Generally, scientists feel that the objectivity of value entails a corresponding duty on the part of man as an agent. They experience this duty as interiorly obligating—so much so that any failure to meet it satisfactorily presents itself as an ethical failure in the precise sense of the word.

Expectably enough, the sense of responsibility generated by science can best be seen by considering the attitude of the scientist toward science itself. There is a gulf between the first bright idea that strikes the mind of the budding investigator and the full-fledged discovery, which constitutes science in its maturity. The intervening steps, which demand much activity and dedication on the part of the investigator, constitute what is called research. It is well known that researchers are normally totally engaged in their work. What is the motivation of their engagement, the absorbing character of which has become proverbial? Should one see it as a psychological compulsion—a sort of uncontrollable urge to satisfy one's curiosity? Or should one explain it as essentially the result of ulterior motives—the desire, for instance, to achieve fame or money? The evidence is that only a genuine sense of responsibility satisfactorily accounts for the dedication of researchers to science.

Psychological investigation of the scientific mind reveals, in effect, as Eiduson put it, that "the scientist is hard on himself." [8] However, this situation does not imply that the scientist operates under a sort of instinctual compulsion. The reason for denying such an interpretation is precisely the psychological finding concerning the spirit that animates the genuine scientist. Such a spirit consists in a system of ethical values. In other words, the scientist who is conscientious operates because of a motivation that deserves to be called an ethical principle or standard. In Eiduson's own explanation, continuing the text cited:

He [the scientist] has built up a judgmental, critical superego which has a built-in, clearly marked scalar system, along which attitudes and kind of performance are measured.

The sense of ethical responsibility animating science is stressed also by sociologists when they speak of the "ethos of science." Indeed, as has been seen (Chapter 4, Section 1a), the scientific community is convinced that science can be created only by adhering to a set of rules and prescriptions, which are internally binding; that is, they operate through the sense of responsibility of the individual. In other words, science is seen as arising not just from doing, but from doing as a response to ethical obligation. In Merton's own terms:

The mores of science possess a methodologic rationale but they are binding, not only because they are procedurally efficient, but because

---

[8] B. T. Eiduson, *Scientists: Their Psychological World* (New York: Basic Books, 1962), p. 189.

they are believed right and good. They are moral as well as technical prescriptions. [9]

In point of fact, as Merton notes in the same context, the scientific community is so serious about the ethical obligation of science that it punishes the transgressors against scientific responsibility. The punishment comes under form of such external sanctions as disapproval, ridicule, and scorn.

Since the scientist is animated by such a strong sense of responsibility in his own work, it should not be surprising that he adopts the same attitude in his entire behavior as a human being. Great scientists have frequently striven to be social reformers as well, at least in the sense of enlightening contemporary public opinion. Why did they not opt for the easy solution of restricting themselves to purely academic discussions rather than go all the way in their efforts to make science relevant to their fellow men? What was the source of their inner courage and consolation in their endless and often socially misunderstood strivings? There is no doubt that they were animated by a sense of mission and responsibility, in the ethical meaning of the term. Galileo and Kepler, for instance, were both thinking of serving God by ministering to their fellow men in their scientific labors. Darwin was convinced of serving truth and mankind. In general, the conviction of science as a source of ethical responsibility is so deeply ingrained in the scientific mind that even the rejection of religion and the absolute does not destroy it. On the contrary, this very rejection is often motivated by a deep sense of personal responsibility. As has been seen (Chapter 6, Section 4bd), Freud's attitude is a typical case in point. He tried to stamp out religion altogether, but he did this as a result of his sense of responsibility aimed at saving the inalienable dignity of man. The psychoanalyst Fromm synthesized the situation aptly:

Freud holds that the aim of human development is the achievement of these ideals: knowledge (reason, truth, *logos*), brotherly love, reduction of suffering, independence, and responsibility.... Freud speaks in the name of the ethical core of religion and criticizes the theistic-supernatural aspects of religion for preventing the full realization of these ethical aims. [10]

To sum up, the experience of doing science stresses a second important aspect of ethical value. Not only should value be recognized as having an immanent foundation in objective reality, but this objectivity should be acknowledged as a source of binding obligation for man. In particular, the experience of the scientific person makes it clear that the very fact of

---

[9] R. K. Merton, *Social Theory and Social Structure* (New York: Free Press, 1957), pp. 552f.

[10] E. Fromm, *Psychoanalysis and Religion* (New Haven: Yale University Press), pp. 18f.

knowing entails a moral responsibility, which man is expected to discharge if he is to live consistently with his dignity.

(bc) Activity, Effectiveness, Progressivity

Developing further the foregoing considerations, we can still speak of a third, comprehensive ethical insight of science. It consists of several interconnected aspects. But all of them derive ultimately from the same basic awareness—namely, the conviction that knowledge demands to be put into *practice*. To express synthetically this manifestation of the scientific experience, one can say with the famous neurophysiologist Ramón y Cajal: "The object of knowledge is to transform, to impart—to know but also to act."[11]

To detail somewhat such an expression of the scientific conscience, we can speak in the first place of *activity*. The term is taken here according to its literal meaning. The scientist feels that man should not adopt a passive attitude vis-à-vis nature but should rather behave with personal initiative relative to the same. What is the origin of such a conviction of the scientific mind? A twofold origin can be mentioned, one subjective and another objective. Subjectively speaking, the scientist feels inclined to adopt an active stand relative to nature because science as such is essentially an undertaking characterized by the activity of man. In fact, as we have seen at length, science is eminently creativity. That is, science can exist only because man engages himself, with the full originality and power of his personality, operating as an active partner of nature. But if this applies as far as the production of science itself is concerned, it is obvious that the same attitude should be adopted by scientific man when his general behavior as a human being is in question. Consequently, the scientist feels that it is incumbent on his dignity as an ethical being to adopt an active stance whenever he has to deal with nature. The second reason leading to the same attitude is the objective information supplied by science about nature. Science discloses that nature is far from perfect under all points of view. Accordingly, scientific man feels obligated to use his knowledge and power to correct and improve nature wherever there is the need.

An example of the active mentality fostered by science is the attitude adopted toward the genetical structure of man himself. To the nonscientist it may appear obvious that in this area, if any, the ethical norm is just that of submitting oneself to nature, with patience and resignation. The scientist generally disagrees with such a posture. The reason is his conviction that man is endowed with creative initiative and that nature is not neces-

---

[11] S. Ramòn y Cajal "Rules and Counsels for the Scientific Investigator" in E. H. Craigie and W. C. Gibson, *The World of Ramòn y Cajal with Selections from His Nonscientific Writings* (Springfield, Ill.: Charles C. Thomas, 1968), p. 192.

sarily perfect in its manifestations. As the great contemporary biologist Medawar put it, perhaps by overstating his point:

It is a profound truth... that nature does *not* know best; that genetical evolution, if we choose to look at it liverishly instead of with fatuous good humour, is a story of waste, makeshift, compromise, and blunder. [12]

Another expression of the ethical conviction engendered by science, also directed toward practice, can be called *effectiveness*. It consists in the persuasion that science should manifest its worth by bringing about some effective results, especially by transforming man and society. To be sure, some scientists—especially because of Marxist influence—tend to go to the extreme in this area. They contend that science can be ethical only if it provides the means for a technological transformation of society. [13] We should beware of such exaggerations. What is at issue here is something far more profound and illuminating. It implies nothing less than the awareness that science, to be faithful to its calling as the search for truth, must commit man to the total service of truth. In other words, if science is humanistic because it aims at knowledge as an end in itself, it must also be humanizing by making man translate consistently into practice the knowledge he has acquired. This, of course, can take place also by means of technology. The emphasis, however, should be placed elsewhere —namely, on humanization as such. After all, socially relevant technological applications of science are just one possible way to live out in practice the ethical inspiration we are considering here. This inspiration deserves to be called effectiveness in the comprehensive meaning of the term because it urges man to become effectively more human by engaging himself energetically in all ethical areas where science can be of any assistance.

Documentation of the ethical effectiveness of science is plentiful, even when we forgo entirely its technological manifestations. To begin with, scientists as a rule judge truly great those among their colleagues who are not just great minds, but who actually contribute to the betterment of other people. To summarize the evidence with the psychologist Eiduson:

The great men are not the great thinkers only.... Scientists have reservations about the men whose achievements have been built through the efforts of others. They have respect for individual talent... they are fond of the men who have used their talents more to inspire others than to further private interests. The teacher is rewarded with a kind of idealization.... [14]

---

[12] P. B. Medawar, *The Future of Man* (London: Methuen, 1960), p. 100.
[13] For an example of such a mentality see J. D. Bernal, *The Social Function of Science* (Cambridge, Mass.: MIT Press, 1967).
[14] Eiduson (note 8), p. 167.

Possibly the most striking manifestation of science as a source of ethical commitment is offered by a peculiar type of engagement that can be detected in the life of great scientists. Scientists, as a rule, make poor politicians. They have neither the time nor the inclinations nor the ability to change the structures of society. And yet, from time to time history records steps of political significance undertaken by scientists—steps that the ordinary person would simply deem foolhardy. The phenomenon in question is genuinely scientific because, as is well known, its tradition goes all the way back to Galileo. Galileo attempted to influence the most powerful sociocultural structure of his time, the Church. He failed, yet his failure did not deter other scientists from striving after similar goals, even though by all appearances they were just as unattainable. Several instances of such a behavior have taken place in recent years. We have already referred briefly to Einstein's political efforts, first to defend democracy, then to prevent new wars through the establishment of a world government (see Chapter 6, Section 4ca). Niels Bohr adopted the same course of action. He contributed to the production of the atom bomb to save free society, but then he did everything possible to have the bomb banned on a worldwide basis. He was not spared humiliation and ridicule by the politicians to whom he appealed in person, yet he remained undaunted in pursuing what he considered his duty as a scientific expert. [15] Another outstanding example of personal engagement, with political connotations, is that of Heisenberg. At the suggestion of Planck, an uncompromising anti-Nazi, he changed his original plan of migrating from Germany when the Nazis came to power. He decided to remain because he hoped, through the prestige of his science, to be able to assist the youth of his nation withstand the trials of tyranny. The risk the famous young physicist was thereby incurring was great, particularly as far as his reputation among foreign colleagues went. Many, at least for a while, were quite bitter about his decision. [16] Leaving aside here any question of merit, Heisenberg's behavior can at least be seen as another instance of that ethical attitude we have called effectiveness. Science, if properly integrated into man's own personality, urges him to engage himself wholly in the service of his fellow men. [17]

The third feature of the practical ethical insight disclosed by science

---

[15] For details see, for instance, R. Moore, *Niels Bohr: The Man, His Science and the World They Changed* (New York: Knopf, 1966).

[16] Evidence of the resentment aroused by Heisenberg's decision among fellow scientists is for instance the book by S. A. Goudsmit, *Alsos* (London: Sigma Books, 1947). A summary of this book is the essay by I. I. Rabi "The Atom Bomb Race" in his *Science: The Center of Culture* (New York: World, 1970), pp. 94-100.

[17] For details about Heisenberg's reasons for not migrating see his *Teil* (note 7), Chapters 12, 14 and 15.

can be called *progressivity*. Science stresses progressivity or developmentalness in both epistemology and ontology (see Chapter 5, Section 2c; Chapter
6, Sections 2a and 4bb). In particular, science emphasizes the inexhaustibility of reality (see Chapter 3, Section 2b). It is not surprising, therefore,
that also in the ethical realm scientific man may feel moved to adopt a
dynamical attitude. The inspiration supplied by science here is specifically
that of considering continual progress an ethical duty. The typical attitude
of the scientific mind in this area has been summarized by the famed
biologist Thorpe:

In short, the scientific conscience is just as imperious and impelling in
its sphere as is the moral conscience—if indeed the two can be separated.
That is to say, it is the *duty* of the scientist to accept no pre-ordained
limits but tirelessly to push his investigations as far as possible in every
field of experience. [18]

For the purpose of our investigation it is important to note the originality of science in making man aware of the dynamical or progressive
aspect of ethical value. Many people tend nowadays to take the idea of
progress for granted. Some even go so far as to sneer at it, as though
the term "progress" is an empty word. Waiving for the moment a comprehensive discussion of the issue, it is clear upon reflection that the
notion of progress is necessary in order to lead a full-fledged ethical life,
particularly on the social scale. In fact, if man had no idea of progress,
what would motivate his social striving? Here, then, the ethical significance of science as an inspiration to progress proves its worth. Surely, one
should not oversimplify the data. Culturally speaking, the genesis of the
idea of progress owes much to the Judeo-Christian conceptions of history
as well as morality. For precisely in such a culture man first began to
think of time as of a unidirectional development, oriented toward a goal
whose attainment depended on the cooperation of man himself. [19] Nevertheless, it is a historical fact that man never became fully aware of the
implications of progress until after science had become widely accepted
by public opinion, in the 18th century. Bury, the famous expert on this
subject, defined the situation as follows:

That idea [of progress] could not insinuate itself into the public mind
and become a living force in civilized societies until the meaning and
value of science had been generally grasped, and the results of scientific
discovery had been more or less diffused. [20]

---

[18] W. H. Thorpe, *Biology and the Nature of Man* (New York: Oxford, 1962), p. 6.
[19] For details see for instance C. Dawson, *Progress and Religion* (New York: Doubleday
Image Books, 1960).
[20] J. B. Bury, *The Idea of Progress: An Inquiry into Its Growth and Origin* (New
York: Dover, 1960), p. 113.

Why was science so decisive in making man aware of progressivity as an essential aspect of ethical value? The reason is clear. Prescientific man may well have had the religious motivation presupposed by the idea of progress. But he lacked the theoretical illumination needed to conceive of progress as a goal both feasible and dutiful. For he was not aware of the intrinsic intelligibility and inherent developmentalness of the world. Thus, on the theoretical level, prescientific man felt he had no choice but to resign himself and accept the world as it was. That is, occultism and phenomenism in epistemology entailed passivity and resignation in ethics. The situation changed totally upon the advent of science. For science proved that the observable world is intrinsically intelligible and developmental. As a consequence, man realized that he was able to contribute creatively to the continual improvement of his world. But ability entails obligation. This is therefore the way the idea of progress, as we know it today, came to assert itself—namely, as an awareness of an ideal goal as well as an ethical obligation.

To sum up, the ethical relevance of science can now be seen more precisely. It is not only that science is permeated by ethics in that the goal of science is eminently ethical—knowledge as an end in itself. Over and above that, science is ethically relevant because it has brought to the consciousness of man several aspects of ethical value that were unknown or at least much too vague in prescientific times. A deeper sense of human dignity is thereby fostered by science (see Chapter 1, Section 3b). This sense of man's own worth can be exaggerated and become unethical. We then have the arrogance that mars many of the manifestations of the scientific culture. But such unethical results are not the fault of science as such. They simply prove that scientists, after all, remain human. As for science itself, however, its eminently ethical character is clear in that, by fostering the sense of human dignity, it inclines man to be more ethically alert. To formulate comprehensively the ethical awareness fostered by science we can speak of a sense of *nondelegable responsibility,* which man experiences as affecting all his relationships with observable reality. We shall discuss the new humanistic perspective that arises therefrom for mankind living in the scientific age. However, before doing that, we must probe still deeper into the ethical implications of science at the social level.

(c) Science as a Socially Lived Ideal

The ethical insights we have surveyed can easily give the impression that science is perceived by many of its practitioners as an embodiment of the ethical ideal. The impression is correct, as has been seen briefly while discussing the motivation of genuine science (see Chapter 4, Section

1a). The great theoretical physicist Louis de Broglie synthesized the
situation suitably by writing:

Removed from all utilitarian preoccupation, solely devoted to the search
for truth, it [pure science] appears to us as one of the noblest activities
of which we are capable. By the wholly ideal nature of the goal it pursues, by the intensity and the disinterested character of the efforts that
it demands, it possesses a moral value which cannot be denied. [21]

To realize the ethical import of science in its fullness we must thus
rapidly consider this theme of science as a socially lived ideal. Let us start
out by defining the term with precision. The ideal, as is well known,
is a conception that is closely allied to that of value, but stands for something even deeper and more comprehensive. Indeed, there are many
values, but only one ideal for man as such. *Ideal,* thus, can be defined
as the all-encompassing value, the supreme goal of man as man. What
are then the reasons for speaking of science in terms of ideal? In particular, what typical features of the human ideal are presented by science?
The reasons that justify speaking of science in terms of ideal are
supplied by the psychological investigation of the scientific mind. In the
first place, science really attracts the entire personality of the scientist.
As Eiduson put it:

Any dichotomy between intellectual and emotional resources in personality
[of scientists] is suspended by the data; for the cognitive styles show
that intellectual abilities became interlocked quickly and early with
emotional and motivational characteristics. All of these, then, operate as
a unit. [22]

In the second place, science is perceived by the scientist as a criterion
of human maturity. This occurs in two main senses. The first sense is
an expectable one. The scientist thinks that he should, as a man, be
molded by the properties characteristic of the scientific researcher. In
Eiduson's words:

Personality features that tend to promote logic, rationality, control, and
intellectual strength are also interpreted as valuable for the scientist. More
than that, they are attributes of the "mature" scientist. By the same
token, scientists reject in themselves and in others such characteristics
as impulsiveness, personal expansiveness, emotional spontaneity, and intense involvement outside of work. [23]

---

[21] L. de Broglie, *Physics and Microphysics,* trans. M. Davidson (New York: Grosset &
Dunlap, 1966), p. 211.
[22] Eiduson (note 8), p. 248.
[23] *Ibid.,* p. 251.

But the second sense with which the scientist experiences science is more specifically to the point concerning his conception of ideal. For the scientist actually comes to look upon science as the embodiment of a comprehensive ethical norm. This is surprising but is proved by psychological data. The scientist considers science as having the right to dictate his entire moral conduct—up to the point that he scorns his colleagues who do not conform with such a norm. To quote Eiduson again:

Some [scientists] are ashamed of their "preadolescent colleagues"; some ridicule the "scientific freaks," the men who are "silly personally." These labels are pasted indiscriminately on the sexually promiscuous, the frequently divorced, the immodest, the crusaders—and even on those who spend what is supposed to be their leisure time working around the clock... as if the rule of the scientific game were somehow being overstepped. [24]

In brief, there can be no doubt that science—at least from the psychological point of view—justifies speaking of itself in terms of human ideal.

If science presents itself to scientific man as an embodiment of human ideal, what are the characteristic traits of such an ideal? There are several of them we can survey, beginning with that of *cosmic reverence*. Indeed, by doing science, man feels moved by a sense of awesome respect for the cosmos and its ultimate meaning. The reason, as has been seen, is the inexhaustible greatness and richness of nature's intelligibility (see Chapter 3, Section 2). To be sure, such an experience is an intellectual one, but it leads unavoidably to the adoption of a genuinely ethical attitude of reverence. As Poincaré put it:

Science keeps us in constant relation with something which is greater than ourselves; it offers us a spectacle which is constantly renewing itself and growing always more vast. Behind the great vision it affords us, it leads us to guess at something greater still; this spectacle is a joy to us, but it is a joy in which we forget ourselves and thus it is morally sound. [25]

Einstein, too, insisted on such an ethical implication of science. In his view, the order of nature that man experiences by doing science is a source of inspiration for overcoming one's own egotistic interests and ordering oneself within the harmony of the whole. In his own words:

The individual feels the futility of human desires and aims and the sublimity and marvelous order which reveal themselves both in nature and in the world of thought. Individual existence impresses him as a sort

---

[24] *Ibid.*, pp. 225f.
[25] H. Poincaré, *Mathematics and Science: Last Essays*, trans. J. W. Bolduc (New York: Dover, 1963), p. 105.

of prison and he wants to experience the universe as a single significant whole. [26]

A second feature of the human ideal experienced by man doing science can be called *active truthfulness*. In fact science, as is obvious, is nothing if not truthfulness. But this term must now be taken with all the active connotations which are typical of the scientific attitude as such. For science is essentially creativity. But such a character of science amounts implicitly—as Poincaré put it—to "an entire code of morality." [27] For, as he goes on to explain, scientific truthfulness should become a habit that, if properly integrated into man's total personality, should affect all aspects of human conduct. As a consequence, the scientist who acts consistently to his principles loves truth for its own sake in all of its manifestations. Above all, he combats all dishonesty, beginning with the instinctive tendency to acquiesce to wishful thinking and consequent self-deception. To cite Poincaré directly:

When we have acquired the habit of scientific methods, of their scrupulous exactitude, of the horror of all attempts to deflect the course of experiment; when we have become accustomed to dread as the height of ignominy the censure of having, even innocently, slightly tampered with our results; when this has become with us an indelible professional habit, a second nature: shall we not then reveal in all our actions this concern for absolute sincerity to the extent of no longer understanding what makes other people lie? And is this not the best means of acquiring the rarest, the most difficult of all sincerities, the one which consists in not deceiving oneself?

A third aspect of the human ideal experienced by the scientist may be termed *freedom as inalienable obligation*. Indeed, as Einstein remarked in a famous text, science is wholly dependent on freedom, in various manifestations of the same. One form of freedom intimately connected with science consists in liberty of expression and discussion; another form is independence from oppressive material needs. Still another form of freedom is "inward freedom." This bears clear ethical connotations because it inspires man to live up to his essential dignity. In Einstein's own words:

It is this freedom of the spirit which consists in the independence of thought from the restrictions of authoritarian and social prejudices as well as from unphilosophical routinizing and habit in general. This inward freedom is an infrequent gift of nature and a worthy objective for the individual. [28]

---

[26] A. Einstein, *Ideas and Opinions*, trans. S. Bargmann (New York: Crown, 1954), p. 38.
[27] Poincaré (note 25), p. 105.
[28] Einstein (note 26), p. 32.

A final feature of the ethical ideal that science discloses to man is a sense of *cooperative solidarity with nature and mankind as a whole.* In fact, science leads the reflective person to feel himself intimately associated with both nature and mankind. This is so because of a twofold ground. First, science makes man realize the universal interconnectedness of nature, including man himself as an observable being. Second, science shows that man can succeed in his enterprise of knowing and molding nature only if he acts in cooperation with other men. For science itself is essentially a communal rather than an individualistic undertaking (see Chapter 2, Section 3c). The ethical importance of science as a source of cooperative spirit has been rightfully stressed by Poincaré. In his own words:

It [science] is a collective task; and it cannot be otherwise. It is like a monument whose construction requires centuries and for which each one must carry a stone, and this stone in some cases requires a lifetime. It therefore imparts to us the feeling of necessary cooperation, of the solidarity of our efforts and those of our contemporaries, and even those of our forefathers and of our successors. [29]

To conclude, we have several reasons for speaking of an intrinsic connection between science and ethics. Science is ethical first of all because of its inherent motivation, which consists in pursuing knowledge as an ethical goal. Then science is ethical because it makes its reflective practitioners become more clearly aware of some basic aspects of ethical value. Finally, science is ethical because it inspires its practitioners to perceive and live the ethical ideal as a commitment of their entire personalities.

2. THE ETHICAL PROBLEM OF SCIENTIFIC MAN

We have realized the ethical originality and overall significance of science. To proceed in our effort of exploring the foundations of scientific humanism, we must now ask the critical question: Does science suffice in providing ethical guidance to man? Many people think so. William K. Clifford, for instance, wrote about a century ago: "Scientific thought is not an accompaniment or condition of human progress, but human progress itself." [30] In our present time B. F. Skinner maintains: "Science supplies its own wisdom." [31] But many other persons, as is well known,

---

[29] Poincaré (note 25), p. 106.
[30] Quoted in K. Pearson, *The Grammar of Science* (London: Dent, 1937), p. 37.
[31] B. F. Skinner, *Science and Human Behavior* (New York: Free Press, 1965), p. 6.

vehemently reject such views. They not only point to ethical weaknesses of science, but frequently go so far as to deny any ethical significance of science at all. Obviously, then, the critical question mentioned is a serious one. In fact, it touches the very core of that phenomenon called the two-culture split. Man fights against man as a result of different interpretations of the ethical significance of science. As a result, it is incumbent upon us to face this question squarely and systematically.

To avoid prejudging the issue, we shall proceed by degrees. Our aim is to assess with precision the ethical competence of science. But ethics has both a practical and a theoretical aspect. Hence we shall begin by testing the ability of science to act as an ethical guide on the practical level. Then we shall test the theoretical ability of science in the ethical realm. Only at this point shall we try to give a precise answer to our question and examine the implications that follow from it.

(a) The Practical Inadequacy of Science Concerning Ethics

The basic criterion for testing science as a possibly self-sufficient source of ethics is plain. A tree is to be judged from its fruit. Thus we must start by examining how science actually performs as an ethical guide in the everyday practice of its practitioners.

Before we begin our discussion, however, an important methodological point should be stressed. What we are interested in here is exclusively a theoretical issue. That is, we want to find out whether science suffices by itself to make man ethical. As a consequence, we must not allow other considerations to divert our attention from the main point. In particular, it is not our business here to pass a favorable or unfavorable judgment on the ethical behavior of scientists as a class. After all, even though we may find that scientists have many professional defects, what is surprising about that? Does not any class of men show its humanity also through the weaknesses of its members? Therefore we should not commit the unfairness—and logical fallacy—of expecting that scientists be ideal people as a result of the experiential fact that science inspires them with a genuine sense of the ideal. Indeed, for instance, no one condemns artists as a class because many of them fail to practice the lofty values of beauty and harmony to which they subscribe in theory. Thus no one should be too critical of the ethical weaknesses of scientists. With this proviso, however, no objection should be raised against our exploring the effective ethical conduct of scientists. For, indeed, this is the nub of our question. If science is a self-sufficient source of ethics, it must prove itself in the ethical behavior of its practitioners—the more so, the more the persons in question devote their lives to science.

The idea of knowledge as an adequate source of practical ethical behavior is a very old one. It can be traced all the way back to the Greek philosophers, especially Socrates. It goes usually under the name of *ethical intellectualism*. Its tenet is simple: Instruct a person sufficiently, and he will certainly choose ethical good. Expectably, the doctrine of ethical intellectualism came into prominence especially after science started to be socially accepted on a wide scale. Indeed, if science is genuine knowledge, and ethics is essentially a question of cognition, why not assume that science suffices to make man ethically good? As a consequence, many people considered ethical intellectualism the obvious corollary of their esteem for science. Already Kepler, writing with enthusiasm of science as a contemplation of God, found it unobjectionable to declare: "If the intellect has agreed to contemplate what God has made, it also agrees to do what God has bid." [32] Later on, when science had established itself with great success, and many had rejected the religious foundations of ethics in the name of science, ethical intellectualism thrived. A classical expression of such a mentality is contained in some statements by Karl Pearson, writing toward the end of the 19th century.

Study and knowledge alone absolve from sin; morality is impossible to the ignorant.... Reason is the only lawgiver, by whom the intellectual forces of the nineteenth century can be ordered and disciplined. The only practical method of making society as a whole approach the freethinker's ideal of morality is to educate it, to teach it to use its reason in guiding the race instincts and social impulses. [33]

In regard to the present, the intellectualistic interpretation of ethics is clearly quite common. A fundamental piece of evidence is the slogan, very operative in our society, that "education is salvation." As for the increasing criticisms of science, uttered by a part of public opinion, the reply of the majority still insists on the same intellectualistic conviction. "The solution to the problems of science is more science," the new slogan runs.

How does ethical intellectualism, as a moral doctrine for scientific man, stand the test of practice? The answer can be obtained only by examining the behavior of the scientific community. The psychologist Eiduson has collected highly revealing data in this regard. To begin with, she notes that science—as an ethical code—presents itself as ambiguous. In her own formulation:

---

[32] Quoted in M. Caspar, *Kepler,* trans. C. D. Hellman (London: Abelard-Schuman, 1959), p. 376.

[33] K. Pearson, *The Ethic of Freethought: A Selection of Essays and Lectures* (London: Fisher Unwin, 1888), pp. 124-126.

The code of science, what is acceptable behavior and what is not, and what are appropriate symbols of recognition, is peculiarly ambiguous. Scientists are not undecided about what should be done, but they are undecided about what the right reasons are for doing what they do. [34]

From Eiduson's findings it is clear that science is not even adequate to act as a practical guide for the ethical life of the scientists. For scientists normally know what the ethical requirements of their work are, but frequently they do not know why they should uphold such requirements. Much less are they able to situate the meaning of scientific research within the overall moral vocation of mankind. To quote Eiduson again, when explaining the foregoing statements:

Whether a reason is right or not has no direct connection with the content of what they [the scientists] do for the most part—the uncertainty about whether to work on the A-bomb project in peacetime may be, for some, an exception. They are concerned more with how much they should be motivated by the tangible rewards that they anticipate from their work, and if "extrascientific considerations" serve as conscious motivations, how much they will conflict with the impersonality, the freedom from bias, and the objectivity to which they are dedicated.

The actual conduct of scientists seems to prove that science, far from leading its average practitioner to ethical betterment, inclines him toward some typical failures of moral kind. In a sense, the situation can be explained on psychological grounds. The enormous concentration demanded by research can easily weaken the human personality (see Chapter 4, Section 3a). In any case, the inability of many scientists to lead a genuine ethical life because of science is a fact. One of their common deficiencies regards their family responsibilities.

For one thing, the interviews reveal that the scientists are not able to give the most significant parts of themselves to home and family. They feel that most of their individuality and personality comes out in their work, and what they as individuals can contribute uniquely is drawn out in studies and laboratories. [35]

Another widely deplored ethical deficiency of scientists is their tendency to ignore the social implications of their work. In particular, scientists are often unable or unwilling to make their knowledge relevant to their fellow men.

They do not identify with the larger intellectual community, nor are they particularly affiliated with the interests of those outside their own scien-

---

[34] Eiduson (note 8), pp. 177f.
[35] *Ibid.*, p. 203.

tific sphere. Few of them explore the continuity that exists between what they do and have been doing, and what other intellectuals do. [36]

The evidence surveyed, though limited, appears sufficient to give an answer of principle to our question. Is science to be considered adequate as a practical source of ethics? Clearly, the answer must be negative. In summary, the reason for such a conclusion is twofold. First, science is not able to make its followers ethically better in proportion to their dedication to science itself. Second, science—if allowed to become the only norm of behavior for man—tends to make him act unethically rather than ethically because of a internal psychological dynamism. It is especially the detection of such a dynamism that justifies rejecting science as the adequate practical source of ethics. Indeed, as has been seen, if science is taken as the only ethical guide, man is dehumanized rather than humanized. For human life is then made into a function of science, rather than the other way around—as truly ethical humanization must demand. The situation was aptly synthesized by Eiduson when she spoke of the tendency of science to produce a psychological conditioning in its followers:

Therefore science has at one and the same time lured the man into an intellectual field, but once there, has dictated and circumscribed a kind of intellectual functioning for him. [37]

In sum, one should not be surprised—much less, scandalized—that science be ethically limited. But, on the other hand, one should acknowledge the lesson, for the good of man and of science itself.

(b) The Theoretical Inadequacy of Science Concerning Ethics

If science does not suffice in practice as an ethical guide, does it at least suffice in theory? After all, we all know from painful experience that ethical practice is often much more difficult than theory. To pursue our investigation, therefore, we must now face the question announced. We have to see whether science can be considered the adequate source of ethical theory. To this end, let us survey several doctrines that assume this to be the case.

(ba) The Naturalistic Opinion

The most widespread doctrine embodying the conviction that ethics is ultimately to be explained in terms of science is the naturalistic view. *Naturalism* can be defined as the conviction that nature, as known by science, is the ultimate or absolute criterion of value. According to this

---

[36] *Ibid.*, p. 257.
[37] *Ibid.*, p. 259.

view, an act is ethical only and insofar as it expresses a spontaneous tendency of man's own nature—as opposed to any deliberate striving after a man-transcending goal or ideal.

Historically speaking, naturalism antedates science, being traceable to the ancient Greek philosophers. But the advent of science, especially the success of the Newtonian synthesis, greatly enhanced the naturalistic position. Why did public opinion come to identify science with naturalism? The basic reason is easy to understand. The public mind, unacquainted with the complexity and limitations of the scientific knowledge of nature, tended to idealize Newton's achievement. Nature was seen as a transparently intelligible and perfectly ordered whole. Therefore, the dynamism of nature, as known by science, was conceived—to put it in Koyré's terms—as "the embodiment of God's will and reason." [38] On this basis ethical naturalism arose most spontaneously. What could man, in fact, do better than conform to nature and its laws? To continue with Koyré:

To follow nature and to accept as highest norm the law of nature was just the same as to conform oneself to the will, and the law, of God.

To go into some detail, what is the conception of man as an ethical being that is typical of naturalism? A twofold answer can be given according to the influence exerted, respectively, by the mentality of classical mechanics and that of biological evolution. To begin with the former, in the heyday of classical mechanics ethical man tended to be conceived in starkly mechanistic terms. It was assumed that to understand him the mechanical categories, which had proved so successful in the study of inanimate nature, would suffice. Thus the thesis was originated in which it is asserted that to discover the total nature of man, it is enough to explore it by means of scientific observation and experimentation. Consequently, the role of ethics began to be interpreted as consisting essentially in making man develop systematically his instinctive tendencies revealed by science. This was substantially the ethical conception of man entertained by many representatives of the 18th-century Enlightenment and also by many social philosophers of the 19th century. It was a mechanistic doctrine compounded with an optimistic and intellectualistic view of man. The optimism of the doctrine consisted in assuming that nature was the self-evident source of ethical goodness for man. That is, it was postulated that man had nothing more to do in order to become fully ethical than to develop the tendencies of his own nature and those of nature in general. The intellectualism of the doctrine consisted in supposing that man, being inherently good by nature, was but in need of instruction if he were to effectively achieve ethical perfection.

---

[38] A. Koyré, "The Significance of the Newtonian Synthesis," in his *Newtonian Studies* (Chicago: University of Chicago Press, 1968), pp. 21f. See the whole essay, pp. 3-24.

The most systematic attempt to develop a mechanistic doctrine of ethical man was undertaken by Auguste Comte. Since mechanical physics had been extraordinarily successful in explaining the material world by means of statics and dynamics, Comte postulated that the same approach sufficed to understand and guide ethical man. He thus constructed his doctrine—which he called sociology or social science—in terms of statics and dynamics. Social statics studied the basic structures of human society. Social dynamics studied the successive stages in the immanent development of social structures. According to Comte's conviction, the overall outcome to be expected of his doctrine was the assurance of the progress of man as such. In his view, in fact, progress was nothing but the unfolding of nature's own intrinsic order. As he put it, aphoristically; "Progress is the development of order." [39]

When the scientific theory of biological evolution was discovered, ethical naturalism as we know it today came into being. In fact, nowadays it is not so much the mechanistic conception of man as his evolutionistic interpretation that prevails in naturalistic circles. Since evolutionistic naturalism is so widely taken as a foregone conclusion of science, we must explore somewhat deeply its presuppositions and implications. Our guide here will be the psychologist Skinner, whose great merit is his frankness and logical consistentcy. What are the fundamental reasons for connecting naturalism so strictly to science? According to Skinner, they are but two, one epistemological and the other ontological. The epistemological reason is the conviction that science is the unique and exclusively reliable form of knowledge. The ontological reason is the conviction that science has demonstrated that the whole of reality, including man himself, has to be understood in deterministic terms (see Chapter 6, Section 3b). Given such presuppositions, Skinner offers a consistent account of ethics in terms of evolution. Indeed, the central category of ethics is value. But what is value, according to Skinner? In his view, value has no connection—as has been traditionally assumed—with choice and effort on the part of man. Rather, he considers it as nothing but the product of the deterministic processes that molded man during the course of his evolution. As he puts it:

The "value" which the individual appears to have chosen with respect to his own future is therefore nothing more than that condition which operated selectively in creating and perpetuating the behavior which now seems to exemplify such a choice. [40]

---

[39] Quoted and discussed in R. Aron, *Main Currents in Sociological Thought,* trans. R. Howard and H. Weaver, vol. I (New York: Basic Books, 1965), p. 86.
[40] Skinner (note 31), p. 433.

In particular, if one accepts Skinner's viewpoint, what should be considered the most important value for man? Skinner replies that, properly speaking, there is only one value possible—*survival*. He defends this point by claiming that it is but the logical consequence of the premises stated above. The premises amount to the assumption that man is totally to be explained in terms of the intrinsic and irresistible dynamisms of nature that account for his evolution. But the highest feat of the theory of evolution is precisely that of showing how species survive by adapting themselves to new environmental conditions. As a consequence, Skinner conceives survival as the unique value for man—that is, as the only goal toward which man can reasonably strive if he takes the scientific theory of evolution into systematic account. In regard to the objection that survival is a newfangled value, unknown to ethical man for uncounted millennia, Skinner has a ready answer. Prescientific people did not think highly of survival as value because they lacked scientific illumination.

Survival arrives late among the so-called values because the effect of a culture upon human behavior, and in turn upon the perpetuation of the culture itself, can be demonstrated only when a science of human behavior has been well developed.

Against all criticisms, Skinner contends that survival must be considered *the* value for man living in the scientific age. Actually, he is so certain of his thesis that he goes to the extreme of spelling out its most unpleasant implications. In particular, he insists that science is entitled to try to suppress all traditional values in order to insure the survival of man. In his own words:

A scientific analysis may lead us to resist the more immediate blandishments of freedom, justice, knowledge, or happiness in considering the long-run consequences of survival. [41]

In the end, we may ask, what is the overall meaning of the dynamism of reality? Can one still speak of progress according to the ethical acceptation of the term in the sense that man becomes more human by acquiring an increasing control over himself and over nature? Skinner rejects this interpretation as deceptive. In his view, it is only an appearance that man can acquire a better control of nature. Actually, he maintains, it is nature that increases its control over man by means of reinforcement mechanisms.

Man is now in much better control of the world than were his ancestors, and this suggests a progress in discovery and invention in which there

---

[41] *Ibid.*, p. 436. Skinner has made this point unmistakably clear in his recent best seller, significantly entitled *Beyond Freedom and Dignity* (New York: Knopf, 1971).

appears to be a strong element of originality. But we could express this fact just as well by saying that the environment is now in better control of man. Reinforcing contingencies shape the behavior of the individual, and novel contingencies generate novel forms of behavior. Here, as anywhere, originality is to be found. [42]

To sum up, ethical naturalism, if taken strictly, must oppose as unacceptable any traditional conception of the freedom of man. However, not all supporters of this doctrine are willing to pay the entire humanistic price involved. Some still prefer to speak of freedom, but according to a radically new interpretation. An example of such a reinterpretation is given by the crystallographer and social philosopher Bernal. As he put it:

In an integrated and conscious society this conception of freedom is bound to be replaced by another—freedom as the *understanding of necessity*. Each man will be free insofar as he realizes that he is taking a conscious and determinate part in the common enterprise. [43]

Scientific critique. Ethical naturalism is widely accepted as scientific on both intrinsic and extrinsic grounds. Intrinsically, because it is regarded as the necessary consequence of the scientific understanding of nature. Extrinsically, because many of its defenders are renowed scientists. The question is now to ascertain whether such a scientific status is objectively grounded or is just conveniently postulated by the supporters of the doctrine in question. In view of our humanistic goal, we must deal briefly with this critical issue. Does naturalism follow of logical necessity from the scientific view of nature? In particular, is naturalism accepted as satisfactory by the vast majority of scientists who have reflected deeply on ethics?

To begin at the beginning, there is no consensus at all among scientists about the scientific status of naturalism. Serious objections, of both theoretical and practical nature, are raised against its acceptance. With regard to theoretical objections, Simpson points out a basic *equivocation in the use of key ethical terms*. In particular, he focuses his criticism on the identification between survival and value. As he puts it: "Certainly it is 'good' to survive, but we fall into a semantic trap if we think that this is an *ethical* good." [44] The objection is substantial. For, of course, no genuine ethics—especially a scientifically satisfactory one—can be founded on faulty logic. But the objection becomes even graver when we notice that the propounders of evolutionistic naturalism seem to take for granted that evolution, and adaptation-survival, are the only appropriate ways through which man can develop and progress in a truly human fashion.

[42] *Ibid.*, p. 255.
[43] Bernal (note 17), pp. 381f.
[44] Simpson (note 1), p. 302.

Such a view, indeed, is by no means self-evident. For, after all, evolution by itself is nothing more than an observationally established fact. But facts, as such, are completely neutral as far as ethics is concerned. Thus, properly speaking, the evolution that science discovers has nothing to do with the ethical category of value or desirable goal. Hence, as far as science is concerned, the ethical question of evolutionistic naturalism remains open. Thus one cannot propound this doctrine merely on the strength of science, but has to prove it on strictly philosophical grounds. To summarize with the famous geneticist Dobzhansky:

No theory of evolutionary ethics can be acceptable unless it gives a satisfactory explanation of just why the promotion of evolutionary development must be regarded as the *summum bonum*. [45]

Another manifestation of the weakness of ethical naturalism comes to the fore when one examines its *dehumanizing practical consequences*. Our best guide here can be a scientist who adheres fully to this doctrine but, precisely because of that, is all the more pessimistic about science. Rostand is so much in favor of the evolutionistic view as to subscribe entirely to its ethical consequences. As he put it in a summary:

The choice of the germ cells and the preparation of the somata—these are the whole province of morality.... Our duty is this: to expedite the future. The first anthropoid that stood upright served morality better than the most affectionate of the four-handed creatures. [46]

The definition of morality given is quite consistent with the general doctrine of evolutionistic naturalism. But, after having drawn this logical conclusion, Rostand himself cannot help feeling puzzled by such a strange morality where value seems to be replaced by crushingly blind might. In his own words:

Morality of the geneticist: evil is dominant, good recessive. The most horrendous of guilty humans is no less innocent than the universe. [47]

As a result, Rostand feels prompted to warn man against science. Science, in his view, is dangerous to man because—through its evolutionistic ethics—it tends to unleash his passions. Hence his warning: "Science had better not free the minds of men too much, before it has tamed their instincts." But, if the message of science is so uninspiring from the ethical point of view, would it not be better to forget it all—science and nature itself? Rostand inclines to adopt this solution:

---

[45] T. Dobzhansky, *The Biological Basis of Human Freedom* (New York: Columbia University Press, 1956), p. 128.

[46] J. Rostand, *The Substance of Man,* trans. I. Brandeis (New York: Doubleday, 1962), p. 18.

[47] *Ibid.*, p. 137.

Man has no recourse but to forget the brute immensity that ignores and crushes him—to work at becoming as "uncosmic" as the universe is inhuman. [48]

The inhuman character of evolutionistic naturalism is so evident that many scientists campaign positively against its being accepted as ethical principle. An outstanding example in this regard is constituted by the behavior of T. H. Huxley. This foremost champion of scientific evolution used to insist that a genuinely human ethics should oppose rather than second the blind tendencies of purely biological evolution. As he put it in a moving page:

The practice of that which is ethically best—what we call goodness or virtue—involves a course of conduct which, in all respects, is opposed to that which leads to success in the cosmic struggle for existence. In place of ruthless self-assertion it demands self-restraint; in place of thrusting aside, or treading down, all competitors, it requires that the individual shall not merely respect, but shall help his fellows; its influence is directed, not so much to the survival of the fittest, as to the fitting of as many as possible to survive. It repudiates the gladiatorial theory of existence.... Let us understand, once and for all, that the ethical progress of society depends, not on imitating the cosmic process, still less in running away from it, but in combating it. [49]

Of course, one could perhaps object that Huxley's ethical stand is weakened by an excessively brutal interpretation of biological evolution. Be it as it may, such a stand deserves serious consideration because of the scientific eminence of its defender. Nor can it be said that subsequent history has proved Huxley's position to be substantially wrong. Rather, his sweeping condemnation of evolutionistic naturalism presents itself to us as remarkably prescient. Indeed, recent developments do point out that naturalism can be quite harmful for man as such. Hence we understand, for instance, why such an outstanding contemporary biologist as Medawar is so bitterly opposed to this doctrine. The reason, as he emphasized, is precisely the *positive harmfulness* of naturalism:

People who brandish naturalistic principles at us are usually up to mischief. Think only of what we have suffered from a belief in the existence and overriding authority of a fighting instinct; from the doctrines of racial superiority and the metaphysics of blood and soil; from the belief that warfare between men or classes of men or nations represents a fulfilment of historical laws. [50]

---

[48] *Ibid.*, p. 68.

[49] In C. Bibby, ed. *The Essence of T. H. Huxley: Selections from His Writings* (New York: St. Martin's, 1967), p. 173.

[50] Medawar (note 12), p. 103.

In summary, it is clear that ethical naturalism is neither science nor a necessary consequence of science. This is so, extrinsically, because many scientists reject it while upholding science itself. But this is also so intrinsically because naturalism effectively amounts to an arbitrary generalization; it simply extends to the realm of values the factual explanations that science discovers while studying nature. Of course, however, it is not scientific to take for granted that facts and values are exactly the same reality. In other terms, it is not scientific to assume as self-evident that science is the uniquely valid way of exploring the intelligibility of reality. To be sure, such an assumption is frequently made, but is based on a philosophical fallacy that is quite unscientific. The fallacy, as has been seen, is that of scientism (see Chapter 5, Section 5bb). But if ethical naturalism is not science, it is itself only a philosophical opinion, which stands or falls according to the philosophical arguments used to support it. As far as we have seen, the arguments of naturalism are not persuasive. The reason is that they do not meet the essential requirements of a typically human ethics. As Simpson put it pregnantly:

There are no ethics but human ethics, and a search that ignores the necessity that ethics be human, relative to man, is bound to fail. Attempts to derive an ethic from evolution as a whole, without particular reference to man, are further examples of the "nothing but" fallacy of the nature of man. [51]

(bb) The Relativistic Opinion

Another ethical doctrine which is widely entertained as a necessary consequence of science, is the cultural-relativistic one. Its perspective is quite different from that of naturalism, with its conviction of basing ethics on the immanent laws of nature. *Cultural relativism* can be defined as the systematic denial of any laws of nature. That is to say, relativism refuses to admit any absolute standards of ethical right and wrong for man. To be sure, relativism continues to speak in terms of values, but the word itself is now interpreted in a way quite incompatible with the traditional one. For values are now supposed to be just the product of the social environment in which a person happens to live. Thus, since environments vary greatly and change continually, values are declared to be completely relative. A typical expression of this mentality is the following:

The so-called eternal values ... the *good*, the *true*, and the *beautiful*, are social constructs and have no meaning apart from society. Not only that, they grow and change with society, and all attempts to fix them

---

[51] Simpson (note 1), p. 308. For the "nothing but" fallacy concerning man see Chapter 6, Section 3c of the present text.

or elevate them into eternal values are simply attempts to fix the particular forms of society, attempts always doomed to failure. [52]

What about the connection between cultural relativism and science? It is easy to detect, although it is not an original one. In fact, the origins of relativism lie hidden in the mists that surround the very origin of philosophy. However, the advent of the social and anthropological sciences contributed greatly to revitalize the doctrine of relativism, because of two main reasons. First, the investigations of these sciences brought to light a great variety of values entertained by man in different cultures. Second, these sciences proved that—in some real sense—man's way of conceiving values is truly a function of his cultural environment and, indirectly, even of his historical as well as geographical environments.

Scientific critique. We must welcome the information that science gives to man, in the cultural as well as any other field of observation. But, of course, we must distinguish sharply between what is scientific and what merely claims to be so. Accordingly, let us discuss in some detail the so-called relativity of culture and its implications for the ethical realm. To do this, we shall follow the guidance of two leading contemporary anthropologists in a celebrated synthethical study. [53]

To begin with, there is no doubt that science has discovered a certain relativity of culture, hence also of values. But, if this is the case, one must in the first place realize what is the aim of science in bringing to light such relativity. Kroeber and Kluckhohn say explicitly that anthropology aims at an eventual unification, not fragmentation, of cultural phenomena. They compare the anthropological research into cultural relativity to the early biological work in taxonomy. In both cases there is relativity, but only as a preliminary stage toward a final unity.

In attaining the recognition of the so-called relativity of culture, we have only begun to do what students of biology have achieved. The "natural classification" of animals and plants, which underlies and supplements evolutionary development, is basically relativistic.... It is evident that the comparative study of cultures is aiming at something similar, a "natural history of culture"; and however imperfectly as yet, is beginning to attain it. [54]

Concerning the ethical consequences to be drawn from the anthropological discoveries in cultural relativity, we should distinguish with our authors between negative and positive results. Starting with the former,

---

[52] J. D. Bernal, *Science in History* (New York: Hawthorn, 1965), p. 742.
[53] A. L. Kroeber and C. Kluckhohn, *Culture: A Critical Review of Concepts and Definitions* (New York: Vintage Books), section entitled: "Values and Relativity."
[54] *Ibid.*, p. 345.

it is true that anthropological sciences have swept away many ethical conceptions that were formerly considered absolute. However, this outcome is only apparently relativistic, because it has effectively contributed to a permanent better understanding of man.

And relativism is not a negative principle except to those who feel that the whole world has lost its values when comparison makes their own private values lose their false absoluteness. Relativism may seem to turn the world fluid; but so did the concepts of evolution and of relativity in physics seem to turn the world fluid when they were new. Like them, cultural and value relativism is a potent instrument of progress in deeper understanding—and not only of the world but of man in the world. [55]

With regard to the positive results of anthropology concerning values, one thing is certain. Cultures are multiple and varied in their ethical manifestations. But beneath such a manifoldness there is a clear convergence on some basic values, which can truly be called universals:

Considering the exuberant variation of cultures in most respects, the circumstance that in some particulars almost identical values prevail throughout mankind is most arresting. No culture tolerates indiscriminate lying, stealing, or violence within the in-group. The essential universality of the incest taboo is well-known. No culture places a value upon suffering as an end in itself.... We know of no culture in either space or time... where the fact of death is not ceremonialized. [56]

Having discovered the universality of certain values, the social scientist as such must refrain from asserting the absoluteness of the same. For social science, being bound to factual observation, is inherently incompetent to establish what constitutes the essential structure of man as unobservable personality. However, as these authors point out, the factual existence of ethical universals strongly hints that the same may actually be absolutes. Indeed, how could one account otherwise for the continuity and persistence of ethical universals in most diverse environments? The inescapable answer seems to be that such universals constitute essential manifestations of the personality of man as such.

The fact that a value is a universal does not, of course, make it an absolute.... However, the mere existence of universals after so many millennia of culture history and in such diverse environments suggests that they correspond to something extremely deep in man's nature and/or are necessary conditions to social life. [57]

The upshot of the analysis undertaken by our authors is that cultural relativity is indeed a fact. But it would be completely misleading and

---

[55] *Ibid.*, p. 347.
[56] *Ibid.*, pp. 349f.
[57] *Ibid.*, p. 351.

unscientific to stop at it. For this fact points beyond itself, namely to what is common and universal in man. Hence cultural relativism as an absolute position cannot claim to rest on any scientific basis. For science starts and ends with the explicit recognition that there is such a thing as a common and permanent nature of man. In our authors' words:

...cultural differences, real and important though they are, are still so many variations on themes supplied by raw human nature and by the limits and conditions of social life. In some ways culturally altered human nature is a comparatively superficial veneer. The common understandings between men of different cultures are very broad, very general, very easily obscured by language and many other observable symbols. True universals or near universals are apparently few in number. But they seem to be as deep-going as they are rare. Relativity exists only within a universal framework. Anthropology's facts attest that the phrase "a common humanity" is in no sense meaningless. [58]

To summarize, two main reasons can be derived from the very nature of anthropological research for not accepting cultural relativism as an ethical doctrine which embodies the spirit of science. The first reason is its *postulational character*. The doctrine of relativism, in fact, simply assumes that the relativity of cultures is an absolute, but does not prove it. The anthropologist and philosopher Bidney, for instance, made this point strongly:

The fallacy of cultural relativism consists in regarding culture as if it were an autonomous, absolute, closed system which determines the relative modes of human experience. [59]

The second reason for rejecting the alleged scientific character of relativism is its *incompatibility with the science of culture* as such. Kroeber and Kluckhohn note this explicitly in the course of the discussion we have surveyed. The cultural relativism that science discovers does not justify philosophical or absolute relativism, for such an absolute relativism would clearly exclude the possibility of a true science of culture as such. In their own formulation:

Cultural relativism does not justify the conclusion that cultures are in all respects utterly disparate monads and hence strictly noncomparable entities. If this were literally true, a comparative science of culture would be *ex hypothesi* impossible. [60]

---

[58] *Ibid.*, p. 352f.

[59] D. Bidney, "The Philosophical Presuppositions of Cultural Relativism and Cultural Absolutism" in L. R. Ward, ed., *Ethics and the Social Sciences* (Notre Dame, Ind.: University of Notre Dame Press, 1959), pp. 51-76; especially p. 67. See also D. Bidney, *Theoretical Anthropology* (New York: Schocken Books, 1967), *passim*.

[60] Kroeber and Kluckhohn (note 53), p. 347.

If we take science as a whole into account, we can still add a third reason for rejecting cultural relativism as the ethical doctrine of scientific man. This reason is its *incompatibility with scientific humanism* as such. Indeed, science humanizes man in that it presents itself to him as an ideal (Chapter 4, Section 1a). This ideal consists basically in the search for knowledge as an end in itself (Chapter 7, Section 1a). In addition, science moves man to become more human by engaging himself entirely toward the achievement of continual progress (Chapter 7, Section 1bc). But, if cultural relativism were justified, all of these humanistic features of science would be reduced to nothing. In fact, if there are no absolute values, is it not illusory even to speak of ideal? In particular, what sense would it still make to strive after cognition as an end in itself? For, indeed, where there is no absolute value there is nothing that can be called an end in itself. Above all, cultural relativism—by definition—makes it impossible to speak of progress in the genuine sense of the term. In brief, the intrinsic incompatibility of cultural relativism with the authentic spirit of science could hardly be more pronounced.

To sum up, we can say that the very existence of science is evidence not only of the falsity of cultural relativism as an ethical doctrine for scientific man, but also of the necessity for overcoming such a doctrine in order to be consistent with the genuine spirit of science. In fact, the glory of science—from Galileo onward—has always been that of spurring man to overcome the cultural conditioning that society tends to impose on its members. Hence one cannot be authentically human and scientific without resolutely breaking the shackles of cultural relativism.

(bc) The Anthropocentric Opinion

Yet a third ethical doctrine is widely connected with science in the contemporary mind. Often its proponents label it "ethical humanism," obviously on the assumption that it is the most adequate doctrine for ethical man as such. To avoid semantic confusion and also to refrain for the moment from considering the question of merit, we shall rather speak of *anthropocentrism*. The reason for such a designation is the very content of the doctrine itself, which insists on the centrality of man in the ethical realm.

The diffusion of anthropocentrism in the scientific age justifies a careful analysis of it. Our guide will be Erich Fromm, in a remarkably clear and consistent presentation. [61] To begin with, Fromm claims to speak scientifically. *"Humanistic ethics is the applied science of the 'art of living'*

---

[61] E. Fromm, *Man for Himself: An Inquiry into the Psychology of Ethics* (Greenwich, Conn.: Fawcett, 1968).

based upon the theoretical 'science of man.' [62] In regard to his conception of science, he tries to be comprehensive. For he belongs to the current of psychology that calls itself humanistic, in opposition to the behavioristic one. Hence, for instance, Fromm not only does not reject, but strongly upholds the objectivity of values and ethical norms. As he summarizes his position in the Foreword:

The value judgments we make determine our actions, and upon their validity rests our mental health and happiness. To consider evaluations only as so many rationalizations of unconscious, irrational desires—although they can be that too—narrows down and distorts our picture of the total personality. Neurosis itself is, in the last analysis, a symptom of moral failure (although "adjustment" is by no means a symptom of moral achievement). In many instances a neurotic symptom is the specific expression of moral conflict, and the success of the therapeutic effort depends on the understanding and solution of the person's moral problem. [63]

The main thesis of ethical anthropocentrism, as presented by Fromm, is exactly that of the centrality of man. As he puts it in the title of the book under consideration: "man [stands] for himself." Hence one can understand why he claims to be speaking in the name of humanism as such. Actually, he urges a

... return to the great tradition of humanistic ethics which looked at man in his physico-spiritual totality, believing that man's aim is *to be himself* and that the condition for attaining this goal is that man be *for himself.* [64]

The motivation of anthropocentrism, according to Fromm, is twofold, negative and positive. Negatively speaking, he inveighs against what he calls "authoritarian ethics" (elsewhere, "authoritarian religion"). [65] He defines such an ethics as follows:

Formally, authoritarian ethics denies man's capacity to know what is good or bad; the norm giver is always an authority transcending the individual. Such a system is based not on reason and knowledge but on awe of the authority and on the subject's feeling of weakness and dependence; the surrender of decision making to the authority results from the latter's magic power; its decisions cannot and must not be questioned. *Materially,* or according to content, authoritarian ethics answers the question of what is good or bad primarily in terms of the

---

[62] *Ibid.,* p. 27.
[63] *Ibid.,* pp. vf.
[64] *Ibid.,* p. 17.
[65] Fromm discusses and rejects "authoritarian religion" in exact parallel to "authoritarian ethics." See his *Psychoanalysis* (note 10), *passim.*

interest of the authority, not the interests of the subject; it is exploitative, although the subject may derive considerable benefits, psychic or material, from it. [66]

On the strength of his rejection of authoritarian ethics, Fromm refuses to admit any self-transcendence of man in the ethical realm. [67] Likewise, he rejects what he calls "absolute ethics" and defines it as follows:

The first meaning in which "absolute" ethics is used holds that ethical propositions are unquestionably and eternally true and neither permit nor warrant revision. This concept of absolute ethics is to be found in authoritarian systems, and it follows logically from the premise that the criterion of validity is the unquestionably superior and omniscient power of the authority. [68]

The positive motivation of anthropocentrism is spelled out by Fromm when he defines what he calls "humanistic ethics" in opposition to the authoritarian one. In his own words:

Formally, it is based on the principle that only man himself can determine the criterion for virtue and sin, and not an authority transcending him. Materially, it is based on the principle that "good" is what is good for man and "evil" what is detrimental to man; the sole criterion of ethical value being man's welfare. [69]

In this connection, Fromm makes the explicit point that, to him, ethics must be anthropocentric in a literal sense.

Humanistic ethics is anthropocentric ... man, indeed, is the "measure of all things." The humanistic position is that there is nothing higher and nothing more dignified than human existence. [70]

The originality of anthropocentrism is clear from the preceding definition. Essentially, it amounts to the contention that man is the source of his own ethics.

The first "duty" of an organism is to be alive. "To be alive" is a dynamic, not a static, concept. ... The aim of man's life, therefore, is to be understood as the unfolding of his powers according to the laws of nature. ... Virtue is responsibility toward his own existence. Evil constitutes the crippling of man's powers; vice is irresponsibility toward himself. [71]

---

[66] Fromm (note 61), p. 20.
[67] Ibid., pp. 23f.
[68] Ibid., p. 239.
[69] Ibid., p. 22.
[70] Ibid., p. 23.
[71] Ibid., p. 29.

In particular, Fromm tries to vindicate the validity of anthropocentrism through his definition of conscience. Conscience, according to him, is just the intimate knowledge that places man in front of himself. But such a knowledge suffices—in Fromm's view—to adequately found ethics. For it not only stimulates man to develop himself, but also persuades him to do so as the fulfilment of an ethical obligation.

Conscience is thus a re-action of ourselves to ourselves. It is the voice of our true selves which summons us back to ourselves, to live productively, to develop fully and harmoniously—that is, *to become what we potentially are.* It is the guardian of our integrity; it is the "ability to guarantee one's self with all due pride, and also at the same time *to say yes* to one's self." If love can be defined as the affirmation of the potentialities and the care for, and the respect of, the uniqueness of the loved person, humanistic conscience can be justly called *the voice of our loving care for ourselves.* [72]

Scientific critique. In the light of Fromm's skillful presentation, it is not difficult to see why the doctrine of ethical anthropocentrism has come to be widely accepted in our times. The desire to attain a comprehensive ethical theory—both scientific and humanistic—is one of the central aspirations of man living in the scientific age. But ethical anthropocentrism presents itself as a fulfilment of this aspiration. For it insists on science while, at the same time, giving prominence to man.

We must acknowledge the noble intentions of ethical anthropocentrism. But, on the other hand, it is obvious that we should not accept it at its face value. For the interpretation of man and his science is too serious a matter to be adopted without a critical investigation. Thus, to satisfy ourselves about the validity of the doctrine under examination, let us ask two basic questions. First, how does ethical anthropocentrism harmonize with the ethical experience that pervades science? Second, how does ethical anthropocentrism effectively succeed in interpreting man as an ethical being?

To begin with the first question, it is clear from our entire research that not only is science permeated by a genuinely ethical experience, but that this experience is characterized by an objective attitude. In fact, to recall briefly some highlights, the objectivity of scientific discovery is so overpowering that the scientist perceives it as a shock of unexpectedness, an awareness of inexhaustibility, a perception of mystery (Chapter 3, Section 2). Also, the sense of attraction exerted by research on the scientist is so objectively oriented that he experiences it as a personal calling (Chapter 4, Section 1b). Above all, the objectivity of values and the corresponding responsibility on the part of man are two of the main ethical insights

---

[72] *Ibid.*, p. 163.

brought to light by science and fostered by it in the conscience of its
practitioners (Chapter 7, Sections 1b and 1c). But if this is the case,
does the attitude underlying ethical anthropocentrism actually harmonize
with that of science? The answer is plainly negative. In fact, it is obviously impossible to unify such disparate outlooks as the reality-centered-
ness of science on the one hand and the man-centeredness of anthropocentrism on the other. In particular, Fromm's definition of conscience
as "re-action of ourselves to ourselves ... the voice of our loving care
for ourselves" documents most clearly the radical incompatibility of the
two ethical attitudes. Indeed, if one recall just for a moment the effective
personal commitment of genuinely ethical scientists—let us say, Galileo
while risking everything in order to illuminate his fellow Catholics or
Bohr while being rebuffed in his attempts to have the atomic bomb
banned—it is hardly possible to defend oneself from the impression that
Fromm's definition is not meant as a caricature. In any case, it contains
such a pallid and misleading reflex of the genuine scientific experience that
we must decide against its adequacy in portraying the scientific experience
itself.

In brief, thus, we have a first reason for rejecting anthropocentrism
as the ethical doctrine of scientific man. This reason is simply that *anthropocentrism fails to take into due account the objectivity of value as experienced by doing science.*

To take up the second question enunciated above, only a few words
are needed to synthesize the interpretation of ethical man entertained by
anthropocentrism. Man is supposed to be so ethically unambiguous as to
possess in himself virtually the totality of ethical perfection. In fact, not
only is he taken to be the uniquely competent source of ethical right and
wrong, but also he is thought to achieve ethical perfection by merely striving to obtain self-fulfilment. Thus, our question now is to see how such
a view of man actually fits the data available about ethical man himself.

To obtain an adequate idea of ethical man let us follow the considerations of a prominent psychologist whose views were tried and found
valid in that "living laboratory and testing ground" which was the extermination camp system of Nazi Germany. Viktor Frankl—the founder
of the so-called Third Viennese School of Psychotherapy—minces no words
in rejecting entirely the theoretical conception of anthropocentrism. Why
does he adopt such an uncompromising position? His reasons are experiential. To begin with, he knows from bitter experience in the camps
that man, far from being ethically unambiguous, is so completely ambiguous
as to be totally unpredictable. As Frankl puts it in colorful language,
man is potentially one of two extremes—either a "saint" or a "swine." [73]

---

[73] V. E. Frankl, *Man's Search for Meaning: An Introduction to Logotherapy,* trans.
I. Lasch (New York: Washington Square Press, 1963), pp. 213f.

What about the anthropocentric tenets considering man as the source of ethical value and considering ethics as the pursuit of self-fulfilment? Frankl rejects both of them. In his conviction, in fact, "the meaning of our existence is not invented by ourselves, but rather detected."[74] As regards self-fulfilment, he warns against the danger of dealing with values "in terms of the mere self-expression of man himself." His reason is that "were [values] nothing but a mere expression of self" they would be "no more than a projection of his wishful thinking." In other words, the meaning of man's life would cease at once to be such: "... it would immediately lose its demanding and challenging character; it could no longer call man forth or summon him."[75] As a consequence, Frankl insists with all the conviction at his command on the need for self-transcendence in ethical life, as opposed to self-actualization. Actually, he sees self-transcendence as so essential that he warns about the impossibility of attaining self-actualization except by way of the self-transcendence itself. His reasons are the fundamental ones just given, namely the objectivity of the ethical meaning and the ethical ambiguity of man as such. In Frankl's own terms:

I wish to stress that the true meaning of life is to be found in the world rather than within man or his own psyche, as though it were a closed system. By the same token, the real aim of human existence cannot be found in what is called self-actualization. Human existence is essentially self-transcendence rather than self-actualization. Self-actualization is not a possible aim at all, for the simple reason that the more a man would strive for it, the more he would miss it. For only to the extent to which man commits himself to the fulfilment of his life's meaning, to this extent he also actualizes himself. In other words, self-actualization cannot be attained if it is made an end in itself, but only as a side effect of self-transcendence.[76]

Above all, Frankl rejects anthropocentrism because of its inability to help man confront successfully suffering and death. Indeed, if self-love were *the* ethical standard, people ought to be commended for going to all possible lengths in order to preserve themselves. For suffering and especially death amount in a genuine sense to the destruction of the self. But, if this attitude were adopted, it would simply dehumanize man as, precisely, the history of the extermination camps documents abundantly. Clearly, then, it is not self-actualization but self-transcendence that humanizes man as an ethical being. This is therefore the reason why Frankl lays such a weight on suffering as the decisive test of ethical theory. It consists in the

---

[74] *Ibid.*, p. 157.
[75] *Ibid.*, p. 156.
[76] *Ibid.*, p. 175.

fact that suffering offers man an opportunity to humanize himself by means of self-transcendence. To quote Frankl once again:

Whenever one is confronted with an inescapable, unavoidable situation, whenever one has to face a fate that cannot be changed ... just then is one given a last chance to actualize the highest value, to fulfil the deepest meaning, the meaning of suffering. For what matters above all is the attitude we take toward suffering, the attitude in which we take our suffering upon ourselves. [77]

In brief, then, we have a second reason for rejecting anthropocentrism as the ethical doctrine of scientific man. *Anthropocentrism* is not humanistically scientific because it *lacks the power to humanize man,* especially when he is confronted with suffering and death.

(c) The Ethical Incompetence of Science

We have examined three of the leading doctrines that are normally propounded to justify the thesis that science suffices by itself to lay the foundations of ethics. We have seen that none of them succeeds in its intent. What conclusion should we draw from this situation? The only plausible conclusion is that the thesis itself is erroneous. That is, it should be acknowledged that science is incompetent as a source of ethics. In fact, if science were competent in the ethical realm, why should the supporters of this thesis fail consistently in their attempts to prove it?

The conclusion we have reached is pivotal for a proper understanding of scientific humanism. Precisely because of its importance, however, it is advisable that we do not accept our result at its face value, but try to test it further according to the experiential criteria on which our entire investigation is based. Thus, let us inquire what the reflective scientists themselves think in this regard.

If there is anything like a consensus on the subject under investigation, this was formulated with trenchant clarity by Max Weber when he wrote:

It can never be the task of an empirical science to provide binding norms and ideals from which directives for immediate practical activity can be derived. [78]

What is the reason that motivated the great sociologist and philosopher to adopt such an uncompromising position? It appears clearly that it was his experience as a creator of science. Indeed, he knew full well that science inspires man from the ethical point of view (see Chapter 7, Section 1a). But he also realized that science—when questioned directly—proves itself unable to justify convincingly the validity of any value and ideal.

---

[77] *Ibid.,* p. 178.
[78] Weber (note 2), p. 52.

The conviction of the ethical incompetence of science has been for-mulated in various ways by reflective scientists. One characteristic ex-pression is to be found in a famous text by Poincaré, who argues strongly for a close connection between science and ethics. In particular, he rejects the idea that science may be seen as unethical; but, in the same breath, he also rejects the possibility of a "scientific morality." As for the reason of his conviction, he expresses it in a way that is typical of his mathematical mind, accustomed to logical rigor. In his own words:

There cannot be a scientific morality: but neither can there be immoral science. And the reason for this is simple; it is a—purely grammatical reason. If the premises of a syllogism are both in the indicative, the conclusion will also be in the indicative. For the conclusion to have been stated in the imperative, at least one of the premises must itself have been stated in the imperative. But scientific principles and geo-metric postulates are and can be only in the indicative. Experimental truths are again in the same mood, and at the basis of the sciences there is and there can be nothing else. [79]

The logical-grammatical formulation of Poincaré's view has attracted much criticism from logicians. A great amount of work has been done to prove it faulty or at least not convincing. These efforts may be justified as far as logic is concerned, but they are beside the point. For, if Poin-caré's formulation is coldly logical, his motivation is not. What motivates him, in fact, is exactly the experience of a person who knows both science and ethics from within. Science manifests its limitations in that it is un-able to go beyond the factual. For science, as a specific form of knowing, starts and ends with observability. But observability is limited to the factual. Thus science is inherently incompetent in the ethical realm precisely because ethics does not deal with the factual, but rather with value and ideal. This is the motive that emboldens Poincaré to feel ab-solutely confident in his position. Actually, he challenges dialecticians to prove him wrong. Thus, in fact, he goes on to explain in the text cited:

That being given, the most subtle dialectician can juggle these principles as he may wish, combine them, and pile them up on one another. All that he will derive from this will be in the indicative. He will never obtain a proposition which will state: do this, or do not do that; that is, a proposition which affirms or contradicts morality.

Another great scientist who resolutely stressed the ethical incom-petence of science was Einstein. To be sure, Einstein was far from blind to the ethical inspiration provided by science. Thus, for instance, in a famous passage he emphasizes what he calls "the heroic efforts of man,"

[79] Poincaré (note 25), p. 103.

which consist in his striving after "objective [scientific] knowledge." He states that scientific work is typical of "the highest of which man is capable." And yet, in the same context, Einstein declares unambiguously that science is completely incompetent in the ethical realm. The reason he gives is the same we have just seen—namely, the inability of science to go beyond the factual in order to attain the valuational. In his own words:

The scientific method can teach us nothing else beyond how facts are related to, and conditioned by, each other ... knowledge of what *is* does not open the door directly to what *should be.* One can have the clearest and most complete knowledge of what *is,* and yet not be able to deduct from that what should be the *goal* of human aspirations. Objective knowledge provides us with powerful instruments for the achievements of certain ends, but the ultimate goal itself and the longing to reach it must come from another source. [80]

In the consistency of his thought, Einstein is not content with merely stating that science is incompetent in the ethical realm. He goes on to explain that science is incompetent to justify the ethical significance of science itself. This he does with great earnestness. To him, in fact, science is knowledge of truth. But knowledge of truth, he contends, does not suffice to motivate ethically any undertaking of man, including the very search for truth. As he puts it, continuing the text cited:

The knowledge of truth as such is wonderful, but it is so little capable of acting as a guide that it cannot prove even the justification and the value of the aspiration toward that very knowledge of truth. Here we face, therefore, the limits of the purely rational conception of our existence.

To sum up, the overall result of our inquiry into the ethical significance of science is clear. On the experiential level, science is ethical because it consists in the striving of man after knowledge as value. Furthermore, science is ethical because, by doing science, man comes to perceive more clearly important aspects of the human ideal as such. On the reflective level, however, science proves itself totally incompetent as an ethical guide. The reason of this situation is that science, as a specific form of knowing, deals only with the factual, whereas ethics deals with the valuational. The inference to be drawn, then, is that science should be greatly respected, but by no means overrated, in regard to ethics. Einstein warns: "We should be on our guard not to overestimate science and scientific methods when it is a question of human problems." [81] Such a warning must be taken very seriously, because otherwise science itself is transformed into a dogma-

---

[80] Einstein (note 26), pp. 41f.
[81] *Ibid.*, p. 152.

tism. [82] But scientific dogmatism, as has been seen, is bound to harm man and science itself (Chapter 4, Section 3c; Chapter 6, Section 5bb). As a final consequence, therefore, it is clear that both respect for science and consistency with its spirit call for the recognition of a philosophical problem of ethics, in the strict sense of the term.

(d) The Ethical Problem of Scientific Man

We can easily detect the philosophical structure of the ethical problem confronting man in the scientific age if we begin by recalling the definition of ethics. Ethics is the reflective study of value—where value is any goal worthy to be striven after by man with all his resources, without reservation or ulterior motive. Accordingly, the philosophical problem envisaged here consists in integrating the ethical insights brought about by science into a comprehensive doctrine of human values. But three main questions can be asked concerning value. First, what are the values that man as such should strive after? Second, why should man strive after values? Third, who is ethical man to be expected in principle to strive after values? The structure of the philosophical problem under investigation consists in examining these questions by taking science into systematic account.

The first question to be faced concerns the "what." *What are the values necessary and sufficient for man* so that he may achieve fullness of humanity while living in the scientific age? In other words, what is the comprehensive set of goals scientific man should strive to realize in practice in order to become fully human? The basic task of the ethical philosopher is to enumerate and clarify such a set of goals by means of a critically systematic reflection. That is to say, the philosopher must take into account the characteristic values stressed by science—for instance, truthfulness and cooperativeness—and insert them into the balanced whole that constitutes the ethical aim of man as such. For, to be genuinely ethical, man must first of all know what goals he is supposed to strive after—and these goals must cohere harmoniously, precisely because genuinely ethical man is a harmoniously developing being.

The second question to be faced when dealing philosophically with the ethical problem of man concerns the "why." *Why is man expected to*

---

[82] A recent example of the dogmatism incurred by the scientist who fails to acknowledge the necessity of philosophy concerning the foundation of ethics is offered by J. Monod's international best seller *Le Hasard et la Nécessité: Essai sur la Philosophie Naturelle de la Biologie Moderne* (Paris: Seuil, 1970). Monod's position is clearly dogmatic. He declares that one must accept as a postulate the proposition that science alone is the source of knowledge. This is what he calls the "postulate of objectivity" (*postulat d'objectivité*), pp. 189-191. But Monod combines dogmatism with ethical intellectualism. In fact, he reduces ethics to cognition. This is what he calls "cognitional ethics" (*éthique de la connaissance*), pp. 191-193.

strive after values at all?   In other words, what are the reasons that should properly motivate man in his behavior, so that he may be ethical in the genuine sense of the term?   The second contribution of the ethical philosopher, thus, consists in helping man behave in an authentically human way —that is, by following principle and conviction instead of superficial impressions and instinctive reactions.   Here, again, one can see the importance of the role of philosophical reflection that takes into systematic account the data of science.   For science has disclosed that man is exposed to many biological, psychological, and cultural urges.   But science itself is unable to make man overcome his instinctive urges.   For science as such deals only with the factual; that is, it is competent to point out only what is there, not what ought to be.   Hence it follows that the ethical philosopher should, indeed, take very seriously into consideration the data of science in order not to misunderstand the complexity of man's attitudes.   But, at the same time, he should also go beyond the mere data of science and motivate man to become fully human by transcending the instinctive conditioning that threatens to impede his development as a person.

The third question facing the ethical philosopher in the age of science confronts him with the very core of the ethical issue.   *Who is man* so as to be taken *as an ethical being*?   That is, how should man conceive his nature as a moral agent when he takes into account all the information that science supplies to him about himself?   This comprehensive question, even more than the two foregoing ones, demonstrates the vital importance of philosophical reflection for the self-humanization of ethical man in the scientific age.   For, indeed, before the advent of science man had little doubt about what it meant to be a moral agent.   To him, in particular, it seemed clear that he knew sufficiently what ethical freedom was.   He spoke confidently of autonomous self-determination as though such a conception were a self-evident one.   But the advent of science, by disclosing the manifold dependence of man on nature, produced such a shock that the very notion of freedom seemed to be called into question.   Here, then, lies one of the most vital tasks for the ethical philosopher in the scientific age.   He has to rediscover the profound meaning of ethical *freedom* while accepting all the data that science offers about man's greatly embodied and dependent autonomy.   That is, the philosopher should assist man to understand himself as an ethical agent, and build this understanding on all the objective information that man can acquire about himself.   For, of course, genuine ethics must be grounded on all objective information available— as opposed to any wishful and sentimental thinking.

But, if ethics has to be objective, it must also be subjective.   For ethical life is, by definition, a characteristic manifestation of man as a personal subject.   Thus we can also formulate otherwise the central question to which the ethical philosopher should address himself.   The question now reads: *What constitutes the ethical dignity of man?* This question is all-important precisely because of the pivotal role played by the notion of dignity in the ethical life of man. Ordinarily, people take for granted that man has a dignity of his own. In particular, they speak of man as the natural subject of inalienable rights as a person. But the dignity of man is far from being a self-evident reality, for obviously it is not enough to exist as a human being in order to possess genuine dignity in the ethical sense of the term. Thus, if man has to do something more than merely exist in order to be dignifiedly human, what does this additional requirement consist in? In particular, what are the objective criteria according to which man can be entitled to be recognized a genuine ethical dignity? The philosopher should assist man to rediscover such criteria and motivate him to follow them conscientiously. This is especially necessary in the age of science because science is ambiguous as regards the dignity of man. On the one hand, in fact, it extols the dignity of man as a creative knower (see Chapter 1, Section 3b). On the other hand, however, it studies man only insofar as he manifests constant regularities in his behavior. Hence science can easily be used to deny and suppress the dignity of man. Consequently, to enable man to become fully ethical while being scientific, the philosopher must face with new urgency the question of man's own personal dignity and the criteria thereof.

To sum up, we can synthesize in a few words the import of the ethical problem of scientific man. The overall ethical experience impressed by science on its practitioners is one of nondelegable responsibility—vis-à-vis truth, nature, and man himself (Chapter 7, Section 1b). But reflection shows that science is inherently unable to motivate and guide man properly in the effective exercise of such a responsibility. Consequently, the ethical problem in the philosophical sense examined arises as a challenge to thoroughgoing consistency. Man is summoned to become more consciously ethical through his being scientific.

3. PERSONAL CORESPONSIBILITY AS THE IDEAL: THE ETHICAL PERSPECTIVE OF SCIENTIFIC HUMANISM

We have explored in some detail the ethical significance of science: the insights and challenges it offers to reflective man. To close our exploration we must now examine briefly the comprehensive humanistic perspective that arises from the preceding analysis. What is the general outlook that should guide man in his effort to be both scientific and ethical? Two main convictions can be considered as reliable guidelines. First, both science and ethics are necessary for the genuine humanization of man.

Second, science and ethics make it possible to speak of authentic human ideal as consisting in personal coresponsibility.

(a) The Humanizing Complementarity of Science and Ethics

What is the reason for asserting that both science and ethics are necessary for genuine humanization in the scientific age? In the light of our foregoing investigation, the answer can be given synthetically. Neither philosophical ethics nor science—if taken separately—know man well enough to guide him adequately in the humanization process. For man is both an observable object and a personal subject. But philosophical ethics by itself does not know man enough as an observable object. Conversely, science does not know him enough as a personal subject. Hence, if science and ethics do not work together, man can hardly humanize himself. In point of fact, he is in positive danger of dehumanization.

An example can clarify the all but desperate humanistic situation to which an overemphasis on the scientific attitude can drive even the most distinguished human beings. It is taken from Darwin's tortured confession in his *Autobiography,* where he discusses the ethical motivation of his total dedication to science. Darwin was sure that—as he put it—he had "acted rightly in steadily devoting my life to science." [83] And yet, in the same context, he admits his complete bafflement concerning a proper justification of his conduct. With frankness bordering on rudeness he speaks of himself in terms of undiluted instinctivism. In fact, he likens the conduct of man to that of a dog—with the only difference that the animal acts without knowing, while man does know. In his own words:

A man ... can have for his rule of life, as far as I can see, only to follow those impulses and instincts which are the strongest or which seem to him the best ones. A dog acts in this manner, but he does so blindly. A man, on the other hand, looks forwards and backwards, and compares his various feelings, desires and recollections. He then finds, in accordance with the verdict of all the wisest men that the highest satisfaction is derived from following certain impulses, namely the social instincts. [84]

To be sure, one should not be unfair toward Darwin by taking his words too literally. After all, to hear his biographers, Darwin achieved a remarkable human maturity despite the theoretical confusion in his mind concerning the significance of human life. Nevertheless, the humanistic confusion and even desperation to which science can drive its practitioners is warningly documented in the passage cited. Indeed, if the ultimate

---

[83] In N. Barlow, ed., *The Autobiography of Charles Darwin (with original omissions restored)* (New York: Norton, 1969), p. 95.
[84] *Ibid.*, p. 94.

justification of man's behavior is instinct—if man is no better than an animal—what room is still there for authentic values? Should one not quit speaking of human dignity and explain it away in terms of compulsion and conditioning? The warning emerging from Darwin's confession, therefore, is clear. Science can seriously imperil the dignity of man. For it can make man think of himself—and, consequently, act—in a less than human fashion. This it can do precisely because, when considered self-sufficient in the ethical realm, it drives thoughtful man to desperation. In Rostand's disconsolate formulation: "Man is as incapable of giving up his noble actions as of finding a satisfactory meaning for them." [85]

Needless to say, however, when the unilateral exaggeration of science is abandoned and a frank cooperation with ethics is accepted, the threat of dehumanization disappears and scientific humanization presents itself as an encouraging goal. For science comes now to be realized for what it really is—a profound, if implicit and vague, awareness of ethical value as such. In other words, science reveals that there is an aspect to reality that is even more important than the search for knowledge as such. This more important aspect is the valuational one. The point in question has been made, for instance, with great force by Eddington, the celebrated astrophycist and philosopher:

The problem of knowledge is an outer shell underneath which lies another philosophical problem—the problem of values.... A scientist should recognize in his philosophy—as he already recognizes in his propaganda—that for the ultimate justification of his activity it is necessary to look, away from the knowledge itself, to a striving in man's nature, not to be justified by science or reason, for it is itself the justification of science, of reason, of art, of conduct. [86]

As a consequence, we can confidently define the proper relationships of science and ethics by speaking in terms of *humanistic complementarity*. It is an encouraging sign of our times that—despite the widely lamented cultural split—numerous scientists as well as philosophers concur in stressing this complementarity. For instance, the renowned physicist Louis de Broglie quotes with approval the famous philosopher Henri Bergson. Their point of agreement is the conviction that, through science and its technological capabilities, mankind has acquired "an increased body [which] awaits a supplement of the soul." [87] The same conviction is expressed by other thinkers. For example, the outstanding geneticist Waddington speaks of what he calls "the most important task of ethical

---

[85] Rostand (note 46), p. 73.
[86] A. Eddington, *The Philosophy of Physical Science* (Ann Arbor: University of Michigan Press, 1958), p. 222.
[87] de Broglie (note 21), p. 258.

value." He sees it as consisting in an achievement that is even superior to the conquest of nature, which constitutes the glory of science. This superior goal is wholly humanistic: "the 'conquest of the conquest of nature'..." [88] In brief, if science and ethics cooperate, humanization of man will ensue because both have some light to shed on man's most fundamental issue: the meaning of life. With the penetration of a creative scientist who is also a profound philosopher, Heisenberg aptly defines this meaning as "the fashioning of our lives in such a way that we may integrate ourselves in the overall interconnectedness of reality." [89] Clearly, genuine humanism begins to appear a realizable as well as attractive goal for man living in the scientific age.

(b) The Ideal of Personal Coresponsibility

If science and ethics are mutually complementary, how should one define the overall humanistic perspective of ethical man illuminated by science? In particular, what are the main practical attitudes that people should adopt in order to become more genuinely ethical in the scientific age? A few rapid considerations on this comprehensive topic may serve to conclude our entire discussion.

To begin with, the basic practical attitude to be adopted in the ethical realm as a result of scientific enlightenment consists in the effort of behaving with an unflagging sense of *personal originality*. The reasons for this statement are obvious. In fact, personal originality—that is, the ability to think creatively and act effectively—constitutes the essential prerequisite of ethical life as such. For, indeed, man can behave ethically only if he is able to think and act for himself. However, if personal originality has always been indispensable to ethical life, we can confidently say that its indispensability has become more obvious as a result of science. For science renders man conscious of his own creative powers in a uniquely persuasive way and enables him to put them effectively to work on an unprecedented scale. It follows that, to be genuinely ethical in the scientific age, man must first of all act by following this fundamental inspiration of science. That is to say, negatively speaking, man should not allow himself to be carried along by any sort of external constraint or psychological conditioning—no matter how imposing or pervasive their sources may be. This consideration applies, of course, to all kinds of pressure that are external to science, such as those due to social, political, and religious ideologies. But it applies just as well to the sources of pressure that are internal to science itself—in particular, to the widespread tendency that consists in transforming science itself into still another kind

---

[88] Waddington (note 6), p. 216.
[89] Translated from *Teil* (note 7), p. 189.

of oppressive ideology, namely scientism. Positively speaking, the attitude of personal originality stressed by science entails the duty of employing all means available to the purpose of making oneself more ethically aware and consistent. In particular, this attitude invites man to investigate with his own mind—through study and personal reflection—the very realm of ethical values. He should inquire about their nature and justification. Then he should consistently put into practice the convictions he has reached in theory. In summary, the first requirement of the ethics of scientific humanism is that man should commit his whole self to the ethical endeavor. In particular, he should commit to this task what constitutes the most specific characteristic of his being—namely, the ability to think and act originally.

It is heartening to notice that the foregoing considerations—although they seem to be remote from the concrete practice of science—are effectively shared by the most thoughtful of the scientific practitioners themselves. A single example may suffice to illustrate this point. The geneticist Waddington, having given much thought to the ethical implications of science, decided to coin a special verb to express them. Thus he speaks in terms of "ethicizing." The reason for employing such an unusual verb is that of stressing the active attitude we have just analyzed. That is to say, he implies that, as a result of the information and inspiration provided by science, man should feel moved to humanize himself in an original and creative manner. Thus, in fact, Waddington defines the new verb:

The function of ethicizing is to mediate the progress of human evolution, a progress which now takes place mainly in the social and psychological sphere. [90]

But a second ethical attitude also arises from the spirit of science and serves to complement the first one. In fact, if science moves its practitioners to adopt a personal posture, science itself warns them not to go to extremes in such a frame of mind. For, indeed, if one were to insist one-sidedly on personal originality, this could easily degenerate into subjectivism and arrogant arbitrariness. But these manifestations would not only be unethical, they would also oppose the genuine spirit of science. In fact, as has been seen, if science is creative, its creativity is essentially a dependent one (see Chapter 1, Section 2). Furthermore, science is permeated by a sense of wonder and awe-inspired respect for reality (see Chapter 3, Sections 1 and 2). Also, the personal commitment to science is experienced by many scientists as a response to a call (Chapter 4, Section 1b). Finally, scientists are led by their very work to insist on ethical values as having an objective validity, immanently inscribed in reality,

---

[90] Waddington (note 6), p. 59.

so that they entail a genuine responsibility on the part of man (Chapter 7, Sections 1ba and 1bb).   Thus it is obvious that the second practical attitude fostered by science in the ethical realm consists in recognizing the objective character of ethics itself.   In short, we can say that science stresses the *objective accountability* of man's personal originality.

At first sight it may well appear that this second contribution of science to the conception of ethics bears no special significance.   For ethics has always been thought to contain an objectively normative element.   In fact, some objectivity of ethical norms and obligations is obviously indispensable if man is to avoid total subjectivity and arbitrariness in his moral behavior.   Yet the insistence of the spirit of science on the objectivity of ethics is not without considerable importance.   For man is not always able to keep a proper balance between the subjective and the objective aspects of ethics.   As is well known, in fact, even ethical philosophers incline sometimes to overemphasize the subjectivity of ethics at the expense of its objectivity.   The spirit of science encourages man to overcome such a temptation.

To be sure, the illumination provided by science with regard to the objectivity of ethics cannot be said to be apodictic.   For science—as has been seen—is inherently unable to prove apodictically anything that belongs to the realm of value.   And yet its contribution in this area is great because of the frame of mind that it originates.   In fact, the practice of science makes man more clearly aware that he, as a person, is essentially a goal-dependent being.   That is to say, science reminds man experientially that he needs to engage himself actively in order to become fully human, and this engagement has to be oriented toward objective and normative goals. In other terms, one can say that the practice of science emphasizes strongly the basic fact that the person who really intends to be genuinely human should never be self-satisfied, but should always strive to reach an objective that lies beyond himself.   This situation has been aptly described by Frankl in terms of tension—"the tension between what one has already achieved and what one still ought to accomplish or the gap between what one is and what one should become." [91]   In short, we can say that science reminds man of the need for tension in a fruitfully human life because the scientific endeavor itself leads to fruitfulness only through continual engagement and striving.

We can also express the humanizing importance of this second ethical attitude inspired by science by speaking in terms of *meaning*.   Generally people agree that man can be fully human only if he achieves genuine meaning in his life.   Nevertheless there is much controversy whether meaning ought to be interpreted in subjective or objective ways.   That

---

[91] Frankl (note 73), pp. 165f.

is to say, the opinions are split whether meaning ought to be taken in a normative and man-transcending way or merely in a suggestive and man-centered manner. Although science cannot settle the matter peremptorily, it can certainly offer valuable indications toward a solution by pointing out that the subjective and objective aspects of meaning do not exclude but rather complement each other. This science can do because, in effect, the practice of science never separates the two aspects of meaning, but rather harmoniously integrates them together. For science is indeed, as a creative enterprise, a very personal and subjective undertaking. However, it is also an undertaking that is essentially dependent on objective norms of right and wrong. The ethical lesson to be derived from the spirit of science in this regard, then, is an invitation to rediscover the central role of meaning when the term is taken in its normative and man-transcending acceptation.

A confirmation of the almost irresistible power with which the spirit of science carries its practitioners beyond a man-centered conception of meaning can be found, surprisingly enough, in Rostand's collection of philosophical aphorisms. Rostand, as we have seen repeatedly, is normally quite consistent with his scientistic philosophy. But scientism must oppose any transcendence because transcendence is the hallmark of metaphysics, and scientism is irreconcilable with metaphysics. Thus one would expect Rostand to share the widespread view that considers man himself the source of meaning. Yet Rostand is so opposed to such a view that he writes defiantly: "Certain famous words to the contrary, I for my part refuse to think that man is a sufficient future for man." [92] What can be the reason that motivates such a surprising position? The reason seems to be just one. Logical consistency with scientism leads necessarily to overall meaninglessness (see Chapter 4, Section 3d). But Rostand is not merely a logically thinking philosopher. Above all, he is a creative scientist. But the practice of science inspires man with an objective and transcendent sense of meaning. Accordingly, Rostand prefers to be inconsistent with his philosophy rather than allow the message of his science to be drowned in the meaninglessness of his philosophical speculations.

To close, we can now take up the first general question enunciated above. How should one define the overall humanistic perspective of ethical man illuminated by science? In the light of the preceding, it is not difficult to formulate an answer to this question. In fact, we have just seen that ethical life should, in the scientific age, compound the features of both personal originality and objective accountability. But, then, one can speak of the overall ethical perspective of scientific humanism as consisting in a sense of personal coresponsibility to be taken as the ideal of man as such.

---

[92] Rostand (note 46), p. 29.

It seems that it is advantageous to synthesize the ethical perspective of scientific humanism as consisting essentially in an attitude of personal *coresponsibility*. For this conception indicates at once that the new ethics is genuinely such—that is, faithful to the traditional definition of ethics—while at the same time marking a notable advance over former views. In fact, the interpretation of moral life in terms of personality and responsibility has always been typical of ethical doctrine as such. For, after all, ethical doctrine is nothing but the theory that should guide man in the discharge of his moral obligations. And yet the responsibility emphasized by the spirit of science is an improvement over past views because it enables man to realize more vividly the role he should play as a person in order to fulfil his moral obligations. Indeed, it is well known that prescientific man—despite his best efforts—was frequently unable to integrate into a harmonious theory the subjective as well as objective aspects of ethics. Thus some ethical philosophers insisted too much on personality whereas others insisted too much on responsibility. But these exaggerations were effectively harmful to ethical man because they tended to make him either too self-confident or too submissive as an agent. Man is genuinely ethical, in fact, only when he succeeds in steering clear of both such extremes. But here is exactly where the spirit of science—if properly assimilated through philosophical reflection—is capable of making man take a decisive step forward in his understanding of ethics. For science gives to man a uniquely convincing awareness of, simultaneously, his creative power as well as the inherent limitation of the same. The creativity of science, in fact, is essentially a dependent one. Thus the person who reflects on science is brought to understand more precisely the creative yet dependent role that man should play in the world in order to fulfil his ethical calling.

The interpretation of morality in terms of coresponsibility leads, finally, to the additional advantage that ethical life can be perceived more persuasively as the pursuit of the human *ideal* as such—where, the term is taken in the entire comprehensiveness of its meaning. Indeed, we have seen that science inspires its practitioners with a genuine, if vague, sense of the human ideal (Chapter 7, Section 1c). Thus it is not difficult to realize how the spirit of science may lead man to perceive more persuasively his ethical life as the pursuit of ideal. The decisive factor is precisely the new awareness of personal coresponsibility engendered by science in the minds of those who reflect upon it. In fact, theoretically speaking, people have always agreed that ethical life and pursuit of ideal amount to the same thing. In practice, however, prescientific humanism has often been unable to harmonize the two concerns properly, especially in regard to man's effective relations with the material world. Thus man frequently has been torn between two conflicting tendencies. On the one hand, he felt he had to flee the world in order to foster the so-called spiritual values.

On the other hand, he felt he had to get himself involved with the world because this was clearly a part of his humanness. The inspiration provided by science serves to reconstitute the unity in the ethical life of man by orienting all his behavior to the practice of coresponsibility as the effective ideal of man as such. Indeed, to act coresponsibly in the world is for man not only the fundamental duty but also the highest achievement—the one that makes him increasingly worthy of his unique dignity as both a creative and dependent agent.

CHAPTER 8

THE HUMANISTIC SIGNIFICANCE OF SCIENCE: A SYNTHESIS

Has science a humanistic significance of its own? In the course of this book we have attempted to find a detailed, if only preliminary, answer to this basic question. This we have done by exploring the relevance of science both for man doing science and for man in general. At this point it is clear that science deserves to be called humanistic in the genuine sense of the term. Yet the very breadth and complexity of the theme explored may well have left us wondering about a clear and comprehensive answer to our basic question. Indeed, if science is truly humanistic, to what does its humanistic message actually amount? This is therefore the final question we must examine in order to bring our investigation to a satisfactory conclusion. That is, we must synthesize what science has to say to man as such—in retrospect and prospect, theoretically and practically.

To obtain a synthetical view about science, we shall begin by situating science itself within the general striving of man after knowledge. Second, we shall summarize the reasons why science has come to be seen as a threat rather than a contribution to man's humanity. Third, we shall survey the main indications that science offers toward self-humanization. Finally, some practical considerations will be added that may enhance the concrete effort of the scientist to become a better man through his doing science.

1. The Self-Discovery of Man

Science is a form of knowledge—in particular, of knowledge about man. Accordingly, we are justified in beginning our conclusive discussion by asserting that science, ultimately, must amount to an original self-discovery of man. Nevertheless, it is clear that such an assertion may easily sound vague and even presumptuous. Therefore, to obtain a synthetical answer to our question about the meaning of science for man, obviously we must start by exploring how science, as a form of knowledge, is relevant to man as man. This will then be our concern in this first section: to situate the meaning of science within the general striving of man after knowledge. As for a thread for our discussion, we can find it by studying the continuity as well as originality of science when contrasted with nonscientific knowledge.

(a) Scientific and Ordinary Knowledge: Advance through Continuity

The relationships between scientific and ordinary knowledge are basic for a humanistic evaluation of science. Indeed, why is science widely accused of being nonhumanistic if not because its views seem to be incompatible with those of the knowledge on which the prevailing humanistic categories are founded? This apparently abstract question, then, is directly relevant to our purpose here. The point at issue is: How does science stand in relation to ordinary knowledge, and its humanistic implications?

The first result that deserves to be stressed is a genuine *continuity* between the scientific and the ordinary ways of knowing. Actually, contemporary research has disproved the position that viewed the cultural manifestations of the prescientific mind as inherently inferior and unacceptable. In particular, concerning the attitude toward nature, anthropology increasingly brings to the fore a basic compatibility between the prescientific and scientific mentalities. As an example, let it suffice to quote here the conclusion reached by the famous anthropologist Malinowski, as a result of his fieldwork among the primitive inhabitants of the Trobriand Islands. In a polemical allusion to Lévy-Bruhl's philosophical hypothesis about the so-called prelogical mentality of savages, Malinowski declares that no such mentality exists. In particular, he makes the convincing point that science itself—albeit in a very rudimentary way—should be acknowledged as present among the primitives. In his own words:

If by science be understood a body of rules and conceptions, based on experience and derived from it by logical inference, embodied in material achievements and in a fixed form of tradition and carried on by some sort of social organization—then there is no doubt that even the lowest savage communities have the beginnings of science, however rudimentary. [1]

If primitive societies have a cultural attitude that is basically compatible with that of science, the same can be asserted with even greater right of the many advanced cultures that flourished before the advent of science proper. It would be enough here to think of such astonishing feats as the founding and administration of the ancient empires. Also one could recall the prodigious technological accomplishment of prescientific man (from the Egyptian pyramids to the medieval cathedrals) as well as the mathematical and logical sophistication of the Ancients, especially the Greeks. In short, we are justified in speaking of continuity between the

---

[1] B. Malinowski, *Magic, Science and Other Essays* (New York: Doubleday Anchor Books, 1954), p. 34.

prescientific and the scientific mentalities. In other words, we can say that that the scientific attitude, in a genuine sense, is anchored in tradition, and should therefore not be conceived as a wanton rebellion against established cultural views. Not without reason, for instance, did Galileo—though full of scorn for the Aristotelians—have a great respect for Aristotle himself. Toward the end of his life he could still defiantly declare to one of those pedants: "I am quite certain that, if Aristotle should come back to the world, he would receive me among his followers." [2]

But, if there is genuine continuity between the scientific and the prescientific mentalities, this fact should not blind us to an equally true datum. In fact, science is not an obvious outgrowth of the prescientific attitude toward nature, but marks a truly novel and creative step in the culture of man. The *novelty* of science is obvious. Evidence of this situation is especially the sense of amazement with which public opinion has kept greeting the great scientific discoveries, from Galileo's time up to the present. We should not belabor this point, but it is necessary to stress that science is such a novel way of conceiving nature that it has taken a long time and much effort on the part of man to bring it about. Historical investigations as well as psychological research—especially that conducted by Jean Piaget—have pointed out why science arose so late in the history of mankind. The reason is the very complex attitude that science presupposes in man when dealing with nature. In fact, many previous cultural factors—ranging from technology to logic and religion—contributed in various ways to the historical making of science. In particular, science could start only when man had succeeded in developing a novel metaphysical conviction. This conviction, as has been seen, consisted in the certainty that, despite all appearances to the contrary, nature is intrinsically intelligible (see Chapter 6, Section 4a).

To realize a bit more the type of cultural novelty that finds expression in science, it may be useful to recall here the chief reason people tend to take things for granted. In psychological terms this reason may be designated with the highly ambiguous and yet illuminating name of *common sense*. Common sense, as the phrase is frequently used, is an ambiguous term because it is a catchall. It stands for all views and convictions that people belonging to a given culture tend to take for granted. Hence common sense is something largely relative. It changes in time and space, according to the changes and diversities of the cultures involved. Nevertheless, the very fact that common sense exerts such a powerful hold on the human mind is greatly illuminating for the purpose of our discussion. Indeed, why are people inclined to take the cultural views of a given

---

[2] Letter to Fortunio Liceti; Sept. 15, 1640. Translated from G. Galilei, *Opere*, F. Flora, ed. (Milan: Ricciardi, 1953), p. 1075.

society so much for granted? The psychological reason is clear. Human knowledge presents a strong social dependency. To put it bluntly with the anthropologist Childe (without agreeing, however, with his materialistic conclusions) one can say: "What I notice is very largely determined for me, not by nature, but by nurture." [3] Hence we can understand why common-sense views appear so obvious. Each individual is born and raised in a given society. Thus he takes as obvious what the society itself considers such. After all, society is older and, being larger, it is also presumably wiser than any of its members. The consequence is that people really come to sense—that is, to experience—things in a way that is common to the cultural society to which they belong. But, given the hold exerted by common sense on people, how is it possible for man to overcome cultural relativism as regards the cognition of nature? Here is precisely where science inserts itself with its fundamental humanistic contribution.

The cultural *advance* brought about by science consists basically in making man perceive nature with a new mind. In other words, the fundamental contribution of science to humanism is the liberation of the mind of man from the tyranny of the so-called common sense. The arduousness as well as the cultural impact of such a contribution can be easily seen if we recall just one example: the establishment of the Copernican hypothesis as the only acceptable interpretation of heavenly phenomena. It is not without justification that people speak proverbially of revolution in this context. In fact, science managed to overturn the all but universally shared conviction of the times, which seemed to be based on the evident testimony of the senses. This example appears to be paradigmatic because the characteristic of science is precisely that of making man continually revise what he tends much too easily to consider as self-evident, without realizing that it is just the fruit of his cultural prejudices. As Born remarked pointedly, while insisting on the significance of the example considered:

Galileo's opponents declared that it was a "necessity of thought" that the earth is at rest at the center of the universe.... "Necessities of thought" are often just habits of thought. [4]

To put it in rigorous terms, we can say that the cultural improvement produced by science consists basically in making man realize the *superiority of critical knowing over instinctive thinking* as far as the cognition of nature is involved. But this is also a great humanistic contribution, for two main reasons. In the first place, man is characterized by the ability to know. Hence an improvement in his knowing entails by itself an improvement in his human condition. In the second place, the

---

[3] V. G. Childe, *Society and Knowledge* (New York: Harper & Row, 1956), p. 56.
[4] M. Born, *The Restless Universe,* trans. W. M. Deans (New York: Dover, 1951), p. 163.

relationship of man with nature is fundamental for the development of a satisfactory humanism. For man depends on nature in countless ways. Thus it is important to stress that science basically humanizes man by enabling him to overcome many forms of instinctive thinking that tend to make him a prisoner of his imagination and prejudice. For one recent example of the way science enables man to overcome imagination in order to achieve better understanding, we need but refer to the results of quantum physics. For difficulties in the interpretation of this branch of science largely disappear if man avoids falling prey to his imagination. As Born, one of the great interpreters of the field, put it forcefully:

Difficulties in the interpretation of quantum physics arise solely if one transcends actual observations and insists on using a special restricted range of intuitive images and corresponding terms. Most physicists prefer to adapt their imagination to observations. [5]

A number of important humanistic corollaries can be easily derived from the preceding discussion. In the first place, we can understand better the sense of *intellectual dignity* that characterizes the scientific mind as such (Chapter 1, Section 3b). This attitude can be synthesized as follows with Wertheimer, in his investigation of the creative thinker:

An attitude is implied on his part, a willingness to face problems straight, a readiness to follow them up courageously and sincerely, a desire for improvement, in contrast with arbitrary, willful, or slavish attitudes. This, I think, is one of the great attributes that constitute the dignity of man. [6]

Another corollary concerns the *public esteem* in which scientific research—notably of the theoretical kind—should be held by people. Unfortunately, this is not the case. But then one can understand the bitter reactions of scientists to the misunderstandings of a public only interested in pragmatic results. For instance, as Campbell put it:

It is the practical man and not the student of pure science who is guilty of relying on extravagant speculation, unchecked by comparison with solid fact. [7]

A third corollary regards the interpretation of the *relationships between science and prescientific knowing of nature*. To the extent that prescientific knowing was genuinely such—that is, objective information based on critical information—there is no justification for opposing it to science. For science is just a refinement of this kind of knowing. In fact, there is a consensus of reflective scientists on this basic point. Ein-

---

[5] M. Born, *Natural Philosophy of Cause and Chance* (New York: Dover, 1964), p. 108.
[6] M. Wertheimer, *Productive Thinking* (London: Social Science Paperbacks, 1966), p. 243.
[7] N. R. Campbell, *What is Science?* (New York: Dover, 1952), p. 174.

stein, for instance, writes: "The whole of science is nothing more than a refinement of everyday thinking." [8]   The same conviction is also stated by Eddington when he says: "Its [science's] intention is to supplement not to supplant the familiar outlook." [9]

A final corollary offers a criterion to remove the ambiguity inherent in the ordinary conception of *common sense*.   Common sense has quite an important role to play, provided it is genuine.   That is to say, its statements must express what people effectively sense in common, on an objective basis—as opposed to the instinct of sensing things under the influence of the prejudices of the cultural environment.   In that case common sense is quite compatible with science.   As T. H. Huxley put it in a famous statement:

Science is, I believe, nothing but trained and organized common sense, differing from the latter only as a veteran may differ from a raw recruit.... [10]

To state it in other terms, one can even say that the basic contribution of science is that of making common sense more adequate.   As Simpson writes:

Science is our principal means of thus changing common sense and bringing it into closer (hence, too, more workable) correspondence with the objective world. [11]

(b) Scientific Quest: Man's Effort Toward Self-Understanding

Having realized the basic continuity between scientific and ordinary knowledge, we must now hasten to explore more directly the humanistic meaning of science as an original form of knowing.   Its generic humanistic meaning has already been indicated.   In fact, since science supplies information about nature, and man is closely connected with nature, it is clear that science as a source of information must have some relevance for man.   But what kind of relevance?

To find an answer to our question let us consider two examples taken from sciences that, directly, have nothing to do with man as such: geodesy and astronomy.   Historians of geodesy record a great controversy that raged during the 18th century about the shape of the earth.   The Newtonians maintained that the earth was flat at the poles as a result of gravitation.   Their French opponents, led by the astronomer Cassini, insisted that the earth was actually elongated.   Since the dispute was scientific, it was

---

[8] A. Einstein, *Ideas and Opinions,* trans. S. Bargmann (New York: Crown, 1954), p. 290.
[9] A. Eddington, *Science and the Unseen World* (London: Allen and Unwin, 1929), p. 51.
[10] In C. Bibby, ed. *The Essence of T. H. Huxley: Selections from His Writings* (New York: St. Martin's, 1967), p. 51.   See pp. 46-55.
[11] G. G. Simpson, *Biology and Man* (New York: Harcourt, 1969), p. 45.

finally decided to settle it by scientific means. Two French expeditions, led respectively by Maupertuis and La Condamine, were charged with undertaking precise geodesic measurements, one in the polar area (Lapland) and one in the equatorial area (Ecuador). The expenses and even dangers involved were great. What was the ultimate justification of it all? Just the satisfaction of curiosity or the desire of acquiring political prestige? Poincaré rightly points out that much more was at stake there. What was at stake was nothing less than testing, in one crucial case, the doctrine of the intelligibility of the universe as put forward by Newton. In his own words:

So when Maupertuis and La Condamine ... braved such diverse climates, it was not only for the sake of knowing the shape of our planet, it was a question of the system of the whole world. If the earth was flattened, Newton was victorious, and with him the doctrine of gravitation and the whole of the modern celestial mechanics. [12]

This example shows the fundamental sense in which science is relevant to man. The profound reason for speaking of science as humanistic is the ability of science itself to reveal the intrinsic intelligibility of the observable world. As a consequence, since man is characterized by intelligence, every advance of science constitutes, at least virtually, a significant advance in his humanization process.

Some considerations about astronomy serve to clarify the point just made. Frequently the relevance of astronomy is reduced by pragmatic thinkers to its utility. Astronomy, we are told, is significant for man because it supplies practical information to be used, for instance, for the purpose of navigation. In actual fact, however, such an interpretation banalizes this science to the point of nonrecognition. To convince ourselves that this is the case, let us recall briefly the influence exerted by astronomy on the cultural development of mankind. Was it not the study of planets and stars that first brought man to the realization of an intrinsic intelligibility in the observable world? Was it not the experience of cosmic orderliness that elevated the mind of man and spurred him to explore and understand the world he lived in? Above all, was not astronomy the factor that gave to man a more adequate standard to judge himself, both regarding his puniness and his greatness? Astronomy has taught humility to man because it has kept disclosing ever-increasing dimensions of the universe. But astronomy has also revealed to man his spiritual dignity because, as Pannekoek has put it strikingly, "The history of astronomy is the growth of man's concept of the world." [13] In other terms, it is true to say that through astronomy man has acquired a better perception of his

---

[12] H. Poincaré, *Science and Method*, trans. F. Maitland (New York: Dover), p. 271.
[13] A. Pannekoek, *A History of Astronomy* (New York: Interscience, 1961), p. 13.

own mental capacity. His mind must be truly powerful if it is able to free itself from the apparently ineluctable limitations of the senses and range sovereignly over the whole observable universe. Laplace expressed this reaction beautifully:

Astronomy, through the dignity of its object and the perfection of its theories, is the highest monument of the human spirit, the most noble evidence of his intelligence. [14]

In brief, it is certainly no exaggeration to say that astronomy has contributed, in a unique way, to the revelation of man to himself. Poincaré asks with a shudder what the fate of man would have been had the earthly atmosphere not been transparent to the light of the stars. To him the usefulness of astronomy is all here: in inspiring man to become better man. His poetical words are a fitting epitome of the humanistic meaning of astronomy—and even of science in general:

Astronomy is useful because it raises us above ourselves; it is useful because it is grand; that is what we should say. It shows us how small is man's body, how great his mind, since his intelligence can embrace the whole of this dazzling immensity, where his body is only an obscure point, and enjoy its silent harmony. Thus we attain the consciousness of our power, and this is something which can not cost too dear, since this consciousness makes us mightier. [15]

In the light of the preceding it is clear why science deserves to be called humanistic. The central reason is that science reveals man to himself. All the sciences contribute to this goal, either directly or indirectly. Even theories that may appear totally abstract have a great significance for the self-knowledge of man. Compton, for instance, notes that Einstein is great because of his relativity theory. But his greatness does not stem from any practical usefulness of the theory itself. The reason is rather that Einstein

... has shown us our world in truer perspective, and has helped us to understand a little more clearly how we are related to the universe around us. [16]

In general, reflective scientists concur with enthusiasm on this central point. Rabi, for instance, asserts emphatically: "The aim of science is to make the universe, including man himself, understandable to mankind." [17] Clif-

---

[14] Translated from P. S. Laplace, *Précis de l'Histoire de l'Astronomie* (Paris: Courcier, 1821), pp. 159f.

[15] H. Poincaré, *The Value of Science,* trans. G. B. Halsted (New York: Dover, 1958), p. 84.

[16] In M. Johnston, ed., *The Cosmos of Arthur Holly Compton* (New York: Knopf, 1967), p. 208.

[17] I. I. Rabi, *Science: The Center of Culture* (New York: World, 1970), p. 130.

ford, for his part, points out that science can actually not do anything but study man. "The subject of science is the human universe; that is to say, everything that is, or has been, or may be related to man." [18]

To sum up, it is clear how humanism constitutes both the profound inspiration and the typical contribution of science. Man, to humanize himself, needs first of all cognition—cognition of other beings, but especially of himself. The basic precept of age-old wisdom has always been that man should strive to know himself. But science is precisely the systematic quest for self-knowledge, as attainable through exploration of observable reality. Hence science should be recognized as inherently humanistic. This conclusion has been best expressed by Schrödinger, the great theoretical physicist of our days:

What, then, is the value of natural science? I answer: Its scope, aim and value is the same as that of any other branch of human knowledge. Nay, none of them alone, only the union of all of them, has any scope or value at all, and that is simply enough described: it is to obey the command of the Delphic deity, get to know yourself. [19]

(c) The New Self-Awareness of Man

We have seen that there are two fundamental reasons that justify speaking of science as humanistically significant. First of all science greatly expands the information of man about the world in which he lives. Second, science supplies much new information to man about himself. In the light of the preceding, we can now formulate synthetically the essential contribution of science to humanism. This contribution consists in nothing less than giving to man a new awareness of himself.

Claude Bernard expressed strikingly the new human awareness brought about by science when, quoting an anonymous poet, he wrote: "Science— it is us. (*La science—c'est nous.*)" [20] This pregnant formulation is illuminating because it states unambiguously what science is in the ultimate analysis. *Science is man.* In fact, science is an unprecedented discovery of man as an observable object. Furthermore, science is an unexampled revelation of man to himself as a personal subject. Hence one can easily realize the new human awareness brought about by science. It consists in the frame of mind that arises as a result of both doing science and reflecting

---

[18] Quoted in J. R. Newman's Introduction to W. K. Clifford, *The Common Sense of the Exact Sciences* (New York: Knopf, 1946), p. xxv.

[19] E. Schrödinger, *Science and Humanism: Physics in Our Time* (London: Cambridge University Press, 1951), p. 4.

[20] C. Bernard, *Introduction à l'Etude de la Médecine expérimentale* (Paris: Delagrave, 1938), p. 127. H. C. Green translates the French words quoted: "Science is ourselves." (New York: Dover, 1957), p. 43.

systematically over it. Such a frame of mind is embodied in a few basic convictions, which can be briefly formulated as follows.

In the first place, the person who made himself familiar with the spirit of science cannot admit any doubt that science itself is *intrinsically humanistic*. That is to say, he is positive that the results and perspectives of science belong essentially to humanism as such. In other words, he does not even conceive the possibility that a genuinely satisfactory doctrine of man may be developed without integrating the data of science into it. To be sure, such a person does not necessarily claim that science suffices to produce a genuine humanism. Much less does he claim that all forms of humanism developed before the advent of science are unworthy of their name. Yet he maintains with all the conviction at his command that, in the scientific age, it is no longer possible—not even in principle—to formulate an acceptable doctrine of man without taking science systematically into account.

In the second place, the person familiar with the spirit of science is sure that science itself is *inherently humanistic*. That is, such a person is not only firmly persuaded that science is needed for the development of genuine humanism in the scientific age, but he is equally positive that science as a whole is needed to that purpose. In other terms, this second conviction differs from the first one in stressing the necessity for taking into account the totality of science—as opposed to selected data and perspectives. Concerning the motive that justifies this explicit insistence on the totality of science, it is but the necessity for avoiding misleading compromises that can easily be accepted by unwary humanists. In particular, it is currently fashionable among many humanists to welcome the data of the so-called human sciences while ignoring or even rejecting the insights provided by the other sciences. Such a position must be denounced because it fails to recognize properly the inherently humanistic spirit of science. Indeed, there is only one scientific spirit, which pervades all of the sciences. And it is this spirit that must be taken systematically into account in order to develop a genuine humanism in our age.

In the third place, the person who is reflectively familiar with science is certain that science is *originally humanistic*. That is, he considers it beyond all question that the appearance of science demands not just a humanism, but a humanism that bears new features when compared with those of its prescientific counterparts. To be sure, the person who has meditated sufficiently on the matter will never claim that scientific humanism must be absolutely different from the various forms of humanism that have been entertained by prescientific man. In point of fact he expects the new humanism to mark an advance rather than a rejection in man's ongoing quest for a comprehensive doctrine of himself. Nevertheless, he is convinced that the humanistic implications of science are so novel and far-reaching that only a total recasting of previous humanistic views will suffice to the task.

To sum up, the attitude of the person permeated by the spirit of science can be defined by speaking in terms of the novelty of man himself. *Scientific man is a new man.* This is true in the immediate sense that science supplies to man an immense increase in information and power when contrasted with those available in prescientific times. This is even truer in the profound sense that science has led man to entertain a new consciousness of himself.

2. THE SELF-CONFRONTATION OF MAN

The humanistic interpretation of science, as is clear from the foregoing synthesis, rests on solidly objective evidence. And yet it is well known that many of our contemporaries take a quite different view of science itself. Numerous people, in fact, think that science and humanism have nothing in common. In particular, some hold science to be nonhumanistic; others view it as antihumanistic. Therefore, to develop our synthetical discussion of the meaning of science, we must now summarize the reasons behind the positions mentioned. Especially, we must synthesize the grounds upon which earnest-minded persons seriously think that they have to be inimical to science in order to serve man. To achieve this end, here are the successive steps of our discussion. First, we shall consider the psychological-historical grounds that explain the emotional rejection of science. Second, we shall survey the instinctive dynamism that can effectively transform science into a dehumanizing pseudoideal. Third, we shall try to define the overall cultural situation of man living in the scientific age.

(a) Science and the Ordinary Worldview: Cultural Shock

To understand the fundamental reason why many people feel moved to hostility toward science, we must start from an elementary consideration. Man is essentially a cultural being, and his culture is originally based on ordinary or nonscientific knowledge. In speaking of man as essentially cultural we mean that to be genuinely human—that is, a mature personal being—he cannot be satisfied to live simply in a world of facts; he also needs an overall, unified worldview that includes principles, values, and ideals. That this is the case, of course, is confirmed by reflection on our own individual experience. But, then, we can immediately realize why the culture of man is inextricably connected with ordinary—that is, nonscientific—knowledge. The reason is obvious in that although man has always been a cultural being, science is a relatively recent arrival on the cultural scene. As a consequence, we can easily detect the basic cause of the widespread hostility toward science. This cause lies in the tension originated by the novelty of science relative to the cognition of nature and its repercussions regarding the overall worldview of man.

From the historical point of view it is indisputable that prescientific man entertained a unified cultural outlook and that he was convinced such an outlook was based on a satisfactory knowledge of nature. Carolus Bovillus, writing shortly before the rise of science, aptly expressed such a prevalent conviction when he called the universe "nothing but an enormous home for man." [21] In fact ordinary knowledge—apparently confirmed by everyday observation and canonized by philosophical speculation—gave to prescientific man the impression that he had attained a quite adequate, even self-evident, worldview. For everything and everyone seemed to have a proper place, to be taken as obvious and unchangeable. The very stability and orderliness of cosmic processes appeared to guarantee the system of certainties and values currently prevailing. Thus man felt instinctively at home in the world of ordinary experience. As Martin Buber put it, while commenting on medieval culture, to nonscientific man the world is:

... a manifold universe, ordered as an image, in which every thing and every being has its place and the being "man" feels himself at home in union with them all. [22]

The advent of science marked a sufficiently sharp break with the past to deserve to be called a cultural revolution. To be sure, objectively speaking, science constitutes a genuine advance in knowledge, which is quite compatible with prescientific certainties (see Chapter 8, Section 1a). Subjectively speaking, however, the rise of science brought about a cultural revolution because it destroyed many widely, if uncritically, accepted convictions and values. Science, in fact, replaced the closed, anthropocentric view of the universe with a boundless conception of cosmic space. It disproved the static interpretation of natural phenomena and substituted for it a conception of perennial dynamism. Above all, science showed that the meaning of the universe could not be visualized or perceived through instinct and imagination, but had to be explored through a continual investigation of an abstract theoretical kind.

It should not be surprising that man felt bewildered when science made its first appearance. For he had the impression that everything was falling apart—that no certainties were left, that he himself had become homeless in the universe. As is well known, the situation was sung by poets such as John Donne in his famous lines:

new Philosophy calls all in doubt...

---

[21] "*hunc mundum haud aliud esse quam amplissimam hominis domum*". Quoted in M. Buber, *Between Man and Man,* trans. R. G. Smith (London: Collins, 1961), p. 161.
[22] *Ibid.,* p. 166.

'Tis all in peeces, all cohaerence gone;
all just supply, and all Relation. [23]

But it was not only poets who felt upset by the new situation. Actually all reflective persons experienced a deep uneasiness. Kepler, for instance, felt a "secret, hidden horror" when reflecting on the infinity of the universe, which seemed required by the new science. [24] Another highly gifted scientist, Pascal, was prompted to pose with new earnest urgency the basic question: "What is man in infinity?" [25] Even Galileo—despite his unshakable conviction of serving truth—could not help sympathizing with the widespread perplexity aroused by science. His understanding of the situation can be gathered from the fact that he made Sagredo, the enthusiastic supporter of science, analyze at length the fears haunting contemporary people. His words may be caustic, but they do not lack compassion. In fact he compares the situation of all those persons who had taken the Aristotelian or common-sense worldview for granted to the horror of a landlord "who having built a magnificent palace at great trouble and expense . . . beholds it threatened with ruin because of poor foundations. . . ." [26]

But, if the humanistic situation was so perplexing when science first arose, should one perhaps not say that it has notably improved in the meantime, since scientific discoveries have become so familiar to modern man? Unfortunately, as everybody knows, the answer to this question must be completely negative. The reason is obvious. In fact, the more science kept expanding the more it revealed the inadequacy of ordinary knowledge regarding the understanding of man in the universe. Hence there has been a spreading of the opinion that science is essentially incompatible with humanism. Today this opinion is widely held not only by scientifically untrained persons but by many scientists as well. Some go so far as to take a perverse pleasure in claiming that nothing humanistically certain and valuable is left—and assert that this thesis manifests a high degree of scientific sophistication. At any event, the sense of weariness and frustration is widely felt among the persons who reflect on this subject. The temptation, therefore, is great for contemporary man to arrive at a desperate conclusion, as though science would have taught him that the striving after a humanistic worldview is a hopeless, self-defeating enterprise.

A typical expression of the mentality outlined is found in what the

---

[23] Quoted in A. Koyré, *From the Closed World to the Infinite Universe* (New York: Harper Torchbooks, 1958), p. 29.

[24] *Ibid., op. cit.,* p. 61.

[25] "*qu'est ce qu'un homme dans l'infini?*" Quoted in Buber (note 21), p. 163.

[26] G. Galilei, *Dialogue Concerning the Two Chief World Systems, Ptolemaic and Copernican,* trans. S. Drake (Berkeley: University of California Press, 1962), pp. 56f.

astronomer Shapley calls the "four adjustments." These are four cosmo-
logical views, which have successively assigned a different place to man
in relation to cosmic space. The first view was, of course, the instinctive
one. Man felt that he was at the center of the universe, and all planets
and stars revolved around his own earth. The second view, brought
about by the Copernican hypothesis, consisted in removing the earth from
the center and replacing it through the sun. Then, the sun itself was removed from its central position. This took place when astronomy proved
(Shapley, 1917) that the solar system is actually relegated to a corner
of the Milky Way—some 25,000 light-years away from the center of the
galaxy. Finally came the most shocking revelation. Not only is man no
longer at the center of the universe, but the universe itself appears to
have no center at all. For the universe grows larger and larger the more
powerful the instruments we use to observe it. Furthermore, contemporary
astronomy has discovered countless galaxies just like ours, and all of them
seem to move increasingly apart from each other. The conclusion that
many persons draw instinctively from such a situation is simply nihilistic.
If there is no center of the universe and this keeps changing in space and
time, they argue, such a situation implies that the universe itself has no
meaning for man. That is, they consider man as completely lost in the
unfathomable ocean of cosmic space and the ceaseless flow of cosmic time.
As Shapley summarized this conclusion:

Man becomes peripheral among the billions of stars of his own Milky
Way; and according to the revelations of paleontology and geochemistry
he is also exposed as a recent, and perhaps ephemeral manifestation in
the unrolling of cosmic time. [27]

To sum up, it is important to formulate explicitly the basic reason
why people have come to oppose each other in the name of science and
humanism. The reason is of a psychological kind. It amounts to a case
of *cultural shock*. The phenomenon in question is well known to social
psychologists. A person who comes in touch with a foreign culture feels
deeply distressed when he realizes that the system of convictions and
values he had hitherto taken for granted are actually quite relative and
questionable. The impression of this person is then that everything
crumbles, nothing is left that is certain and valuable. The same phenomenon has occurred repeatedly to modern man, ever since science arose.
This being the case, it should not be surprising that people became polariz-
ed because of science. The cause of their polarization is basically the
psychological instinct at work in cultural shock. In fact, when people
undergo such an experience, they have the impression that—as a consequence of it—they are confronted with an unavoidable choice between

[27] H. Shapley, *Of Stars and Men* (New York: Washington Square Press, 1960), p. 98.

two mutually exclusive systems of certainties and values. Hence it is quite understandable that some persons feel they have to opt for humanism in the traditional sense of the term and thus reject science, while others feel that science is so valuable that they have to reject humanism itself because of it.

However, if psychology clarifies the instinctive polarization of man in the scientific age, what profound lesson does this insight contain? That is, why is the cultural shock due to science so divisive that man feels justified to attack man in the name of the highest values? It seems clear that, properly speaking, the source of division cannot lie in science or humanism as such. In fact, science by itself is not necessarily anthi-humanistic, nor is humanism necessarily antiscientific. As a consequence, we must admit that the ultimate source of the upsetting effects produced by science has to be sought in man himself. What is it? None other can be thought of than the inherent complexity of man—a complexity that is far greater than people had been able to realize by means of traditional humanism. Accordingly, we can understand basically why science has proved so shocking to man. The shock arises from the fact that science has confronted man with himself in a quite unexpected manner by disclosing to him that he does not really know himself as he had traditionally assumed. This being the case, we notice here a humanistic aspect of science that we had no occasion to consider so far. This aspect consists in the *self-confrontation of man*. In other words, science is not only a self-discovery of man, but it also confronts man with himself. Such a cultural role of science is obviously quite important from our humanistic point of view. Accordingly, we intend to explore it further by examining more deeply the reasons that induce people to oppose science as antihumanistic.

(b) Scientistic Technicalism: The Fallibility of Man

The cultural shock brought about by science does not account entirely for the deep-seated humanistic malaise experienced by contemporary man. In fact, this malaise is so profound that many persons—even though not affected by the shock in question—conclude by branding science as inherently inhuman and dehumanizing. What is the further cause of such a situation? A frequently given answer is that there is indeed an evil factor at large in our scientifically dominated civilization. But the factor in question is said to be technology or applied science, rather than science itself. Clearly, such an answer cannot be deemed satisfactory. In fact, technology is by itself quite neutral as far as humanistic significance is concerned. Even more, technology must be considered a necessary component of any conceivable culture. For man cannot survive and thrive without some means of changing the environment according to his needs.

Thus technology cannot be inherently evil. Besides, many who know what they are talking about insist that science, not technology, is to blame (see Chapter 4, Section 3c). For instance, Heitler, one of the leading theoretical physicists of our days, writes unambiguously: "So we ought properly to speak not only of the demonic nature of technology, but of science as well." [28] As a consequence, we can expect to find the ultimate cause of the current cultural malaise only by fixing our attention on science itself. To be sure, we must not start our inquiry by supposing that science is inherently evil. But we must explore the reasons intrinsic to science that may give people some justification for considering science inherently evil. In other words, since we are studying a subject with psychological connotations, we must survey the dynamism that, if unchecked, can make science degenerate into a dehumanizing agent.

One typical manifestation of science that arouses humanistic suspicion is its proneness toward *fragmentation*. This phenomenon deserves a careful analysis. To begin with, fragmentation is not the same as specialization, although the two are frequently present simultaneously in the same person. In fact, specialization is by itself something quite positive from the humanistic point of view. For it is an attitude that arises out of a sense of respect for the intelligibility of reality—and, as such, is capable of making an individual develop genuine "personality," as Max Weber put it (see Chapter 1, Section 2b). But fragmentation is fundamentally negative as far as humanism is concerned. For it consists in the tendency of science to become more and more subdivided. Or, to link it to specialization, fragmentation is specialization run wild. But then it is obvious that fragmentation must be suspect from the humanistic viewpoint. The reason is that the more fragmentation advances, the less it becomes feasible for man to acquire the holistic view of reality, which is the necessary prerequisite of humanism.

With regard to the extent to which science tends toward fragmentation, it can be easily documented by some statistical data concerning the accelerating proliferation of science itself. Such a proliferation is evidenced both by the growth of scientific literature and that of the number of scientific workers. The growth of literature has been so dramatic that there is no exaggeration in calling it exponential. In fact, for 100 journals published in 1800 there were 1000 of them in 1850, some 10,000 in 1900, and about 100,000 in 1960. Currently no less than 300 abstract journals try to keep people abreast of new developments in their fields of specialization. [29] The number of scientific workers has also increased immensely.

---

[28] W. Heitler, *Man and Science* (New York: Basic Books, 1963), p. 3.

[29] Data taken from D. J. de Solla Price, *Science Since Babylon* (New Haven: Yale University Press), pp. 95-98.

We have reached the point that "80 to 90 percent of the scientists that have ever been, are alive now." [30]

How should the proliferation of science and its ensuing fragmentation be assessed humanistically? Some observers are quite pessimistic. They go so far as to consider the phenomenon a "disease." [31] Even though we may not be willing to share their pessimism, it is nevertheless clear that the phenomenon in question gives rise to a psychological dynamism whose consequences can be extremely serious from the humanistic standpoint. In fact, the accelerating development of science gives the impression that everything, in the intellectual field, has turned fluid. By contrast, the impression is also aroused that intellectual pursuits of nonscientific kind are hopelessly out of date. For, as one leading sociologist of science put it, "Science has been growing so rapidly that all else, by comparison, has been almost stationary." [32] It is precisely such a lack of harmonious balance in the intellectual development of man that causes concern from the humanistic point of view because of the psychological consequences that it entails on the social level.

To begin with, although the fragmentation of science can be no more than a humanistic embarrassment, in actual fact it often amounts to an exaggerated emphasis laid on scientific specialization. The reason for such an emphasis is obvious in the light of the preceding. In fact, if scientific knowledge is fragmented, and in continual expansion, it is instinctive to infer that scientific specialization alone may be able to provide trustworthy information. Yet once this inference is drawn, it is not difficult to see that it entails quite serious humanistic implications. The first is the very overemphasis of scientific specialization—*specialism*, for short. Specialism is humanistically ominous because it leads man to the conviction that, for all practical purposes, any effort to develop a comprehensive humanism is doomed to failure. In fact, specialism amounts to the conviction that man can know reality only by way of ever-progressing specialization. But specialization will never be able to provide the synthethical view of reality that is essential for genuine humanism. Consequently, specialism makes the attainment of genuine humanism appear practically impossible.

Once the close connection between science and specialism is realized, it is not difficult to perceive that an instinctively dehumanizing trend is likely to assert itself in a society where science reigns supreme. Another manifestation of this trend, which is but a consequence of specialism, is constituted by scientific *pragmatism*. In fact, the pragmatization of science arises out of specialism by an almost unavoidable psychological dynamism. The reason is obvious. Specialism, in principle, tends to make science

---

[30] See Price, *op. cit.*, p. 107.
[31] *Ibid.*, pp. 92-124: "Diseases of Science."
[32] *Ibid.*, p. 108.

socially unproductive by fostering the so-called ivory-tower mentality. But many scientists feel deeply uneasy about such a situation. Hence they convince themselves that scientific specialization is socially relevant, and that its relevancy is automatic because it gives rise to the technological know-how so essential to the development of society. Frequently these scientists actually preach research and dedicate themselves to it with the zeal of a crusader. From the humanistic point of view, however, it is clear that their position is quite dangerous—not merely in theory, but also in practice. For a scientist who embraces specialized research as the overriding goal of his life cannot avoid becoming dehumanized. In actual fact, such persons tend to behave less and less as persons but increasingly more like thinking machines, totally obsessed with the efficiency of their work. Unfortunately, the threat to humanity contained in this situation is currently quite widespread in our Western society as psychology of science proves. Eiduson, for one, has pilloried in indignant terms the person whom she calls "the Ph.D.-research-scientist-turned-technician" and whose behavior she describes as follows:

...the scientist who has become merely a cog in the wheel, the wheel which is so tremendous and intricate that neither he nor any of his "spokemates" know where the vehicle is driving, nor why, nor exactly where his skills or contributions fit. More important, he has no say about how the journey should proceed in the light of what he does. The most valuable scientific man is not the thoughtful intellectual of "old science," who was sensitive to the discontinuities as well as the continuities of the data, and adjusted his problem accordingly; but the superficial extremely competitive man who recognizes and accepts the fact that neither he nor any man can perform the new technical job alone. [33]

Although it is not possible here to analyze in full detail the dehumanizing trend instinctively connected with science, it is still necessary to mention explicitly the final outcome of such a trend. This consists in an ideological position that can be termed *scientistic technicalism*. The terms have to be taken according to the acceptations we have already examined (Chapter 5, Section 5bb; Chapter 4, Section 2c). Thus, the final result of science is often not merely the dogmatic conviction that science is the unique and exclusive form of knowing. But, over and above that, science is made into a sort of psychological conditioning that compels its devotees to look upon science itself and its technological applications as the unique ways of achieving man's ideal.

We need not insist on the fact that scientistic technicalism is inimical to genuine humanism. For this has become increasingly obvious in our

---

[33] B. T. Eiduson, *Scientists: Their Psychological World* (New York: Basic Books, 1962), pp. 151f.

own time, on both the practical and the theoretical levels. Practically speaking, in fact, it is enough to think of the ravages wrought on the ecological environment to realize that technicalism threatens to destroy man. Theoretically speaking, too, we have plenty of evidence readily available. Indeed, as is well known, some contemporary technicalists have gone to the length of popularizing what they consider the necessary reasons for opposing man's freedom and dignity in the name of science. However, for the sake of our discussion, there is one point worth noting in this connection. The point is that, besides opposing humanism, *scientistic technicalism is inimical to genuine science,* too. By realizing this point, we can understand in much greater depth the reasons why contemporary man is so upset by science. By the same token, we can also perceive more clearly the sense in which science amounts to a confrontation of man with himself.

The irreconcilability of scientistic technicalism with genuine science is so pronounced that it is possible to find a widespread agreement on this point among both scientists and humanists in the traditional meaning of the term. To cite some evidence, we can begin with the warning sounded by Max Weber, sociologist and philosopher, early in this century. When public opinion in Germany was permeated by an exaggerated enthusiasm for science, Weber pointed out that science is a culture-dependent phenomenon, hence it should never be taken for granted. To summarize in his own words:

It should be remembered that the belief in the value of scientific truth is the product of certain cultures and it is not a product of man's original nature. [34]

Another piece of evidence that shows how one should be cautious in evaluating the meaning of science for man is contained in an important passage of Poincaré's philosophical writings. This great scientist and enthusiastic promoter of science could not help pointing out that science is not automatically good for man. In fact, as he explains at some length, there is a science that man should fear—this science bearing all the characteristics of what we have called scientistic technicalism. To cite him directly:

We must fear only that science which is incomplete, the one which is in error, the one which lures us with vain appearance and thus incites us to destroy that which we would want to reconstruct later, when we are better informed and when it is too late. There are people who become

---

[34] M. Weber, *The Methodology of the Social Sciences,* trans. E. A. Shils and H. A. Finch (New York: Free Press, 1949), p. 110. For a penetrating commentary on this view see R. K. Merton "Science and the Social Order" in his *Social Theory and Social Structure* (New York: Free Press, 1957), pp. 537-549.

infatuated with an idea, not because it is sound but because it is new, because it is fashionable. These people are terrible spoilers, but they are not ... I was about to say that they are not scientists, but I notice that many of them have rendered great services to science; they are, therefore, scientists, but they are scientists not because of, but in spite of that. [35]

As regards the conviction of humanists in the traditional meaning of the term concerning the incompatibility of genuine science with scientistic technicalism, it is enlightening, for instance, to compare among themselves the three best-selling novels, *Brave New World, 1984,* and *Walden Two.* The writers of these works could not be more sharply divided against each other in condemning or extolling the consequences of technicalism for man. And yet, all of them effectively agree that in a consistently technicalized society there is no room left for genuine science as such.

As a consequence of our discussion, we can now synthesize the most profound reason why contemporary man feels so upset about science up to the point of judging science itself inherently evil. This reason is, in a true sense, immanent in science as such. It consists in the fact that science, if it comes to exert an exclusive influence on man, tends to progressively dehumanize man himself. This is an essential result that must be accepted as definitively established. No amount of enthusiasm or rhetoric should be allowed to obfuscate it. For it would clearly be a self-delusion on the part of man were he to willfully ignore the lessons of history as well as psychology. In point of fact, the dehumanizing power of an exaggerated trust in science is so great that—as we have just seen—it tends to involve science itself by making it impossible for man to cultivate the genuineness of the scientific spirit.

However, if science can be dehumanizing, what light does this fact cast on man as such? After all, science is not a self-standing entity, but rather a product of man. Accordingly, what does the dehumanizing potential of science reveal about man the maker of science? The answer is plain. Science reveals, in a new and unprecedented way, the inherent *fallibility* of man as an ethical being. That is to say, properly speaking, what science reveals is not an inherent evilness of science as such, but rather a deep-seated tendency of man to behave evilly; this tendency is so strong that he can easily transform science itself into an instrument of dehumanization. In other words, what science does in this connection is merely serve as a magnifier of man's own ethical nature. A particular manifestation of science that exemplifies this point cogently is the experience of nuclear power. Robert Oppenheimer, after having watched at close range the first nuclear explosion in the desert of New Mexico, is

---

[35] H. Poincaré, *Mathematics and Science: Last Essays,* trans. J. W. Bolduc (New York: Dover, 1963), p. 110.

reported to have exclaimed that that event had made man acquire a new awareness of sin. We can consider this remark as typical. For science has lent new urgency to man's need for facing the issue of his ethical fallibility with total honesty. In fact, while living in the prescientific age, man could still delude himself about his genuine character as a moral agent. Especially, he could ascribe his moral fallibility or sinfulness to the lack of sufficient enlightenment and power. But science—precisely by conferring on man enormous enlightenment and power—has definitively exposed the inanity of such views.

To sum up, it should not be surprising that science is so much at the center of controversies in contemporary society. The reasons, as we have seen, are many. But the most illuminating is also the most likely to arouse bitter opposition. For such a reason consists in the fact that science confronts man with himself in the most unflattering light possible by disclosing his inherent fallibility. Obviously, no one likes instictively to be reminded of such a situation. Hence, it is most natural for contemporary man to take sides for or against science instead of facing the issue where it really lies. Even so, it is no small humanistic merit of science itself to have touched man at the bottom of his heart. For it is only from there that genuine humanization can finally originate.

(c) The Crisis of Identity and Growth

We have explored the two main grounds that make science a source of cultural upheaval. How should we now define comprehensively the situation of man living in the scientific age? Phenomenologically speaking, we have already seen that this situation can be characterized as drama (see Chapter 4, Section 4). But now, after our long philosophical analysis, we are in a position to synthesize its significance more precisely. Accordingly, we propose to employ some basic psychological categories whose applicability to contemporary cultural man becomes increasingly evident. In brief, we intend to define the situation of man living in the scientific age as one of crisis—a crisis of both identity and growth.

The key concept here is that of man's own *identity*. The reason for employing this concept in connection with our issue emerges from the definition of identity itself, as formulated by one of its leading theoreticians, the psychologist Erik Erikson:

The conscious feeling of having a personal identity is based on two simultaneous observations: the perception of the selfsameness and continuity of one's existence in time and space and the perception of the fact that others recognize one's sameness and continuity. [36]

---

[36] E. H. Erikson, *Identity: Youth and Crisis* (New York: Norton, 1968), p. 50.

Erikson himself indicates in one of his essays why science can affect man's sense of identity. This he does while discussing Charles Darwin's tendency to hypersensitivity. As he puts it:

A peculiar malaise can befall those who have seen too much, who, in ascertaining new facts in a spirit seemingly as innocent as that of a child building with blocks, begin to perceive the place of these facts in the moral climate of their day. [37]

In brief, one can say that science shakes man's sense of identity because it gives the impression that there is an unbridgeable break between the prescientific and the scientific conceptions of man—and that the scientific conception of man is itself continually changing. This state of affairs is increasingly stressed by experts on the subject. Bronowski, for instance, explains as follows the reason for the split between the so-called two cultures:

The breach between them [the two cultures] is not merely a gap in contemporary education, but it is the visible sign of a loss of confidence in the identity of man. [38]

Scientists and philosophers readily agree that the main issue of modern science revolves around the identity of man. Thus, for instance, Max Scheler writes emphatically:

We are the first epoch in which man has become fully and thoroughly "problematic" to himself; in which he no longer knows what he essentially is, but at the same time also *knows* that he does not know. [39]

Loren Eiseley, the well-known anthropologist, compares the development of science to a stirring and yet dangerous voyage, something like a new Odyssey: "The epic of modern science is a story at once of tremendous achievement, loneliness and terror." [40] Repeating the words of Odysseus, Eiseley stresses: "There is nothing worse for men than wandering."

However, if science calls into question man's sense of identity, one should not view this situation as an unwarranted attack of science on man's need for cultural harmony. Rather, one should acknowledge that science has just brought to the fore a deep-lying feature of man as such, namely his own problematicity. Man is problematic simply because he is man— because he must strive after self-humanization by coming in touch with

---

[37] E. H. Erikson, *Insight and Responsibility: Lectures on the Ethical Implications of Psychoanalytic Insight* (London: Faber, 1964), p. 22.

[38] J. Bronowski, *The Identity of Man* (Garden City, N.Y.: Natural History Press, 1965), p. 94.

[39] Quoted in Buber (note 21), p. 220.

[40] L. Eiseley, *The Unexpected Universe* (New York: Harcourt, 1969), p. 4.

the world of observation and experience. This point is stressed, for example, by the philosopher Martin Buber when he writes:

But the real man, man who faces a being that is not human, and is time and again overpowered by it as by an inhuman fate, yet dares to know that being and this fact, is not unproblematic: rather, he is the beginning of all problematic. [41]

Man living in the scientific age is so much upset by the unexpected revelation of his own problematicity that he feels at a loss for a sense of self-identity. This being the case, how should this cultural situation be designated with precision? Erikson, for one, suggests that we use here the term *crisis*, which he employs according to its accurate clinical definition. He writes in fact:

Man's socio-genetic evolution is about to reach a crisis in the full sense of the word, a crossroads offering one path to fatality, and one to recovery and further growth. [42]

In the light of our entire investigation it seems clear that we have reason to accept the foregoing diagnosis as valid. In fact, contemporary cultural man finds himself in a state of serious psychological sickness. Hence we cannot hypothesize that such a situation may continue for long without resulting in either of only two possible solutions—namely, fatality or recovery. In other terms, it is to be expected that science may eventually dehumanize man entirely. Yet this outcome is not necessary, because science can also lead to a greater humanization of man.

Erikson, for his part, adopts an optimistic position. In the context just cited, he elaborates as follows on the nature of the present-day crisis:

But the processes of socio-genetic evolution also seem to promise a new humanism, the acceptance by man—as an evolved product as well as a producer, and a self-conscious tool of further evolution—of the obligation to be guided in his planned actions and his chosen self-restraints by his knowledge and insights.

History appears to justify a guardedly optimistic attitude. Indeed, if science has brought about a profound crisis of man, this very fact should be considered a manifestation of the ability of man to grow, rather than a symptom of tendency toward failure. For, after all, science is clearly a quite positive conquest, one of the greatest achievements that man ever attained. Accordingly, the current cultural situation of man should indeed be assessed with all the earnestness that it deserves—namely, as a genuine crisis. Yet no one should feel justified in implying that such a crisis must be a catastrophic one. In fact, just the opposite outcome is likely to take

[41] Buber (note 21), p. 181.
[42] Erikson (note 37), p. 227.

place—that is, a marked increase in humanization—provided, of course, man is willing to act consistently with the inspiration provided by science. For science does inspire man to be more human, even though it is not able to motivate him adequately in this regard. As a consequence, the comprehensive evaluation of the human situation in the scientific age should not just be one of a crisis of identity. More specifically and more profoundly, it is necessary to speak of a *crisis of growth*. To be sure, man may or may not grow more human because of science. Yet this is the direction to which the development of civilization points.

To sum up, it can be sair that if science is humanistic, its humanism is of a peculiar kind: it is only *partial* and, as a consequence, inescapably *ambiguous*. The partial humanistic character of science needs emphasis in order to remove a widespread misunderstanding. In fact, since science is humanistic, many people tend instinctively to identify science with humanism itself. But such an identification could only lead to disaster—for both humanism and science. Actually, a comprehensive doctrine of man that rested exclusively on science could hardly avoid to foster the antihuman, as well as antiscientific, ideology that we have called scientistic technicalism (see Chapter 8, Section 26). Thus the intrinsic limitation of science from the humanistic standpoint must be accepted as a fundamental datum. But, by the same token, it is also necessary to admit that science is inherently ambiguous as far as humanism is concerned. For, indeed, since the humanism of science is only partial, it can easily lend itself to misunderstandings and misapplications.

And yet, if science is humanistically ambiguous, this feature should not be deemed a negative trait of science itself. For, after all, what science does in this connection is just to reveal man to himself in a realistic way. Indeed, the ambiguity of science simply discloses that—in the final analysis—it is man himself, the maker of science, who is inherently ambiguous. But such a disclosure is a vital contribution to genuine humanism—all the more so the greater the inclination of contemporary man to feel self-complacent because of his intellectual sophistication and technological power. In fact, the point has unfortunately already been reached that some —dazzled by the achievements of science—do not blush to advocate that the very humanity of man should be considered a function of science. Man is exhorted to forget about his own dignity and to entrust his salvation to omniscient and almighty science. Against such a mythologization, science itself—through its inherent ambiguity—warns man to come to his senses by having the courage of acknowledging his own basic ambiguity. In brief, then, it can be said that science lends new urgency to the need for self-humanization on the part of man, at least in a negative sense. For, truly, if man living in the scientific age does not determinately strive after self-humanization, he is bound to effectively dehumanize himself.

3. Indications Toward Self-Humanization

As has been seen, the essential contribution of science to humanism is one of enlightenment. Science is humanistic because it amounts to a discovery as well as a confrontation of self. However, we would not be justified were we to stop our humanistic assessment of science at this point. For science has an important message to offer also in the realm of values— albeit only an indicative, rather than a cogent, one (see Chapter 7, Sections 1 and 3). As a consequence, in our effort to synthesize the meaning of science for man we must now survey the main indications that science itself contributes toward self-humanization. That is, we must discuss how a person should behave who intends to be consistent with the spirit of science.

(a) Authentic Understanding

The basic humanistic indication originating from the spirit of science is obvious. Why, indeed, does man dedicate himself to science in the first place? All the passionate motivation of the searching scientist is beautifully summarized, for instance, in this personal confession by Jean Rostand:

My heart, like most others, desires the survival of the whole man; my mind, however, longs to escape from incomprehension. [43]

As a consequence, we may say that the first obligation incumbent on the person who intends to be consistent with the spirit of science is that of seeking authentic understanding.

It is important to recall that scientific man may fail to attain such an understanding, and thus effectively fail in his self-humanizing task. The reason for this failure is, of course, the tendency of man to overrate the cognitive importance of science itself. In fact, many scientific enthusiasts absolutize science as though it were the uniquely and exclusively valid form of knowing. But, by so doing, they effectively overlook the contradiction in their own intellectual attitude. Indeed, as scientific researchers they insist on the necessity of keeping an open mind, they claim that people should do away with all purely "natural" or spontaneous ways of thinking about reality. And yet as philosophers they demand that science be taken so much for granted as to exclude any reflective way of exploring the significance of science itself. As a result, these people emprison the understanding of man in another kind of "naturalness"—the new natural outlook being now, by definition, the scientific one. We need

---

[43] J. Rostand, *The Substance of Man,* trans. I. Brandeis (New York: Doubleday, 1962), p. 65.

not press the point that such a lack of authentic understanding is incompatible with the spirit of science. Yet we must recall that it has devastating humanistic consequences. For, indeed, if science explains everything, if nothing can be understood but by way of the objective-interpersonal approach of science—if, in other words, the world of observation has no message for man as a reflective being—then one must well conclude that nothing can be understood in a genuinely human way (see Chapter 4, Section 3d). A distressing confirmation of this inescapable conclusion is, for instance, a disconsolate statement by Rostand himself. To him, indeed, it seems clear that man's passionate search for comprehension leads nowhere but to total meaninglessness.

All is tragically simple for the individual human: nothing to understand, nothing to expect, and there would not even be anything to suffer from, were man able to do otherwise. [44]

The immediate implication of the preceding is that man, to be consistent with the spirit of science, must learn not to overvalue science itself. In other words, he must learn not to take anything as self-evident simply because science explains it. On the contrary, man should—consistently with the spirit of science—make himself more open to wonder and awe. For these attitudes are the roots and the rewards of science. But, above all, they are what characterizes man as man. The anthropologist Eiseley has expressed forcefully the cognitive situation typical of man by writing:

No longer, as with the animal, can the world be accepted as given. It has to be perceived and consciously thought about, abstracted and considered. The moment one does so, one is outside of the natural; objects are each one surrounded with an aura radiating meaning to man alone. [45]

But, if man has to understand more than scientifically in order to understand authentically, what other form of understanding must he pursue besides the scientific one? If we take into account the results of our research in the second part of this book, the answer is plain. Man can expect to attain authentic understanding only if he tries to assimilate the import of his science by means of philosophical reflection. In fact, the task of philosophy is to enable man to become reflectively aware of what he knows factually and of the humanistic consequences that arise from his knowledge. In other terms, philosophy enables man to discover an overall meaningful message in the manifold information that reaches him from the observable world. As stated by Meyerson:

Philosophy is an attempt to make us agree with ourselves. Or, if one likes it better—since our mind is what it is—philosophy is an attempt

---

[44] *Ibid.*, p. 69.
[45] Eiseley (note 40), p. 32.

to make the "realities" which assault us on every side agree with themselves. [46]

To go into some detail, the contribution of philosophy to the authenticity of understanding manifests itself in several forms. First, philosophy makes man reappraise the significance of his own knowing. Buber has stressed the necessity of philosophical discovery. "All philosophical discovery is the uncovering of what is covered by the veil woven from the threads of a thousand theories." [47] Second, philosophy alone enables man to overcome the dangerous tendency to identify science with knowledge as such. This point needs to be emphasized because sometimes even would-be humanists continue to overstress science. A case in point seems to be the so-called "science of science." [48] Third, philosophy enables the mind of man to be more judicious as well as penetrating when handling factual information. In fact, if all other conditions are the same, a person is a far better knower if he is also a philosopher than if he is merely a scientist (see Chapter 5, Section 5aa). A typical example in this regard is that of Bohr. Indeed, why was Bohr such a great knower as to be not only an outstanding scientific creator but also a leader of creators? The reason has to be sought in his practice of philosophical reflection. Such a reflection made him see better the complexity of truth. This attitude of his is expressed in one of his most revealing mottoes, which he was fond of repeating: "The opposite of a correct statement is a false statement. But the opposite of a profound truth can again be a profound truth." [49] Sometimes, to be sure, the effort to harmonize science and philosophy may well not be too successful, and produce tension. Even then, however, the humanistic importance of philosophy cannot be denied. An example is that of the opposition exerted by Socrates and Protagoras against the Ionian cosmologists. A contemporary historian of science rightly emphasizes the humanistic worth of their criticisms. "These enemies of science were the first to perceive that observations depended upon the existence of an observer—who tended to forget his own existence." [50]

Finally, the most valuable form with which philosophy contributes

---

[46] Translated from E. Meyerson, *De l'Explication dans les Sciences* (Paris: Payot, 1921), pp. 10f.

[47] Buber (note 21), p. 221.

[48] See D. J. de Solla Price, "The Science of Science," in M. Goldsmith and A. Mackay, eds., *Society and Science* (New York: Simon and Schuster, 1964), pp. 195-208. What makes me uneasy about such a view is the apparent lack of a proper distinction between science and philosophy in Price's position. See especially pp. 200f.

[49] Translated from quotation in W. Heisenberg, *Der Teil und das Ganze: Gespräche im Umkreis der Atomphysik* (Munich: Piper, 1969), p. 141.

[50] P.-H. Michel, in R. Taton, ed., *A General History of Science*, trans. A. J. Pomerans (London: Thames and Hudson, 1963), vol. I, pp. 218f.

to authentic understanding consists in its ability to make man effectively understand himself. Surely, philosophy does not suffice by itself to that task, but needs science. Even so, philosophy alone is able to make man understand himself because such an understanding has to be both personal and reflective. As Buber put it rightly:

The investigator cannot content himself ... with considering man as another part of nature and with ignoring the fact that he, the investigator, is himself a man and experiences his humanity in his inner experience in a way that he simply cannot experience any part of nature. Philosophical knowledge of man is essentially man's self-reflection, and man can reflect about himself only when the cognizing person, that is, the philosopher pursuing anthropology, first of all reflects about himself as a person.... He can know the wholeness of the person and through it the wholeness of man only when he does not leave his subjectivity out and does not remain an untouched observer. [51]

We can summarize the main traits of the basic indication toward self-humanization provided by science. In the first place, man must strive to *know more, not less*. In fact, the very existence of science shows that man can achieve the fullness of his humanity only by way of knowledge. But the requirement is that man must pursue a truly comprehensive knowledge. Holton has put this requirement well while discussing science and the intellectual tradition of mankind.

Salvation can hardly be thought of as the reward for ignorance. Man has been given his mind in order that he may find out where he is, what he is, who he is, and how he may assume the responsibility for himself which is the only obligation incurred in gaining knowledge. [52]

In the second place, the spirit of science moves man to strive after effective *understanding*. That is to say, he must not rest satisfied with superficial explanations, no matter how scientific. On the contrary, he must look for the entire truth, namely for the ultimate message of reality. For only then can he say that he authentically understands what reality means. In Eiseley's appropriate words, man is man precisely because of his ability to perceive and interpret messages.

We live by messages—all true scientists, all lovers of the arts, indeed, all true men of any stamp. Some of the messages cannot be read, but man will always try. He hungers for messages, and when he ceases to seek and interpret them he will be no longer man. [53]

---

[51] Buber (note 21), pp. 154f.

[52] G. Holton, "Modern Science and the Intellectual Tradition" in P. C. Obler and H. A. Estrin, eds., *The New Scientist: Essays on the Methods and Values of Modern Science* (New York: Doubleday Anchor Books, 1962) pp. 19-38; see especially p. 29.

[53] Eiseley (note 40), p. 146.

In the third place, the spirit of science moves man to seek an understanding that is really *humanistic*. That is to say, man must strive to become aware of himself in the concreteness and wholeness of his nature, within the framework of certainties and values that make up a comprehensively human culture. C. F. von Weizsäcker has rightfully spoken of "insight" in this connection.

Insight ... I would call the knowledge which considers the coherence of the whole. Insight must be especially concerned with man himself, his motives and his aims, and with the inner and outer conditions of his existence. Insight may not separate subject and object fundamentally, but must recognize their essential kinship, their mutual dependence and, consequently, their inseparable coherence.... As we are trying to achieve such insight, the notion of responsibility for the whole acquires a specific, concrete meaning. It now means the responsibility for the whole which is the totality of all sciences.... [54]

(b) Authentic Creativity

We can realize another important humanizing indication inspired by science if we consider the most characteristic feature of scientific knowing—namely, creativity. Self-humanization, obviously, implies creativity because man becomes genuinely human through creation. Accordingly, precious light on the process of humanization can be obtained by reflecting on creativity, especially as embodied in science.

The first feature of authentic creativity is a *positive, though critical, evaluation of tradition*. Psychological investigation proves that this is truly a characteristic mark of the genuinely creative person. One researcher specializing in this subject writes for instance:

The attitude that appears most readily to favour creative thinking combines receptivity towards what is valuable, in traditional and new ideas alike, with discriminating criticisms of both. [55]

It is important to stress the positive though critical position of the genuine creator with regard to the past, because only too frequently the public mind assumes that originality amounts to iconoclasm or destruction of the past. Nothing could be farther from the truth as the psychologist just cited insists: "Creativity ... seems to be distinguished by serious receptivity towards previous thought products and unwillingness to accept them as final." [56]

---

[54] C. F. von Weizsäcker, *The History of Nature,* trans. F. D. Wieck (Chicago: University of Chicago Press, 1949), p. 4.
[55] P. McKellar, *Imagination and Thinking: A Psychological Analysis* (London: Cohen and West, 1957), p. 113.
[56] *Ibid.,* p. 116.

As regards the contribution of science to the understanding of this first feature of creativity, a quick look at its history may suffice. In fact, great scientific creators were never iconoclasts or destroyers—on the contrary, they were careful evaluators of past achievements as well as tireless improvers over them. Galileo, the father of science, stands out especially in this regard. Undoubtedly, he was quite impatient of any worshiping of traditional learning. Nonetheless, he always showed genuine respect for such learning and was proud to be considered an inheritor of tradition. Thus he viewed himself as a disciple of the ancient Greeks—not only of mathematicians such as Archimedes, but also of philosophers such as Aristotle. As a result, we can realize the first contribution of the spirit of science to authentically human creativity. It consists in the exhortation to overcome the dichotomy that is so instinctive in man between tradition or conservation, on the one hand, and advance or progress, on the other. Indeed, the very existence of science shows that genuine advance implies continuity with tradition, and genuine continuity with tradition implies a continual effort to improve over past performance (see Chapter 8, Section 1a; Chapter 1, Section 3e).

Another distinctive feature of authentically human creativity, as is well known, is the *total personal engagement* required of the person involved. In fact, creation is what expresses at its best the whole personality of man. But if this is the case, we can notice here another important aspect of the inspiration that science provides to man to humanize himself through creativity. The reason is precisely the fact that science offers an outstanding example of the necessity as well as productiveness of personal engagement in humanizing endeavors. Indeed, science succeeds in humanizing man in the area of knowledge simply because it manages to harness all the resources of his personality to the task. But, then, the spirit of science must necessarily exhort man to extend the same attitude of total personal involvement to the remaining endeavors that are demanded by full-fledged humanization.

Finally, the creativity which is authentically human is defined by still another feature which can be called an *originally personal communion with reality*. In fact, in order to create, man cannot simply start out of nothing; he must have a grounding in preexistent reality. Nevertheless, genuinely human creation is precisely that: an originally personal reinterpretation and remolding of reality itself. As a consequence, creativity humanizes man by establishing an enriching communion between himself and reality. However, if these considerations apply to creativity in general, it is clear that they apply more specifically to scientific creativity as such. For, precisely, more perhaps than other forms of creativity, science stresses the need for creative man to keep continuously in touch with objective reality. As a consequence, we have here another manifestation of the inspiration that science offers toward the self-humanization of man. Science exhorts man to be creative in an authentically human way by overcoming a widespread temptation that tends to mar the creative person by making him either too subjectivistic or too objectivistic.

To sum up, it is certainly inspiring to think of science in terms of creativity. For science, no less than other typical endeavors of man, is truly creative in the proper acceptation of the term. "Science, like art, is not a copy of nature, but a re-creation of her. We remake nature by the act of discovery, in the poem or in the theorem." [57] But, if creativity is so typical of science, the inspiration that it provides must clearly extend to the humanization of man in his entirety. Hence we have a second major reason to speak of the possibility of, as well as necessity for, developing a full-fledged humanism that incorporates science and its spirit as one of its integral parts.

(c) Holistic Culture

To close our theoretical survey of the humanistic significance of science we must still mention a third indication of science itself toward self-humanization. This indication refers to man in the totality of his cultural being. It stresses the need for striving after a holistic or unified culture as an essential requirement of humanization. Already T. H. Huxley was making this point while criticizing Matthew Arnold's overemphasis on the classics as a source of culture:

For culture certainly means something quite different from learning or technical skill. It implies the possession of an ideal and the habit of critically estimating the value of things by comparison with a theoretical standard. Perfect culture should supply a theory of life, based upon a clear knowledge alike of its possibilities and of its limitations. [58]

Reflection on science and its spirit stresses the necessity for a unified culture, on both negative and positive grounds. Negatively speaking, in fact, it is clear why man living in the scientific age should strive to develop a unified culture. The reason is the very real threat to his humanity that is arising from the progressive cultural breakdown typical of current civilization. Indeed, man can become genuinely human only if he possesses a holistic worldview that gives to him a sense of direction and value. But contemporary people have come to take for granted that there are two—or even many—noncommunicating cultures in our midst. However, the more persuasive motivation that science offers toward holistic culture is a positive one. In fact, not only is science one of the most excel-

---

57 J. Bronowski, *Science and Human Values* (New York: Harper Torchbooks, 1965), p. 20.

58 In Bibby (note 8), p. 201. See the whole section "Education," pp. 189-230.

lent products of human culture, but its contributions are of immense cultural value because they give to man a better awareness of himself. Thus it is but consistent with the spirit of science that man living in the scientific age should strive to develop a new holistic culture. This culture should precisely consist in a comprehensive doctrine that takes into systematic account the whole of man—both as revealed by science and as explored by other cultural approaches such as philosophy and art.

Can it be said that the holistic message of the spirit of science is effectively heeded in our times? Despite the very real and increasing polarization of contemporary people, some positive signs should be noted. For, even though they are scattered and merely inchoative, they point in the right direction and give new courage to the committed humanist. The thinkers involved are both scientists and philosophers. As regards scientists, it is currently no longer altogether exceptional to hear them declare, in conversation and in writing, that they are "concerned with the teaching of science itself as a humanistic subject." [59] As for philosophers, they too tend to show a progressive awareness of the indispensable role played by science concerning authentic humanism. A particularly telling example in this connection is that of Buber. He seems to have been largely uninformed about science—up to the point of employing occasional expressions that may hurt the scientist. And yet, moved by his sincere concern for the problem of man, Buber came to realize—at least implicitly—the indispensability of science for a genuine humanism or "philosophical anthropology," as he calls it.

Buber's general thesis is that man cannot be understood except in his "wholeness." This he defines as follows:

... man's special place in the cosmos, his connection with destiny, his relation to the world of things, his understanding of his fellow-men, his existence as a being that knows it must die, his attitude in all the ordinary and extraordinary encounters with the mystery with which his life is shot through.... [60]

As a consequence of his general standpoint, Buber insists that man must be studied by a philosophy that considers him in the totality of his environment. Thus he implicitly refers to science when he says that philosophical anthropology

... must put man in all seriousness into nature, it must compare him with other things, other living creatures, other bearers of consciousness, in order to define his special place reliably for him. [61]

---

[59] E. Rabinowitch, "Science and Humanities in Education," in Obler and Estrin, eds. (note 52), pp. 155-177; see especially p. 170.

[60] Buber (note 21), p. 150.

[61] *Ibid.*, p. 154.

In particular, Buber rejects as totally inadequate the widespread philo-sophical tendency to study man by pure speculation, as though man himself would be essentially reason. In this connection he declares pointedly:

Rather, the depth of the anthropological question is first touched when we also recognize as specifically human that which is not reason. [62]

To synthesize, we can now express in a few words the significance of science for man that we have detected in the course of our long investiga-tion. This significance does not lie in such outward and superficial reasons as the obvious technological usefulness of scientific discoveries or the emotional excitement that they entail. Much less does the significance of science consist in an implicit invitation to man to consider himself superior to the limitations and obligations of the human condition. In other words, science is not human because, by satisfying man's cognitive and technical instincts, it can give him the impression of being a new Prometheus capable of fashioning the universe as he pleases by means of his engineering skills. Actually, science is human—chiefly and essential-ly—because it is humanistic, that is, because it enables man to become more consciously aware of his human condition, both as a fact and as an ideal.

In other words, the central result of our investigation is that the significance of science lies especially in the inspiration it gives. Actually the spirit of science does not allow man any self-complacency but invites him to transcend himself. In fact, if rightly understood, *science exceeds science*. For it points beyond and beneath itself, toward a fully humaniz-ing ideal. This ideal consists, surely, also in knowing, more and more. But, above all, it consists in living. For science invites man to live as a coresponsible creator of the universe entrusted to his care.

Doubtlessly, the humanistic significance of science outlined can be easily ignored and even ridiculed. But, if this happens, the fault is not science's, rather man's. For man can only too naturally be inconsistent with the spirit of his science. However, if he heeds this spirit and puts it into practice, he will hardly fail to humanize himself. For, after all, the very existence of science shows that man is able to transcend the de-humanizing conditioning of his self-centeredness. In the light of the inter-pretation attained, therefore, we are justified in looking forward to a more genuinely human civilization in the age of science without indulging thereby in naïve and self-deceptive expectations. In actual fact, science shines already for many as an attractive and rewarding goal of their humanizing endeavors. Such an inspirational character of science was formulated, for instance, with unique charm by the great contemporary theoretical physicist Wigner when he wrote:

---

[62] *Ibid.*, p. 195.

The promise of future science is to furnish a unifying goal to mankind rather than merely the means to an easy life, to provide some of what the human soul needs in addition to bread alone. [63]

4. SCIENCE AS AN EFFECTIVELY HUMANIZING PROCESS: PRACTICAL CONCLUSIONS

We have brought to a close—albeit only introductorily—the research we had set to ourselves as a goal in the course of the present book. May we say at this point that our task is accomplished? Theoretically speaking, a positive answer is undoubtedly justified because we have attained our original aim of discovering the significance of science for man. Practically speaking, however, it is clear that our results are still far from satisfactory. In fact, the central insight we have obtained is that science is humanistic. But, if science is humanistic in theory, it must also be humanizing in practice. However, how can man effectively humanize himself by doing science? The outcome of our preceding investigation may appear rather disheartening than encouraging from this point of view. For we have dealt with issues that are so manifold and complex that they seem hardly accessible, let alone implementable, to the majority of working scientists. As a consequence, it is clear that we have still to face one obligation in our effort to comprehend the message of science for man. This obligation consists in considering how the practice of science can effectively humanize the working scientist as such.

Our starting position here must be realistic as well as genuinely human. We must acknowledge the seriousness of the objection that views the humanistic consideration of science as an elitist enterprise—open only to a selected few, exceptionally gifted scientists. However, if we must grant that the theoretical understanding of scientific humanism demands some rather special conditions of both leisure and reflective talents, we must insist on the equally true fact that science is and should be a humanizing undertaking for all of its practitioners. For, after all, it is not only a matter of theoretical principle that man is called to be more human through his dedication to science. In practice, too, as experience shows, the average scientist is far from disinclined to think of his science in humanistic terms. Indeed, it is a cliché, and a misleading one, that views the scientist as being opposed, in general, to reflective efforts aimed at the humanization of his work. Rather, it should be said that the vast majority of scientific workers simply feel disheartened and almost cynical as far as humanization of science is involved. Yet this attitude arises

---

[63] E. P. Wigner, *Symmetries and Reflections: Scientific Essays* (Bloomington: Indiana University Press, 1967), p. 280.

normally not out of theoretical conviction, but simply out of a sense of personal hopelessness. As a consequence, therefore, the discussion to follow will start by examining the humanistic difficulties typical of the scientific career. Then we shall proceed to examine the two main components of any humanizing process, namely wisdom and dialogue. Finally we shall apply the insights obtained to the everyday life of the ordinary scientist.

(a) The Humanistic Difficulties of a Scientific Career

The difficulties impeding the humanistic development of scientific workers fall into a few principal classes. Beginning with the most obvious, we encounter the class that can be labeled *environmental pressure*. Rare is the scientist who is able to work in a relaxed atmosphere. No matter how gifted and successful, he is ordinarily exposed to many strains that bear down on him because of the very fact that he belongs to the scientific community. A quite widespread strain arises because of *financial worry*. Some scientists—actually the largest number of them—worry about money simply because they are paid employees. Their employers want as quick a return as possible on the capital they have invested in them. It is obvious that such scientists can hardly think of science as a humanizing search for truth. Their employers, in fact, are not interested in truth at all, but simply in quickly marketable products. But even the relatively few scientists who do not depend for their sustenance on commercially minded employers seldom find themselves free of financial worry. They too, in fact, are under the pressure of justifying to their sponsors and the public the considerable sums of money invested in their projects.

Another form of environmental pressure which strains the energies of scientists to their limits is *professional competition*. Every researcher, even the most humanistically inclined, feels himself confronted with the iron law formulated in the brutal alternative: "publish or perish." Thus, if a scientist wants to enjoy the esteem of his colleagues and make a decent living, he feels compelled to publish at least as much as the others. But if he is ambitious and intends to get ahead in his profession, the pressure of his environment stimulates him to surpass the others as a matter of quasi-ideal. Naturally, then, such a person can hardly have time or interest left for any strictly humanistic pursuit.

A second class of humanistic difficulties besetting the working scientist comes under the heading of *psychological exhaustion*. This phenomenon, closely connected with the pressures of the scientific environment, presents a number of symptoms. Chief among them is a widespread sense of *restlessness*. The ordinary scientist finds it quite hard to pause and reflect philosophically. The psychological root of this situation is obvious. Scientific work is done largely in haste. The worker must hurry to come up with some discovery that may justify the money invested in his salary and equipment. He must rush into print to forestall the competition. He must establish a reputation and keep defending it tirelessly against the continual attacks of his challengers. Obviously, then, no psychological room or justification is left for the recollected quiet demanded by humanistic maturation.

The psychological exhaustion resulting from scientific work easily entails ominous humanistic consequences. One such consequence is what can be called *routinization*. It consists in the fact that the working scientist tends to adopt a uniform approach in all intellectual endeavors—that of his scientific routine. But then, of course, such a person inclines to view humanism as meaningless because it demands an approach so different from that of science. Another quite serious consequence is what can be called the pseudoideal of *mental confinement*. It consists in the fact that the scientist not only tends to forget any other pursuit because of his dedication to science but inclines to condemn all nonscientific pursuits as unworthy of consideration. The psychological root of such a situation is evident. Science demands the whole man in more and more specialized areas of research. But overemphasized specialization not only narrows the mental outlook; it also produces the impression that an immense amount of specialized research has to be completed before man can even dream of dealing with the universal humanistic issues connected with science. For, indeed, specialization fosters specialization: the more one investigates, the more he finds that many additional details have to be investigated before attaining a comprehensive understanding. As a result, the person caught in this process is likely to emerge with an express contempt for humanism and a proselytizing zeal for specialized scientific research as the only justified occupation—at least for the foreseeable future—of any sensible person.

A third class of humanistic difficulties confronting the working scientist is *emotional pessimism*. The reasons leading toward such a position are not difficult to realize in the light of the preceding. The scientist, for all his professionalism, remains human. Hence, if science—to which he sacrifices everything—proves to be itself incapable of satisfying his inmost expectations as a human being, he tends to react by branding as self-deceptive the very hypothesis of humanistic significance in the life of man.

To go into some detail, a rather frequent manifestation of pessimism detectable among scientists is a sense of *disenchantment with science* as such. Normally, of course, the people in question continue to do science, but their ideal motivation is gone. What originally presented itself to them as a goal worth all their enthusiasm and dedication is now "just a job." A second manifestation of scientific pessimism, strictly connected with the foregoing one, is a sense of *estrangement from nature*. The persons who share this attitude have no use for what they call sentimentalism—the consideration of the concreteness, the beauty, the wholeness of nature. They rather contend that what really counts in science are its pragmatic manifestations. Thus they insist on "joblikeness"—the importance of finding formulas and experiments that really work, to the exclusion of every other consideration. A third form of pessimism to be found among scientific workers manifests even more explicitly the pessimistic mentality itself. It consists in a bitter sense of *universal meaninglessness*. Not too rarely, unfortunately, those who entertain this feeling go so far as to boast about it, as though it constituted a great enlightenment due to science. Finally, a fourth form of emotional pessimism can still be found among scientific workers. It is more comprehensive and thoroughgoing than the forms already listed since it represents a veritable *alienation from self*. The persons who entertain such a feeling are not numerous and, frequently, not very vocal. And yet, such a frame of mind really exists and it tends to cause terrible havoc. One of its manifestations is the suicide of the successful scientist. Another manifestation is the boundless arrogance of some would-be social reformers who, in the name of science, declare a total war against everything that is human and humanistic.

To complete our survey of the humanistic difficulties encountered by scientists in the course of their work we must still add a couple of comments. The first regards the earnestness of the situation examined. We have painted a starkly depressing picture because the data warrant it. In this connection, however, two equally misleading, if opposite, reactions must be avoided. One such reaction consists in viewing science as an inherently dehumanizing agent. This view is wrong because the data do not prove that science necessarily dehumanizes man, but simply that man can rather easily allow himself to be dehumanized by science. The other reaction consists in going to the opposite extreme of belittling or even denying the humanistic risks involved in the practice of science. This reaction is dangerous because it amounts to an illusion that flies in the face of the evidence. In fact, great scientists themselves—those certainly humanized by their science—cannot help uttering a concerned cry of alarm when they reflect on the situation we have surveyed (see Chapter 4, Sections 3 and 4).

The second comment we must make on the theme discussed regards the source of the threatened dehumanization of the scientific practitioner. This source consists in the instinctive tendency to overrate human intelligence, particularly as employed in scientific research. Thus, to find a way out of the humanistic threat outlined, it is first of all necessary to recognize that man's intelligence does not suffice alone to the task.

Rostand's rather discomfited remark applies here: "It appears that human intelligence faces man with problems he will certainly be prevented from resolving by intelligence alone." [64]   However, if intelligence is not sufficient to humanize the scientist, one should not jump to the opposite extreme of claiming that man can dispense with intellectual effort in this regard.   On the contrary, the humanization of scientific man will never take place unless he learns to use his intelligence more and more sharply. But he must learn to use his intelligence reflectively also, in the philosophical sense of the term, in order to penetrate beyond the factual data of science toward those realities that Heisenberg is fond of calling "the great interconnections."   If such an attitude is adopted, there is a well-founded reason to expect that science may effectively contribute to the humanization of its practitioners.   In fact Heisenberg himself holds out the intellectual search mentioned as an ideal pursuit to be undertaken by young people striving after the fullness of their humanity.   As he put it:

There will always be young people who will think also about the great interconnections, at least because they want to be honest to the end. And then it does not matter how many they are. [65]

(b) Wisdom and Dialogue: The General Structure of Humanization

Humanization is, by definition, the process through which man strives positively to mature as a person—that is, to attain to the fullness of his humanity.   The reason why man is confronted with this process is the personal character of man himself.   In fact, it is not sufficient to be human in order to be genuinely so.   Rather, one must strive to grow as a person— namely, as a being characterized chiefly by intelligence and will.   As regards these traits of man, they can be easily defined as follows.   Intelligence is the power through which man is able to perceive reality as existent and meaningful.   Will is the power though which man is able to welcome reality and contribute creatively to it.   But intelligence and will, in turn, are not disconnected from the other features of man himself.   On the contrary, they are firmly embedded in the totality that constitutes his nature. Consequently, the process of humanization has to be seen as the undertaking through which man—consciously and creatively—strives to become genuinely human in the harmony of his whole being.

What, in practice, constitutes the structure of man's self-humanizing process?   If we turn for enlightenment to history, the answer is plain. The universal tradition of mankind—present with minor variations in all cultures—has always conceived humanization as identical with the acquisition of wisdom.   As regards the acquisition of wisdom, the widespread

---

[64] Rostand (note 43), p. 240.
[65] Translated from Heisenberg (note 49), p. 333.

conviction has been that dialogue offered the most promising approach. As a consequence, therefore, we are now going to examine these two central humanistic categories in order to apply them subsequently to the situation of man living in the scientific age.

Wisdom. The term "wisdom" is still present in contemporary parlance. But its ordinary usage is more confusing than illuminating for the purpose of our discussion. For people frequently take wisdom as synonymous with prudence or even simply as a vague subjective belief that offers some general guidance for practical life. In view of our aim, therefore, we must strive to understand the original profundity and richness of the idea of wisdom, as contained especially in the sources of our Western culture. [66]

The clue to the understanding of wisdom is to be found by reflecting on the meaning of orderliness. According to the ancient adage: *Sapientis est ordinare*. That is, what characterizes the sage is the ability to organize things in an orderly fashion. But this ability, of course, must be manifold since man himself is manifold and the world in which he lives is quite various. Accordingly, the following are the main features that, in the light of tradition, can be considered typical of the sage as such.

The first characteristic refers to the intelligence of man. It may be termed *comprehensive intrinsic understanding.* ⌐To be wise, then, first of all means to understand. That is, it means to become aware of reality not only as a fact, but also as a source of meaning.⌐ This remark explains the fundamental reason why there is a strict connection between wisdom and orderliness. The connection stems from the conception of understanding just mentioned. In fact, a person understands or becomes aware of meaning only when he perceives that many beings are not just many—in their scattered multiplicity—but actually constitute a unity, through mutual interrelationship. But orderliness is precisely the general term used to designate such a mutual interrelationship of the many converging toward unity. However, what characterizes the sage from the point of view of intelligence is not merely understanding as such, but rather a specific form

---

[66] For a summary idea of wisdom see, for instance, "Wisdom," in P. Edwards, ed., *The Encyclopedia of Philosophy* (New York: Macmillan and Free Press, 1967); "Sapienza" and "Sophia" in Centro di Studi Filosofici di Gallarate, eds., *Enciclopedia Filosofica* (Florence: Sansoni, 1967/69). Useful pieces of information concerning the sapiential views of, respectively, the Primitives and the Orientals can be found in P. Radin, *Primitive Man as Philosopher* (New York: Dover, 1957) and R. G. H. Siu, *The Tao of Science: An Essay on Western Knowledge and Eastern Wisdom* (Cambridge, Mass.: MIT Press, 1957). An exhaustive discussion of the Greek and Hebraic conceptions of wisdom is "Sophia" by U. Wilckens and G. Föhrer, in G. Friedrich, ed., *Theologisches Wörterbuch zum Neuen Testament,* vol. VII (Stuttgart: Kohlhammer, 1964), pp. 465-529. I myself discussed the biblical idea of wisdom in three articles published in *Rivista Biblica* **8** (1960), 1-9, 129-143, 193-205.

of understanding. These specifications of sapiential understanding, as has been indicated, are chiefly two: comprehensiveness and intrinsicness. Comprehensiveness entails that the sage is the person who is able to realize that there is a meaning that encompasses reality as a whole. Intrinsicness entails that the sage perceives the meaning of the whole of reality as objectively present at the very core of reality itself. In short, then, the first characteristic of wisdom is the ability to perceive that reality makes sense to the full extent of the term.

The second characteristic of wisdom refers to the will of man. It may be called *responsive original activity.* To be wise, then, means to act. That is, it means that the awareness of meaning that one has attained through understanding becomes itself the source of initiative on the part of the person involved. Thus, the sage is not a mere knower or contemplator. On the other hand, however, he is also not a mere doer or activist. Rather, the sage is both a knower and a doer as far as the meaning of reality he has perceived elicits from him a personally original response. This response takes the form of welcoming the meaning perceived while striving to perceive it increasingly better. But it also takes the form of engaging oneself creatively toward an effective increase of the meaning of reality. In short, therefore, the second characteristic of wisdom is the willingness to foster the overall sense of reality by engaging one's whole being in the task.

In the light of the two characteristics considered it is easy to see why the search for wisdom has been consistently identified by people with the humanizing process itself. The essential reason is that the search for wisdom amounts to the *effective personalization of man.* Indeed man is man—as has been indicated—essentially because of his personal traits, which consist in intelligence and will. Thus the humanizing process is the same as personalization. But this is precisely what the search for wisdom amounts to—a progressive development of man's intelligence and will, that is, a continual development of man's own personality.

A second reason why the search for wisdom has been traditionally identified with humanization consists in the *dynamical character* of the sapiential striving itself. In fact, the more a person strives to be wise, the more he finds himself in a position to become still wiser—without any limit. This is the case because there is a continual, mutually enriching interaction between intelligence and will in the sapiential striving; that is, understanding generates response and response generates understanding.

To sum up it is clear, at least in principle, how man should proceed in order to achieve the fullness of his humanity in the scientific age. No less than in foregoing ages, he should strive to acquire wisdom. The only qualification is that scientific man should strive after a kind of wisdom that takes into full account the new perspectives of understanding and response disclosed by science. But here, we are faced with a major objection from the practical standpoint. How can the ordinary working scientist effectively attain wisdom? From the analysis made, would it not rather appear that he is debarred in principle from such an attainment? For, after all, we have been speaking of comprehensive intrinsic understanding as the fundamental element of wisdom. But, how is it possible to attain such an understanding in the age of science? In particular, is not the ordinary working scientist committed to increasingly specialized research—the very antithesis of sapiential understanding?

We must take quite seriously the objection outlined. For it uncovers the roots of the humanistic predicament of scientific man. It clarifies why many scientists tend to feel positively antagonistic toward the very idea of wisdom. The reason is precisely what appears to be the inherently tantalizing character of wisdom itself. In fact, wisdom keeps attracting man, even in the age of science, as the very ideal of his life. But then science, through its ever developing specialization, seems to prove that the attainment of wisdom is impossible in principle. Thus contemporary man, especially the practicing scientist, tends to feel frustrated in his most profound aspirations as a human being. And yet if the objection is serious, we need not feel discouraged by it. In fact, traditional sapiential reflection has anticipated its thrust and supplied a principle of solution in the concept of dialogue. As as consequence, we undertake now a rapid survey of this concept. The author, whose considerations we are going to summarize, is the well-known contemporary expert in this area, the philosopher Martin Buber. [67]

Dialogue. The fundamental idea of dialogue is simple. In fact, as the etymology itself says, there is dialogue wherever there is an exchange of a word or message between two beings. In other terms, dialogue is nothing but *mutual communication.* However, if the nature of dialogue is simple in theory, it can be quite complicated in practice. The obvious reason is that communication can only too easily be merely apparent instead of being real—or it can be real without being adequate. Thus, to obtain a comprehensive view of dialogue, we begin by listing various forms of pseudodialogue. Then we are going to outline the typical features of dialogue itself and its humanizing implications.

As is well known from everyday experience, there are various forms of exchanging words or messages that fall short of the definition of dialogue. One blatantly obvious case is polemic. There is no dialogue here simply because there is no mutual communication worth speaking of. In fact,

---

[67] Buber is certainly one of the leading contemporary thinkers—probably the most prominent one—in the area of dialogue. The only problem with his doctrine is its lack of systematicity. The systematization offered in the text is entirely mine.

neither of the polemizing parties has any genuine desire to learn what the
other really means to say. Rather, each party has made up his mind in
advance to confute the other and thus make his own thesis triumph. But,
if polemic is an extreme form of nondialogue, many other less pronounced
manifestations of the same phenomenon can easily be detected in ordinary
life. Specious or counterfeit dialogue ranges from the purely pragmatic
to the determinedly egotistic. An example of the former is the request
for a specific piece of information without any mutual interest of the talkers
in each other. An example of the latter is any conversation conducted with
the sole aim of seeking one's own satisfaction—be it to relieve boredom
or to show friendliness or to relish the togetherness of the moment. In
short, therefore, it must be admitted that dialogue is far from being a
normal—much less, automatic—occurrence in man's life. As for the root
of this state of affairs, it can be readily identified in man's own tendency
to avoid genuine communication. In other words, man inclines strongly
to suppress dialogue in favor of *monologue*. The latter can be defined,
with Buber, as the situation in which

... a man withdraws from accepting with his essential being another
person in his particularity ... and lets the other exist only as his own
experience, only as a "part of myself." [68]

To characterize genuine dialogue, we must first of all endeavor to
bring out the basic personal attitude that it presupposes. In this regard,
it is necessary to exclude several opinions which are rather widespread.
In particular, dialogue should not be viewed as resting on pathos, be it
under form of empathy or sympathy. In fact, to be dialogical one needs
not to project himself into the emotions of another person or even share
his feelings. For dialogue, according to the definition given, is basically
not an emotional but an intellectual enterprise—the willingness to entertain
communication. Thus we can say that the essential feature of dialogue,
its key presupposition, is substantially an attitude of *attention and honesty*.
Attention, in the sense that the person has to give heed to the message
that comes to his notice. Honesty, in the sense that the person has to
perceive the message as it is, without indulging in subjectivistic interpretations of the same. To summarize with Buber, this is the conviction that
constitutes the foundation of dialogue:

What occurs to me addresses me. In what occurs to me the world-happening addresses me. It will, then, be expected of the attentive man
that he faces creation as it happens. It happens as speech, and not as
speech rushing out over his head but as speech directed precisely
at him. [69]

---

[68] Buber, (note 21), p. 42.
[69] *Ibid.*, pp. 28-34.

On the basis of the attitude described, we can now easily detail the principal features of dialogue. To begin with, dialogue implies a relation to the other as other. Accordingly, the first feature of dialogue consists in *accepting the other in his otherness*. That is to say, a person is dialogical when he does not try to reduce the being that confronts him to anything abstract or noncommittal such as a concept, a number, or a mere instrument. On the contrary, the dialogical person welcomes the other in his concreteness, and acknowledges the message that arises from the very fact that the other is there. Buber describes this first feature of dialogue as follows:

This man is not my object; I have got to do with him. Perhaps I have to accomplish something about him; but perhaps I have only to learn something, and it is only a matter of my "accepting." [70]

The second feature of dialogue can be detected by reflecting on the first. In fact, if dialogue consists in accepting the other in his otherness, this cannot take place without an act of *positive volition* on the part of the persons involved. For, to accept the other as such, one must turn to the other with one's whole being, in complete sincerity: "from one open-hearted person to another open-hearted person." [71] In other words, positive volition is an essential feature of dialogue because, without it, the other could not be recognized—let alone accepted—in his concreteness. Actually, it is a psychologically obvious fact that other beings remain incomprehensible and meaningless to an individual as long as he does not decide to take an active interest in their concreteness.

The basic movement of the life of dialogue is the turning towards the other ... with the essential being. In this way ... out of the incomprehensibility of what lies to hand this one person steps forth and becomes a presence. Now to our perception the world ceases to be an insignificant multiplicity of points to one of which we pay momentary attention. [72]

The third feature of dialogue is but a specification of the second one. In fact, if dialogue demands an act of positive volition, this does not entail that one should take the initiative. On the contrary, a person becomes dialogical only if he is willing to respond to the message that arises from the very fact that the other is there, in his concreteness. But, of course, the dialogical person must respond without prejudices or reservations. Accordingly, the third feature of dialogue can be called *open-hearted responsiveness*. As for the form of one's own response, it can be manifold.

The words of our response are spoken in the speech, untranslatable like

---

[70] *Ibid.*, p. 27.
[71] *Ibid.*, p. 24.
[72] *Ibid.*, p. 40.

the address, of doing and letting—whereby the doing may behave like a letting and the letting like a doing. [73]

A complementary feature of dialogue emerges if one take simultaneously into account the three features examined. In fact, even though we have so far spoken of dialogue with the implicit assumption that it is an interpersonal relationship, such a restrictive interpretation is not required as a matter of principle. The reason is simply that the acceptance of the other as other, through positive volition and open-hearted responsiveness, is an attitude which can just as well apply to other beings besides people. The fourth feature of dialogue, therefore, is to make oneself open to the message conveyed by the concrete existence of other beings, *whatever they may be*. Buber rightly emphasizes the importance of this feature of dialogue.

It by no means needs to be a man of whom I become aware. It can be an animal, a plant, a stone. No kind of appearance or event is fundamentally excluded from the series of the things through which from time to time something is said to me.... The limits of the possibility of dialogue are the limits of awareness. [74]

After the foregoing brief analysis, it is not difficult to realize why the dialogical attitude is important from the humanistic point of view. The principal reason consists in the fact that dialogical considerations give back to knowledge its genuinely human dimension as a personal relationship with concretely existing reality. As Buber points out, knowledge is by itself always so human, even the knowledge of entities that people tend easily to take for granted and consider meaningless.

That which will eventually play as an accustomed object around the man who is fully developed, must be wooed and won by the developing man in strenuous action. For no thing is ready-made part of an experience; only in the strength, acting and being acted upon, of what is over against men, is anything made accessible. [75]

As regards the humanistic difference made by the dialogical attitude, let us examine a simple *example*. Buber discusses the various angles from which one can study a tree. [76] The tree can be investigated in many abstract manners: measurement of its height and volume, taxonomic classification of its species, explanation of its specific features by tracing them back to evolution and to the physico-chemical properties of the materials that make up the tree itself. All of these ways of approaching the tree are

---

[73] *Ibid.*, p. 35.
[74] *Ibid.*, p. 27.
[75] M. Buber, *I and Thou*, trans. R. G. Smith (Edinburgh: T. Clark, 1958), p. 26.
[76] *Ibid.*, pp. 7f.

important and even necessary in order to know it.   And yet it goes without saying that such cognitive approaches, either singly or collectively, are insufficient to provide a genuinely human knowledge of the tree itself.   The reason is obvious in the sense that, were we to be satisfied with the results these approaches offer, the real tree would disappear and be replaced by a man-made representation, the collection of abstract sets of observational and theoretical data.   In other terms it can be said that the approaches enumerated are important but inadequate simply because they provide much information *about* the tree, yet they are not *the* cognition of the tree itself.   What attitude does lead to the cognition in question so that man may feel satisfied in his cognitive desire?   Obviously, only the dialogical attitude.   For, indeed, the tree I want to know is not an abstract entity, but a very concrete one.   Hence I can know it satisfactorily only if I turn to it in all its concreteness.   To summarize with Buber:

The tree is no impression, no play of my imagination, no value depending on my mood; but it is bodied over against me and has to do with me, as I with it—only in a different way.

The dialogical attitude is vital for man as a knower because it makes knowledge into an awareness not only of truth but of living truth.   "I know as a living truth only concrete world reality which is constantly, in every moment, reached out to me." [77]   We can now formulate in a few corollaries the overall humanistic significance of dialogue itself.

Dialogue is essentially a personalizing process.   Man, as Buber stresses, is twofold.   He can be an object among objects or can make himself into a mature person.   He is an object when he refuses to be dialogical—when he treats every other being as an "It."   He becomes a mature person when he adopts the dialogical attitude—that is, when he addresses every other being as a "Thou."

The attitude of man is twofold, in accordance with the twofold nature of the primary words which he speaks.... The primary word *I-Thou* can only be spoken with the whole being.   The primary word *I-It* can never be spoken with the whole being. [78]

I become through my relation to the *Thou*; as I become *I*, I say *Thou*. All real living is meeting. [79]

Dialogue makes humanization accessible to everybody.   Dialogue, by personalizing man, humanizes him.   But, by the same token, it brings humanization within the reach of every human being.   Indeed, to be dialogical, man needs nothing more than an earnest desire to be human.   For, after

[77] Buber, *Between Man* (note 21), p. 30.
[78] Buber, *I and Thou* (note 75), p. 3.
[79] *Ibid.*, p. 11.

all, dialogue does not require any special gift besides good will and serious dedication.

The life of dialogue is no privilege of intellectual activity like dialectic.... There are no gifted and ungifted here, only those who give themselves and those who withhold themselves. [80]

Dialogue humanizes man by integrating him in the totality of the human community. Man, owing to his unlimited social nature, can be fully human only if he participates in the universal community of men as such. But dialogue makes each individual attain such a universal participation, at least virtually. For dialogue is precisely the factor that gives rise to community among people, a community virtually universal. Indeed, dialogue joins people to each other because it discloses the much they have in common—especially the values and ideals that constitute the very center of their lives.

The true community does not arise through people having feelings for one another (though indeed not without it), but through, first, their taking their stand in living mutual relation with a living Center, and, second, their being in living mutual relation with one another. [81]

Dialogue humanizes man by integrating him in the totality of the cosmos. Man, owing to his all-encompassing nature, can be fully human only if he is united with the cosmos in its entirety. But dialogue establishes a direct relationship, at least virtual, between each individual and the cosmos as a whole. For dialogue discloses the unitary character of the cosmos itself. It manifests that each component of the world is a source of message and meaning as a constitutive part of the world in its totality.

But when a man draws a lifeless thing into his passionate longing for dialogue, lending it independence and as it were a soul, then there may dawn in him the presentiment of a world-wide dialogue, a dialogue with the world-happening that steps up to him even in his environment, which consists partly of things. [82]

Dialogue humanizes man by making him attain unconditional meaning. Man, owing to his deep-seated aspiration after the nonrelative and ultimate, can become fully human only if he succeeds in overcoming any relativity or conditionality in his striving after meaning. But dialogue is the factor that takes man beyond all relativity and conditionality of meaning. For the dialogical attitude is precisely based on the conviction that not only is meaning present in reality, but this meaning is ultimately nonrelative and nonconditional. Indeed, to be dialogical, a person must welcome

[80] Buber, *Between Man* (note 21), p. 54.
[81] Buber, *I and Thou* (note 75), p. 45.
[82] Buber, *Between Man* (note 21), p. 57.

the other as something unconditional—as the bearer of a message which is objectively valid and commands unrestricted recognition. Thus, by making man transcend the relativeness of his subjectivity, dialogue enables him to become fully human.

Human life possesses absolute meaning through transcending in practice its own conditioned nature, that is, through man's seeing that which he confronts, and with which he can enter into a real relation of being to being, as not less real than himself, and through taking it not less seriously than himself. Human life touches on absoluteness in virtue of its dialogical character. [83]

Dialogue humanizes man by establishing a personal communion with the absolute itself. The innermost aspiration of man as a person is even more far-reaching than the attainment of unconditional meaning. For man feels in his heart of hearts that he can be fully human only if he enters a personal relationship with the ultimate objective source of such a meaning. But dialogue assists man in achieving also this most arduous and sublime goal. Indeed dialogue—if the term is taken in the totality of the concrete commitment that it involves—consists in not just attaining unconditional meaning, but in personally engaging oneself toward such a meaning, at its very source or ultimate origin. This source is what deserves to be called the absolute. Hence dialogue is capable of humanizing man by establishing a personal communion between the absolute and himself. In practice, to be sure , one should not expect that dialogue will always succeed in establishing the communion in question explicitly. Nevertheless, such a communion is assured every time a person decides to be dialogical to the end. In fact, the dialogical attitude consists in taking the unconditional meaning of reality so seriously as to feel bound by it. But such an engagement has always been considered typical of the authentic communion with the absolute, as witnessed by the conviction of all religions. Hence, the person who is consistently dialogical becomes genuinely human in the comprehensive meaning of the term—including the humanization that is ordinarily connected with religion. Conversely, however, the person who refuses to be dialogical can hardly be human, despite all appearances or labels.

He who knows the world as something by which he is to profit knows God also in the same way. His prayer is a procedure of exoneration heard by the ear of the void. He—not the "atheist," who addresses the Nameless out of the night and yearning of his garret-window—is the godless man. [84]

---

[83] *Ibid.*, p. 204.
[84] Buber, *I and Thou* (note 75), p. 107.

Dialogical wisdom: a summary.   In the light of the preceding it is now clear, at least in principle, why—despite all the complexity brought about by science—wisdom justifiably remains the ideal for man striving after self-humanization.   The reason consists in the fact that the dialogical attitude enables each person to consider wisdom as an effectively attainable goal, both theoretically and practically.

Theoretically speaking, the dialogical attitude removes the main obstacle that seems to invalidate the sapiential ideal in the age of science. In fact, wisdom invites man to strive after comprehensive intrinsic understanding.   But how is it possible to view this as an attainable goal, since science has disclosed the inexhaustible character of the intelligibility of reality?   The answer is that this is possible through dialogue. ⌈Actually, dialogue makes each individual capable of overcoming the inherent limitations of his individuality.   To be sure, dialogue does not transform any individual into a superman capable of grasping explicitly the totality of the intelligibility of reality.   The individual continues to be quite limited as far as his particular abilities go.   And yet he is freed of the suffocating sense of hopelessness that seems to be entailed by his individuality.   Science itself is the most convincing evidence of the limitless resourcefulness resulting from the dialogical attitude.   In fact, no individual scientist can explore the entire intelligibility of observable reality.   Nevertheless, if he adopts a dialogical attitude toward all his colleagues, he can really know such a universal intelligibility.   Consequently, if a person adopts in general the dialogical attitude toward all the possible ways with which mankind can ascertain meaning in reality—that is, besides science, art, philosophy, religion, and the like—it is clear that the ideal goal of intrinsic comprehensive understanding remains valid in the age of science.   In a way, perhaps, one could even say that it remains more valid than ever because, through science, man has learned in an unprecedented way to overcome the limitations of his individuality by seeking understanding on a corporate basis.

Practically speaking, dialogue gives new encouragement to man to strive after wisdom simply because it points out the humanness as well as facility of such a striving.   Indeed, without the insights brought by reflection on dialogue, a person could be easily scared by the earnestness of the sapiential endeavor.   He could think that wisdom would demand an exclusive dedication and asceticism that would make it an impractical pursuit for anyone who must live in the tensions and worries of ordinary existence.   But the dialogical attitude removes such a disheartening obstacle by insisting on the responsive character of wisdom itself.   To be sure, such a responsiveness can entail great demands.   For it requires that man be originally creative.   And yet, reflection on dialogue clarifies that the demands of wisdom are not really onerous, much less unbearable.   The reason is precisely the spirit of dialogue.   This spirit invites each individual to meet reality as it presents itself to him, and to welcome it with humble confidence, without fear. For, after all, each manifestation of reality is a carrier of meaning. Thus a person moved by the spirit of dialogue is enabled to perceive a quite practical opportunity to achieve wisdom in any encounter that he happens to make, in any undertaking that lies at hand. As for the outstanding exploits that sometimes seem to be required in order to achieve wisdom the way the celebrated sages of history did, even then the dialogical spirit removes all worries. For it shows that the acquisition of wisdom is essentially and exclusively a matter of response. To be sure, this response must be wholehearted and consistent to the end. Yet a person who sincerely intends to become wise can do nothing better than strive to be dialogically responsive.

To sum up, even in the age of science the effective approach to humanization continues to remain the one embodied in the traditional idea of wisdom. Only, people should now acknowledge more explicitly than in the past that they cannot actually attain wisdom unless they strive after it on a dialogical basis. Thus, the humanizing ideal of contemporary man should be defined more precisely. Instead of speaking merely of wisdom, it is now necessary to speak specifically of dialogical wisdom.

(c) Sapiential Openness: The Humanization of Scientific Knowing

After having ascertained the overall structure of humanization and its enduring relevance, we revert to our central question. How can man effectively humanize himself by doing science? The general answer is that this person must be realistic enough to acknowledge the very serious humanistic difficulties engendered by his career and be courageously determined to overcome them by striving after dialogical wisdom. As for specific indications, they can be easily obtained by reflecting on the two main practical attitudes demanded by the sapiential ideal, namely openness and engagement. To complete our discussion, therefore, we are now going to survey rapidly these two remaining topics as relating to the everyday life of the scientific researcher.

The basic attitude of the sage is that of *sapiential openness*. What is meant by this term? A positive effort to make oneself available to reality and its meaning as far as accessible to the thoroughly and consistently open-minded investigator of truth. In other words, the sage is the person who strives, by means of reflection, to acknowledge explicitly the entire message that reality is able to convey to a genuinely thoughtful person. In particular, the sage strives to perceive the deep meaning of reality as a matter of habit; that is, he does not seek for meaning only occasionally or in selected areas, but constantly and comprehensively and in increasing depth.

Is the attitude of sapiential openness effectively practicable while remaining within the scientific profession? The answer is positive on three counts, as can be readily seen by considering the threefold posture required by scientific creativity. Indeed, creative science demands three types of openness: to things, to persons, and to meaning. But this threefold openness is, in turn, the attitude that can and should humanize scientific knowing from within.

To begin from the very beginning, the fundamental prerequisite of science is obviously that of *openness to things*. The scientist, in fact, can become such only if allows himself to be moved by the message brought to his notice by some of the observable things that surround him. Actually, if we look back at history, why did Galileo and Kepler and Newton give rise to science? Simply because they took seriously the message conveyed to their minds by familiar objects and events, such as the sun and moon and planets, the falling apple, the swinging pendulum, the cannonballs. Likewise, in recent times, the starting point of the creative work of Bohr and Heisenberg was the fascination with such commonplace phenomena as the permanency of the properties of water and the stability of chemical compounds.[85] But, if this is the case, one should not look upon this very first manifestation of science as something that is humanistically trivial. For the achievement that it implies is great and the indication that it contains is even greater. The achievement implied is, of course, that of being authentically factual while overcoming the instinctive tendency to look at things according to the view that is considered orthodox in one's own cultural environment. For, as Rostand remarked caustically, "Facts laugh at orthodoxies. Nature is unorthodox."[86] But this first manifestation of science contains a very great humanistic indication because it points to an openness to things that effectively belongs in the sapiential realm. In fact, if one is consistent with the spirit of science, should one not try to understand the entire message conveyed by the things themselves—even that part of the message that science as such cannot decipher? For, indeed, there is such a message—the one, namely, that man can grasp but only by means of philosophical reflection. Hence it would be quite unscientific as well as unsapiential to stop at the orthodox interpretation that sees science as the uniquely important and exclusively respectable form of knowing.

The spontaneous objection, however, is that the average scientist has neither the training nor the time to perform the philosophical reflection indicated. This objection is valid, although it should never be used as an excuse for avoiding reflection. For reflection is indispensable for humanization. However, if the objection is valid, science itself provides at least a principle of solution. This consists in *openness to persons*, which is typical of the scientific practice as such. In fact, the cognitive limitation of the individual is openly acknowledged by scientists even in scientific

---

[85] For details see Heisenberg, *Teil* (note 49), pp. 37 and 60f.
[86] Rostand (note 43), p. 255.

research. But they do not see in such a situation a reason for discouragement. On the contrary, they successfully overcome this obstacle by sharing
mutually their expertise on a dialogical basis. This practice has been
aptly summarized, for instance, in the preface of Heisenberg's memoirs:

Science is made by men.... Science rests on experiments. It arrives
at its results through the conversations of those who are active in it,
in that they consult each other about the interpretation of the experiments.... Science comes into existence in conversation. [87]

But if openness to persons is so proper to science, who could deny that
the spirit of science exhorts its practitioners to adopt the same dialogical
attitude also with regard to the exploration of the all-important philosophical questions for which they have neither the training nor the leisure?
The great scientists, in point of fact, normally heed this exhortation. An
outstanding example of this conduct is the entire book by Heisenberg, from
which we just quoted.

The third form of scientific openness with sapiential implications is
the most comprehensive one: *openness to universal meaning*. To be sure,
many scientists manage to ignore this kind of openness and even boast
about their doing so. However, they act this way not because of science,
but rather despite it. In fact, the genuinely creative scientist can hardly
fail to be interested in universal meaning—both as a presupposition and
a conclusion of his creative endeavors. An example is contained in a revealing answer by Einstein. When asked how he discovered his relativity
theory, he replied that "he was so strongly convinced of the harmony of
the universe." [88] A more detailed exemplification of the way science leads
its practitioners to the openness to universal meaning is contained in an
enthusiastic passage penned by the famous astronomer and philosopher
John Herschel:

A mind which has once imbibed a taste for scientific enquiry, and has
learned the habit of applying its principles readily to the cases which
occur, has within itself an inexhaustible source of pure and exciting
contemplations—one would think that Shakespeare had such a mind
in view when he describes a contemplative man as finding
    "Tongues in trees—books in the running brooks
        sermons in stones—and good in every thing."
[Such a person] walks in the midst of wonders: every object which falls
in his way elucidates some principle, affords some instruction, and impresses him with a sense of harmony and order. [89]

---

[87] Translated from Heisenberg, *Teil* (note 49), p. 9.

[88] Statement reported by H. Reichenbach in P. A. Schilpp, ed., *Albert Einstein, Philosopher-Scientist* (New York: Harper Torchbooks, 1959), p. 292.

[89] J. F. W. Herschel, *Preliminary Discourse on the Study of Natural Philosophy* (London:
Longman, Rees, Orme, Brown and Taylor, 1830), pp. 14f.

At the present time, when the idea is widespread that science is merely a technical undertaking, it would be surprising if many a practicing scientist would not feel tempted to dismiss the foregoing passage as "sentimentalism." And yet, were he to do so, he would not only wrong humanism, but science itself, as history shows. In fact, the scientific fecundity of Herschel's position proved supreme, notably because of the decisive influence it exerted on the young Darwin. Shortly before participating in the *Beagle* expedition, Darwin read Herschel's work along with Alexander von Humboldt's *Personal Narrative of Travels,* another no less "sentimental" piece of writing. Of both works the discoverer of evolution was to write in his old age:

[These books] stirred up in me a burning zeal to add even the most humble contribution to the noble structure of Natural Science. No one or a dozen other books influenced me nearly so much as these two. [90]

To sum up, it is not only desirable but possible to humanize scientific knowing and thereby attain in practice the fundamental goal of wisdom, namely sapiential understanding. This is possible precisely because the spirit of science itself exhorts the researcher not to consider himself satisfied until he has succeeded in practicing consistently the attitude of openness that is so typical of the scientific endeavor. To be sure, science does not suffice alone to this end, not even to motivate the necessity of total openness. And yet, science points the way because it makes man aware that the universe has a truly comprehensive message and meaning—a message and meaning, in turn, that far transcend what can be seen and touched. At this point, then, science begins to be authentically humanizing—when it succeeds in persuading man to seek for meaning not just with his eyes but, as it were, with his heart. For, as Heisenberg put it charmingly, quoting with approval the "Little Prince" of Antoine de Saint-Exupéry: "One cannot see well except with the heart; the essential is invisible to the eyes." [91]

It is inspiring that, for all the pressures of their work, the sapiential interpretation of scientific knowing is not ignored by great researchers but rather insisted upon increasingly. An example is the exhortation by Synge, a contemporary Nobel laureate in chemistry. He invites investigators to do science "for the good of the soul."

In essence, this natural historian's plea is for the restoration of something which has tended to become lost as civilization has become more

---

[90] In N. Barlow, ed., *The Autobiography of Charles Darwin (with original omissions restored)* (New York: Norton, 1969), p. 68.

[91] Translated from W. Heisenberg, "Naturwissenschaft und Technik im politischen Geschehen unserer Zeit," in his *Schritte über Grenzen: Gesammelte Reden und Aufsätze* (Munich: Piper, 1971), pp. 182-186; especially p. 186.

complicated—that is, a view of humanity in perspective against our natural background. This should inculcate a proper sympathy, respect, fear, and love for the natural universe of which we are a part. [92]

(d) Sapiential Engagement: The Humanization of Scientific Doing

The second, complementary attitude of the sage is that of *sapiential engagement*. We can define this term readily by recalling that the sage is not only a knower, but also a doer—simultaneously, a knower and a doer. Accordingly, the attitude of sapiential engagement consists in putting oneself entirely at the disposal of the deep meaning of reality that one has perceived. In other words, the sage is the person who strives, by means of practice, to express in his life the message of truth that has come to his notice. In particular, the sage strives to translate cognition into edifying action as a matter of habit; that is, he does not respond to meaning only occasionally or in selected areas, but constantly and comprehensively and with increasing creativeness.

Is the attitude of sapiential engagement effectively implementable while remaining within the scientific profession? The answer is definitely positive if one takes into account the spirit of science in its entirety as well as the humanizing indications that arise therefrom. For, after all, science is essentially a matter of engaged response to meaning—a response in which the person dedicates himself wholly with increasing creativeness, from the first tentative observations to the mighty theoretical syntheses. In other words, science—just as wisdom—does not consist merely in knowing, but also very much in doing. Genuine scientific doing consists, precisely, in one's own total dedication to the task. But, if this is so, it is only consistent with the spirit of science that a scientist extend this same attitude of engagement to all other areas affecting his personality as a concretely existing human being.

However, even if the attitude of sapiential engagement is possible to a person who does science, is it really desirable? After all, a person can be quite productive as a scientist without bothering about such an attitude. Even worse, if a person engages himself sapientially, is not this superfluous dedication disadvantageous to the scientific endeavor itself? For, of course, each human being has only a limited amount of time and energy. Thus, if one is a scientist, why should one not concentrate his engagement exclusively on science as such?

The objection is impressive. In fact, numerous scientists seem to dispense more or less completely with the sapiential engagement, and be all the better off for it. Nonetheless, we must deny the validity of the

---

[92] R. L. M. Synge, "Science for the Good of Your Soul," in Goldsmith and Mackay (note 48), pp. 170-178; especially p. 178.

objection. For, of course, it may be true that a person who does not care for wisdom can be more scientifically productive than another who does. However, it would be hardly justified to make this into a causal explanation. In fact, after all, science demands the whole man. But man is a unitary being. Hence any form of behavior that introduces a split in the human personality rather hinders than favors science itself. Actually, psychological and sociological surveys of the scientific community show that science is generally considered by its practitioners as the embodiment of the ideal of the whole man, in the sapiential acceptation of the term (see Chapter 7, Section 1c). As a consequence it must be admitted that the sapiential engagement arising from the spirit of science is far from detrimental to science itself.

A conspicuous example of the way the sapiential engagement can humanize scientific doing is offered by the life of Otto Hahn. This famous discoverer of nuclear fission had to go through most trying experiences because of his achievement. After having constantly compaigned against the military use of nuclear power, he reached the brink of suicide upon hearing about the Hiroshima holocaust. And yet he recovered and for many years, inspired other scientists with a renewed dedication to a peaceful reconstruction of mankind. The secret of his unquenchable courage and optimism? Heisenberg, a long-time friend and collaborator, can only point to a frame of mind that deserves to be called sapiential engagement. As he stresses, this frame of mind seems to account for the entire greatness of Hahn, both as a scientist and as a man. In Heisenberg's own words:

Perhaps his outstanding human and scientific success was deeply rooted in his unconditional "Yes" to his co-workers and friends. [93]

At this point it is clear that the practice of science and the search for wisdom are far from incompatible. Thus we could close our discussion but for a major objection that threatens to invalidate our entire effort. In fact, if the scientist is supposed to practice sapiential openness and engagement in his entire life, toward whom or what has he, in the last analysis, to be so totally dedicated? In other terms, to refer to Heisenberg's words, if the scientist is expected to say "Yes," to what being, if any, is this personal word to be addressed? In former times, one would have normally replied that the ultimate term of reference of the sapiential endeavor was the absolute or God himself, as the primal source of meaning and ideal. And yet, nowadays, such an answer cannot be taken as automatically persuasive. Or, at the very least, if it is to be deemed persuasive, a person would demand to know all the reasons why. But such

---

[93] *Physics Today* **21** (1968), p. 102.

an investigation certainly exceeds both the time and the training of the average scientist. As a consequence it might appear that, to the working scientist as such, the search for wisdom presents itself as not sufficiently motivated. Perhaps it can even give the impression of being a monumental self-delusion. Indeed, if there were no one to acknowledge the "Yes" of man—if this word would be shouted into the void—would not the ideal of wisdom be a cruel mockery? Would not man annihilate himself by chasing nothingness?

The dramatic earnestness of the situation confronting the person who aspires after humanization in the practice of science should be acknowledged. In particular, it should be admitted that the root of this situation lies in the religious question. For such a question is not an artificial or secondary problem, but affects the very center of the personality of man in his search for meaning. Nevertheless, there is no justification for nihilist pessimism here—just the opposite. Indeed, if the situation is so dramatic, the sensible person has no alternative but to face it with total consistency and dedication. This very determination will enable him to find a solution which is in keeping with both the spirit of science and that of wisdom.

To realize how much the spirit of science has in common with that of wisdom, it is sufficient to envision the general attitude of the scientist while pursuing scientific research. The scientist is not a disembodied mind fueled by clear and distinct ideas. He is not, in other words, a person who refuses to commit himself to a message of meaning until he has exhaustively checked its objectivity by means of geometric reasonings. On the contrary, the scientist is a person who operates on the strength of a meaning that is experientially perceived, but whose certainty is as yet logically unproved and even unprovable. In fact, the whole of scientific research is truly a search—the search for something whose presence has been felt in nature, but whose critical justification is still missing. The justification itself will be apodictically obtained only when the discovery is made, often after many years of painstaking searching. But, if this is the case, should one say that the scientist who dedicates all his life to research acts foolishly simply because he cannot give apodictic evidence of the validity of his original insight? Certainly not. In fact, from the very beginning, the researcher knows in his heart of hearts that the goal of his search is not a figment of his imagination, but rather an objective reality that is truly present—though, as yet, in an inexplicit way. Hence the genuine scientist cannot accept at all the view that looks upon his research as an insufficiently motivated effort. On the contrary, he is certain that his effort is supremely motivated because it aims at incarnating in the life of man a message without which man himself would not be sufficiently human. The outcome of scientific research, in fact, is that of humanizing man a little more by disclosing to him some hitherto unsuspected aspects of the overall intelligibility of nature. Thus the genuine scientist would feel guilty, rather than justified, were he to refuse his total engagement in scientific research. For, without his engagement, the message of meaning he had originally perceived would be evacuated, and man, through his fault, would fail to reach the ideal to which he had been called.

We can now apply the foregoing considerations to man's sapiential engagement. If scientific research justifies the dedication of man, the search for wisdom justifies his dedication even more, on two main counts: first, because the goal of science is only limitedly humanizing, whereas that of wisdom is totally so; second, because the experiential foundation of the search for wisdom is more solidly established than even that which underlies scientific research. In fact, the awareness of an intrinsic intelligibility in nature is granted to relatively few people and, as a cultural phenomenon, it has appeared only recently in the history of mankind. But the awareness of an ultimate source of meaning in reality as a whole is an experience that is shared by man as such, throughout space and time. Indeed, it is remarkable that all religious and moral systems of mankind—though differing in everything else—have always agreed on the one point that man must dedicate himself wholly to the ultimate source of objective meaning if he is to effectively humanize himself.

To reply directly to the objection under discussion, we can now say that the scientific practitioner should not fear to adopt the attitude of sapiential engagement. He should not hesitate to say "Yes" with his entire being. For, by so doing, he will experience—even more rewardingly than in science itself—that his "Yes" will not be dissipated into the void of meaninglessness. On the contrary, this personal word will come back to him in the form of a greater awareness of meaning. As a consequence, the scientist himself will be more fully human and the humanistic significance of science will be entirely vindicated.

In brief, it is clear in what sense science can and should humanize the working scientist effectively. There is no question here of great intellectual gifts or other resources—such a plentiful leisure. What is necessary and sufficient is nothing more, but also nothing less, than a consistent fidelity to the spirit of science itself. To put it in other words, the humanistic significance of science is a fact. But it operates like a moral summons rather than a statement of factual evidence. Hence the humanism of science is essentially a question of one's own personal life rather than an issue to be faced mainly on the theoretical level or, even less, by means of a technological approach. For science is humanistic only to the extent that man develops the potentialities of the spirit that gave birth to science itself and keeps it thriving, today. In this sense, then, it is clear that scientific man should be authentically human—not despite his science, but because of it.

BIBLIOGRAPHY

The following is a list of books I found helpful toward understanding the history, philosophy, psychology, and sociology of science as well as the attending problems.

I. Science in General

Baker, J. R., *Science and the Planned State* (New York: Macmillan, 1945).

Barber, B., *Science and the Social Order* (New York: Free Press, 1952).

——, and Hirsch, W., eds., *The Sociology of Science* (New York: Free Press, 1962).

Bartlett, F., *Thinking: An Experimental and Social Study* (New York: Basic Books, 1958).

Bavink, B., *The Natural Sciences: An Introduction to the Scientific Philosophy of Today,* trans. H. S. Hatfield (New York: Appleton, 1932).

Bernal, J. D., *Science in History* (New York: Hawthorn, 1965).

Bernard, C., *An Introduction to the Study of Experimental Medicine,* trans. H. C. Green (New York: Dover, 1957).

——, *Philosophie: Manuscrit Inédit* (Paris: Boivin, 1937).

Bertholet, E., *La Philosophie des Sciences de Ferdinand Gonseth* (Lausanne: L'Age d'Homme, 1968).

Beveridge, W. I. B., *The Art of Scientific Investigation* (New York: Vintage, 1957).

Bibby, C., ed., *The Essence of T. H. Huxley: Selections from His Writings* (New York: St. Martin's, 1967).

Blake, R. M., Ducasse, C. J., and Madden, E. H., *Theories of Scientific Method: The Renaissance through the Nineteenth Century* (Seattle: University of Washington Press, 1960).

Boas, G., *The Challenge of Science* (Seattle: University of Washington Press, 1965).

Boas, M., *The Scientific Renaissance: 1450-1630* (New York: Harper & Row, 1962).

Bronowski, J., *The Common Sense of Science* (Cambridge, Mass.: Harvard University Press, 1963).

——, *The Identity of Man* (Garden City, N. Y.: Natural History Press, 1965).

——, *Science and Human Values* (New York: Harper Torchbooks, 1965).

Bury, J. B., *The Idea of Progress: An Inquiry into Its Growth and Origin* (New York: Dover, 1960).

Bush, V., *Science Is Not Enough* (New York: Morrow, 1967).

Butterfield, H., *The Origins of Modern Science: 1300-1800* (New York: Collier, 1962).

Campbell, N. R., *What is Science?* (New York: Dover, 1962).

Childe, V. G., *Society and Knowledge* (New York: Harper & Row, 1956).

Compton, A. H., *The Human Meaning of Science* (Chapel Hill: University of North Carolina Press, 1940).

Compton, K. T., *A Scientist Speaks* (Cambridge, Mass.: MIT Undergraduate Association, 1955).

Conant, J. B., *Modern Science and Modern Man* (New York: Columbia University Press, 1952).

——, *Scientific Principles and Moral Conduct* (London: Cambridge University Press, 1967).

Crombie, A. C., *Medieval and Early Modern Science,* 2 vols. (New York: Doubleday Anchor Books, 1959).

——, *Robert Grosseteste and the Origins of Experimental Science* (New York: Oxford, 1962).

Dampier, W. C., *A History of Science and Its Relations with Philosophy and Religion* (London: Cambridge University Press, 1966).

de la Vaissière, J., *Méthodologie Scientifique: Methodologie Dynamique Interne* (Paris: Beauchesne, 1933).

de Santillana, G., *The Origins of Scientific Thought* (New York: Mentor Books, 1961).

——, *Reflections on Men and Ideas* (Cambridge, Mass.: MIT Press, 1968).

Eddington, A., *Science and the Unseen World* (London: G. Allen, 1929).

Eiduson, B. T., *Scientists: Their Psychological World* (New York: Basic Books, 1962).

Eiseley, L., *The Unexpected Universe* (New York: Harcourt, 1969).

Enriques, F., *Signification de l'Histoire de la Pensée Scientifique* (Paris: Hermann, 1934).

——, *Problems of Science,* trans. K. Royce (La Salle, Ill.: Open Court, 1943).

Freedman, P., *The Principles of Scientific Research* (New York: Pergamon, 1960).

Gemant, A., *The Nature of the Genius* (Springfield, Ill.: Charles Thomas, 1961).

George, W. H., *The Scientist in Action: A Scientific Study of His Methods* (New York: Emerson, 1938).

Geymonat, L., *Filosofia e Filosofia della Scienza* (Milan: Feltrinelli, 1960).

Gregory, R., *Discovery: On the Spirit and Service of Science* (New York: Macmillan, 1916).

Hadamard, J., *An Essay on the Psychology of Invention in the Mathematical Field* (New York: Dover, 1954).

Hagstrom, W. O., *The Scientific Community* (New York: Basic Books, 1965).

Hall, A. R., *The Scientific Revolution (1500-1800): The Formation of the Modern Scientific Attitude* (Boston: Beacon, 1956).

Hanson, N. R., *Patterns of Discovery: An Inquiry into the Conceptual Foundations of Science* (London: Cambridge University Press, 1958).

Haskings, C. P., ed., *The Search for Understanding* (Washington D.C.: Carnegie Institution, 1967).

Heitler, W., *Man and Science* (New York: Basic Books, 1963).

Hennemann, G., *Naturphilosophie im 19. Jahrhundert* (Freiburg: Alber, 1959).

Herschel, J. F. W., *Preliminary Discourse on the Study of Natural Philosophy* (London: Longman, Rees, Orme, Brown, and Taylor, 1830).

Hildebrand, J. H., *Science in the Making* (New York: Columbia University Press, 1962).

Jordan, P., *Science and the Course of History,* trans. R. Manheim (New Haven: Yale University Press, 1955).

Kuhn, T., *The Structure of Scientific Revolutions* (Chicago: University of Chicago Press, 1962).

Lindsay, R. B., *The Role of Science in Civilization* (New York: Harper & Row, 1963).

Lorenz, K., *Gestaltwahrnehmung als Quelle wissenschaftlicher Erkenntnis* (Darmstadt: Wissenschaftliche Buchgesellschaft, 1964).

McKellar, P., *Imagination and Thinking: A Psychological Analysis* (London: Cohen and West, 1957).

Margenau, H., *Open Vistas: Philosophical Perspectives of Modern Science* (New Haven: Yale University Press, 1961).

Medawar, P. B., *Induction and Intuition in Scientific Thought* (Philadelphia: American Philosophical Society, 1969).

Merz, J. T., *A History of European Scientific Thought in the Nineteenth Century,* 2 vols. (New York: Dover, 1965).

Métraux, G. S., and Crouzet, F., eds., *The Evolution of Science: Readings from the History of Mankind* (New York: Mentor Books, 1963).

————, and ————, *The Nineteenth-Century World*: *Readings from the History of Mankind* (New York: Mentor Books, 1963).

Meyerson, E., *De l'Explication dans les Sciences* (Paris: Payot, 1921).

————, *Du Cheminement de la Pensée* (Paris: Alcan, 1931).

————, *Identity and Reality*, trans. K. Loewenberg (New York: Dover, 1962).

Nagel, E., *The Structure of Science*: *Problems in the Logic of Scientific Explanation* (New York: Harcourt, 1961).

Nash, L. K., *The Nature of the Natural Sciences* (Boston: Little, Brown, 1963).

Needham, J., ed., *Science, Religion and Reality* (New York: Braziller, 1955).

Neugebauer, O., *The Exact Sciences in Antiquity* (New York: Harper Torchbooks, 1962).

Obler, P. C. and Estrin, H. A., eds., *The New Scientist*: *Essays on the Methods and Values of Modern Science* (New York: Doubleday Anchor Books, 1962).

Ostwald, W., *Grosse Männer*: *Studien zur Biologie des Genies* (Leipzig: Akademische Verlagsgesellschaft, 1910).

Pearson, K., *The Grammar of Science* (London: Dent, 1937).

Peirce, C. S., *Essays in the Philosophy of Science* (New York: Liberal Arts Press, 1957).

Pelz, D. C., and Andrews, F. M., *Scientists in Organizations*: *Productive Climates for Research and Development* (New York: Wiley, 1966).

Piaget, J., *Sagesse et Illusions de la Philosophie* (Paris: Presses Universitaires de France, 1965).

Poincaré, H., *Science and Method*, trans. F. Maitland (New York: Dover).

————, *Science and Hypothesis*, trans. W. J. G. (New York: Dover, 1952).

————, *The Value of Science*, trans. G. B. Halsted (New York: Dover, 1958).

————, *Mathematics and Science*: *Last Essays*, trans. J. W. Bolduc (New York: Dover, 1963).

Polanyi, M., *The Study of Man* (Chicago: University of Chicago Press, 1963).

————, *Science, Faith and Society* (Chicago: University of Chicago Press, 1964).

————, *Personal Knowledge*: *Towards a Post-Critical Philosophy* (New York: Harper Torchbooks, 1964).

————, *The Tacit Dimension* (New York: Doubleday Anchor Books, 1966).

Popper, K. R., *The Logic of Scientific Discovery* (New York: Basic Books, 1959).

————, *Conjectures and Refutations*: *The Growth of Scientific Knowledge* (New York: Basic Books, 1962).

Price, D. J. de S., *Science since Babylon* (New Haven: Yale University Press, 1962).

————, *Little Science, Big Science* (New York: Columbia University Press, 1963).

Purver, M., *The Royal Society*: *Concept and Creation* (Cambridge, Mass.: MIT Press, 1967).

Rabi, I. I., *Science*: *The Center of Culture* (New York: Harcourt, 1970).

Randall, J. H., Jr., *The School of Padua and the Emergence of Modern Science* (Padua: Antenore, 1961).

Roe, A., *The Making of a Scientist* (New York: Dodd, Mead, 1953).

Rostand, J., *The Substance of Man*, trans. I. Brandeis (New York: Doubleday, 1962).

Russell, B., *The Scientific Outlook* (New York: Norton, 1959).

————, *Religion and Science* (New York: Oxford, 1961).

Sambursky, S., *The Physical World of the Greeks*, trans. M. Dagut (New York: Collier, 1962).

Sarton, G., *The Life of Science*: *Essays in the History of Civilization* (New York: Abelard-Schuman, 1948).

————, *The Study of the History of Mathematics and the Study of the History of Science* (New York: Dover, 1957).

————, *On the History of Science*: *Essays* (Cambridge, Mass.: Harvard University Press, 1962).

————, *The History of Science and the New Humanism* (Bloomington: Indiana University Press, 1962).

Selye, H., *From Dream to Discovery*: *On Being a Scientist* (New York: McGraw-Hill, 1964).

Shapley, H., ed., *Science Ponders Religion* (New York: Appleton, 1960).

Sullivan, J. W. N., *The Limitations of Science* (New York: Viking, 1933).

Taton, R., *Reason and Chance in Scientific Discovery* (New York: Science, 1962).

——, ed., *A General History of Science,* trans. A. J. Pomerans, 4 vols. (London: Thames and Hudson, 1963-1966).

Thomson, G. P., *The Strategy of Research* (Southampton, England: University of Southampton Press, 1957).

——, *The Inspiration of Science* (New York: Oxford, 1961).

Toulmin, S., *Foresight and Understanding*: *An Enquiry into the Aims of Science* (Bloomington: Indiana University Press, 1961).

——, and Goodfield, J., *Fabric of the Heavens*: *The Development of Astronomy and Dynamics* (New York: Harper & Row, 1965).

——, and ——, *The Architecture of Matter* (New York: Harper & Row, 1962).

——, and ——, *The Discovery of Time* (New York: Harper & Row, 1965).

Trinklein, F. E., *The God of Science* (Grand Rapids, Mich.: Eerdmans, 1971).

van der Waerden, B. L., *Science Awakening,* trans. A. Dresden (New York: Science, 1963).

Watson, D. L., *Scientists are Human* (London: Watts, 1938).

Weaver, W., *Science and Imagination*: *Selected Papers* (New York: Basic Books, 1967).

Weizsäcker, C. F. von, *The History of Nature,* trans. F. D. Wieck (Chicago: University of Chicago Press, 1949).

Wertheimer, M., *Productive Thinking* (London: Social Science Paperbacks, 1966).

Whewell, W., *The Philosophy of the Inductive Sciences Founded upon their History* 2 vols. (London: Parker, 1840).

——, *History of the Inductive Sciences from the Earliest to the Present Times,* 3 vols. (London: Parker, 1857).

——, *On the Philosophy of Discovery*: *Chapters Historical and Critical* (London: Parker, 1860).

Whitehead, A. N., *Science and the Modern World* (New York: Mentor Books, 1948).

——, *Adventures of Ideas* (New York: Free Press, 1967).

Wilson, E. B., *An Introduction to Scientific Investigation* (New York: McGraw-Hill, 1952).

Woolf, H., ed., *Quantification*: *A History of the Meaning of Measurement in the Natural and Social Sciences* (Indianapolis: Bobbs-Merrill, 1961).

Ziman, J. M., *Public Knowledge*: *An Essay Concerning the Social Dimension of Science* (London: Cambridge University Press, 1968).

Frontiers of Modern Scientific Philosophy and Humanism: *The Athens Meeting 1964* (New York: Elsevier, 1966).

The Scientific Endeavor: *Centennial Celebration of the National Academy of Sciences* (New York: Rockefeller University Press, 1965).

II. Physical Sciences

Auger, P., Born, M., Heisenberg, W., Schrödinger, E., *On Modern Physics* (New York: Collier, 1962).

Beer, A., ed., *Vistas in Astronomy*: *New Aspects in the History and Philosophy of Astronomy* (New York: Pergamon, 1967).

Bloch, L., *La Philosophie de Newton* (Paris: Alcan, 1908).

Bohm, D., *Causality and Chance in Modern Physics* (New York: Harper Torchbooks, 1961).

Bohr, N., *Atomic Theory and the Description of Nature* (London: Cambridge University Press, 1934).

——, *Atomic Physics and Human Knowledge* (New York: Science, 1961).

——, *Essays 1958-1962 on Atomic Physics and Human Knowledge* (New York: Interscience, 1963).

Bondi, H., *The Universe at Large* (New York: Doubleday Anchor Books, 1960).

————, *Assumptions and Myths in Physical Theory* (London: Cambridge University Press, 1967).

Bopp, F., ed., *Werner Heisenberg und die Physik unserer Zeit* (Braunschweig: Vieweg, 1961).

Born, M., *The Restless Universe,* trans. W. M. Deans (New York: Dover, 1951).

————, *Experiment and Theory in Physics* (New York: Dover, 1956).

————, *Physics in My Generation: A Selection of Papers* (New York: Pergamon, 1956).

————, *Physics and Politics* (New York: Basic Books, 1962).

————, *Natural Philosophy of Cause and Chance* (New York: Dover, 1964).

————, *My Life and Views* (New York: Scribner, 1968).

Scientific Papers Presented to Max Born (London: Oliver & Boyd, 1953).

Bridgman, P. W., *The Nature of Physical Theory* (New York: Dover, 1936).

————, *The Logic of Modern Physics* (New York: Macmillan, 1946).

————, *The Nature of Some of Our Physical Concepts* (New York: Philosophical Library, 1952).

————, *Reflections of a Physicist* (New York: Philosophical Library, 1955).

————, *The Way Things Are* (New York: Viking Press, 1961).

Brillouin, L., *Scientific Uncertainty and Information* (New York: Academic Press, 1964).

Broda, E., *Ludwig Boltzmann: Mensch, Physiker, Philosoph* (Vienna: Deuticke, 1955).

Burtt, E. A., *The Metaphysical Foundations of Modern Physical Science* (New York: Doubleday Anchor Books).

Campbell, N. R., *Foundations of Physics: The Philosophy of Theory and Experiment* (formerly *Physics: The Elements*) (New York: Dover, 1957).

Cantore, E., *Atomic Order: An Introduction to the Philosophy of Microphysics* (Cambridge, Mass.: MIT Press, 1969).

Caspar, M., *Kepler,* trans. C. D. Hellman (New York: Abelard-Schuman, 1959).

Cohen, I. B., *Franklin and Newton: An Inquiry into Speculative Newtonian Experimental Science and Franklin's Work in Electricity as an Example Thereof* (Philadelphia: American Philosophical Society, 1956).

————, *The Birth of a New Physics* (New York: Doubleday Anchor Books, 1960).

Compton, A. H., *Atomic Quest: A Personal Narrative* (New York: Oxford, 1956).

Crombie, A. C., ed., *Turning Points in Physics* (New York: Harper Torchbooks, 1961).

de Broglie, L., *Matter and Light: The New Physics,* trans. W. H. Johnston (New York: Dover, 1939).

————, *Continu et Discontinu en Physique Moderne* (Paris: A. Michel, 1941).

————, *Savants et Découvertes* (Paris: A. Michel, 1951).

————, *The Revolution in Physics,* trans. R. W. Niemeyer (New York: Noonday, 1953).

————, *Sur les Sentiers de la Science* (Paris: A. Michel, 1960).

————, *New Perspectives in Physics,* trans. A. J. Pomerans (New York: Basic Books, 1962).

————, *Certitudes et Incertitudes de la Science* (Paris: A. Michel, 1956).

————, *Physics and Microphysics,* trans. M. Davidson (New York: Grosset & Dunlap, 1966).

Louis de Broglie, Physicien et Penseur [Festschrift] (Paris: A. Michel, 1953).

D'Elia, P. M., *Galileo in China,* trans. R. Suter and M. Sciascia (Cambridge, Mass.: Harvard University Press, 1960).

de Santillana, G., *The Crime of Galileo* (Chicago: University of Chicago Press, 1955).

de Sitter, W., *Kosmos: Development of Our Insight into the Structure of the Universe* (Cambridge, Mass.: Harvard University Press, 1932).

Deutsch and Shea, Inc., *A Profile of the Physicist* (New York: Industrial Relations News, 1962).

de Vaucouleurs, G., *Discovery of the Universe: An Outline of the History of Astronomy from the Origins to 1956* (London: Faber, 1957).

Dijksterhuis, E. J., *The Mechanization of the World Picture,* trans. C. Dikshoorn (Oxford: Clarendon Press, 1961).

452 BIBLIOGRAPHY

Dingle, H., *Through Science to Philosophy* (Oxford: Clarendon Press, 1937).

———, *The Scientific Adventure: Essays in the History and Philosophy of Science* (New York: Philosophical Library, 1953).

Dingler, H., *Das Experiment: Sein Wesen und seine Geschichte* (Munich: Reinhardt, 1928).

———, *Die Methode der Physik* (Munich: Reinhardt, 1938).

Dockx, S., and Bernays, P., eds., *Information and Prediction in Science* (New York: Academic Press, 1965).

Dugas, R., *Histoire de la Mécanique* (Neuchâtel: Griffon, 1950).

———, *Mechanics in the Seventeenth Century: From the Scholastic Antecedents to Classical Thought,* trans. F. Jacquot (Neuchâtel: Griffon, 1958).

———, *La Théorie Physique au Sens de Boltzmann et ses Prolongements Modernes* (Neuchâtel: Griffon, 1959).

Duhem, P., *Sozein Ta Phainomena: Essai sur la Notion de Théorie Physique de Platon à Galilée* (Paris: Hermann, 1908).

———, *The Aim and Structure of Physical Theory,* trans. P. P. Wiener (New York: Atheneum, 1962).

Eddington, A., *The Nature of the Physical World* (London: Cambridge University Press, 1931).

———, *The Philosophy of Physical Science* (Ann Arbor: University of Michigan Press, 1958).

———, *New Pathways in Science* (Ann Arbor: University of Michigan Press, 1959).

Einstein, A., *The World as I See It,* trans. A. Harris (New York: Covici, Friede, 1934).

———, *Out of My Later Years* (New York: Philosophical Library, 1950).

———, *Ideas and Opinions,* trans. S. Bargmann (New York: Crown, 1954).

———, *Lettres à Maurice Salovine* (Paris: Gauthier-Villars, 1956).

———, and Infeld, L., *The Evolution of Physics: The Growth of Ideas from Early Concepts to Relativity and Quanta* (New York: Simon and Schuster, 1951).

Feinman, R., *The Character of Physical Law* (Cambridge, Mass.: MIT Press, 1965).

Frank, P., *Einstein: His Life and Times,* trans. G. Rosen (New York: Knopf, 1947).

G. Galilei, *Opere,* F. Flora, ed. (Milan: Ricciardi, 1953).

———, *Discorsi e Dimostrazioni Matematiche intorno a Due Nuove Scienze,* A. Carugo and L. Geymonat, eds. (Turin: Boringhieri, 1958).

———, *Dialogue concerning the Two Chief World Systems, Ptolemaic and Copernican,* trans. S. Drake (Berkeley: University of California Press, 1962).

Discoveries and Opinions of Galileo, S. Drake, trans. and ed. (New York: Doubleday Anchor Books, 1957).

Gerlach, W., *Humanität und naturwissenschaftliche Forschung: Sammlung von gelegentlichen Vorträgen* (Braunschweig: Vieweg, 1962).

Geymonat, L., *Galileo Galilei: A Biography and Inquiry into His Philosophy of Science,* trans. S. Drake (New York: McGraw-Hill, 1965).

Glasser, O., *Wilhelm Conrad Roentgen* (Springfield, Ill.: Charles C. Thomas, 1958).

Gonseth, F., *Le Problème du Temps: Essai sur la Méthodologie de la Recherche* (Neuchâtel: Griffon, 1964).

Hahn, O., *A Scientific Autobiography,* trans. W. Ley (New York: Scribner, 1966).

Hall, A. R., *From Galileo to Newton: 1630-1720* (New York: Collins, 1963).

Heelan, P. A., *Quantum Mechanics and Objectivity: A Study of the Physical Philosophy of Werner Heisenberg* (The Hague: Nijhoff, 1965).

Heisenberg, W., *The Physical Principles of the Quantum Theory,* trans. C. Eckart and F. C. Hoyt (New York: Dover).

———, *Philosophic Problems of Nuclear Science,* trans. F. C. Hayes (London: Faber, 1952).

———, *The Physicist's Conception of Nature,* trans. A. J. Pomerans (New York: Harcourt, 1958).

——, *Physics and Philosophy: The Revolution in Modern Science* (New York: Harper Torchbooks, 1962).

——, *Der Teil und das Ganze: Gespräche im Umkreis der Atomphysik* (Munich: Piper, 1969). English translation: *Physics and Beyond: Encounters and Conversations,* trans. A. J. Pomerans (New York: Harper & Row, 1971).

——, *Die Bedeutung des Schönen in der exakten Naturwissenschaft* (Stuttgart: Belser, 1971) [Bilingual edition: English trans. by E. Cantore].

——, *Schritte über Grenzen: Gesammelte Reden und Aufsätze* (Munich: Piper, 1971).

Heller, K. D., *Ernst Mach: Wegbereiter der modernen Physik* (Vienna: Springer, 1964).

Holton, G., and Roller, D. H. D., *Foundations of Modern Physical Science* (Reading, Mass.: Addison-Wesley, 1958).

Hoyle, F., *Man and Materialism* (New York: Harper & Row, 1956).

——, *The Nature of the Universe* (New York: Harper & Row, 1960).

——, *Of Men and Galaxies* (Seattle: University of Washington Press, 1964).

——, *Encounter with the Future* (New York: Simon & Schuster, 1965).

——, *Man in the Universe* (New York: Columbia University Press, 1966).

Ihde, A. J., *The Development of Modern Chemistry* (New York: Harper & Row, 1964).

Jaki, S. L., *The Relevance of Physics* (Chicago: University of Chicago Press, 1966).

Jeans, J., *Physics and Philosophy* (London: Cambridge University Press, 1943).

Johnston, M., ed., *The Cosmos of Arthur Holly Compton* (New York: Knopf, 1967).

Jordan, P., *Das Bild der modernen Physik* (Hamburg: Stromverlag, 1947).

——, *Atom und Weltall: Einführung in den Gedankeninhalt der modernen Physik* (Braunschweig: Vieweg, 1960).

Kaplon, M. F., ed., *Homage to Galileo* (Cambridge, Mass.: MIT Press, 1965).

Kockelmans, J. J., *Phenomenology and Physical Science: An Introduction to the Philosophy of Physical Science* (Pittsburgh: Duquesne University Press, 1966).

Koestler, A., *The Sleepwalkers: A History of Man's Changing Vision of the Universe* (London: Hutchinson, 1959).

Koyré, A., *From the Closed World to the Infinite Universe* (New York: Harper Torchbooks, 1958).

——, *Newtonian Studies* (Chicago: University of Chicago Press, 1968).

——, *Metaphysics and Measurement: Essays in Scientific Revolution* (London: Chapman & Hall, 1968).

Kuhn, T. S., *The Copernican Revolution: Planetary Astronomy in the Development of Western Thought* (New York: Vintage, 1962).

——, Heilbron, J. L., Forman, P., Allen, L., eds., *Sources for History of Quantum Physics* (Oral Interviews), copyright The American Philosophical Society, Philadelphia, 1967.

Laue, M. von, *History of Physics,* trans. R. Oesper (New York: Academic Press, 1950).

Leprince-Ringuet, L., *Atoms and Men,* trans. E. P. Halperin (Chicago: University of Chicago Press, 1961).

Lewis, G. N., *The Anatomy of Science* (New Haven: Yale University Press, 1926).

Lovell, A. C. B., *The Individual and the Universe* (New York: Harper & Row, 1959).

——, *Our Present Knowledge of the Universe* (Cambridge, Mass.: Harvard University Press, 1967).

Mach, E., *The Science of Mechanics: A Critical and Historical Account of Its Development,* trans. T. J. McCormack (La Salle, Ill.: Open Court, 1960).

——, *Popular Scientific Lectures,* trans. T. J. McCormack (La Salle, Ill.: Open Court, 1943).

McKie, D., *Antoine Lavoisier: Scientist, Economist, Social Reformer* (New York: Collier, 1962).

McMullin, E., ed., *Galileo Man of Science* (New York: Basic Books, 1967).

Manuel, F. E., *A Portrait of Isaac Newton* (Cambridge, Mass.: Harvard University Press, 1968).

Millikan, R. A., *Science and Life* (Boston: Pilgrim Press, 1924).
———, *Evolution in Science and Religion* (New Haven: Yale University Press, 1927).
———, *Science and the New Civilization* (New York: Scribner, 1930).
———, *Time, Matter and Values* (Chapel Hill: University of North Carolina Press, 1932).
———, *Autobiography* (Englewood Cliffs, N. J.: Prentice-Hall, 1950).
Moore, R., *Niels Bohr: The Man, His Science and the World They Changed* (New York: Knopf, 1966).
More, L. T., *Isaac Newton: A Biography* (New York: Dover, 1962).
Moszkowski, A., *Einstein, the Searcher: His Work Explained from Dialogues with Einstein,* trans. H. L. Brose (New York: Dutton).
North, J. D., *The Measure of the Universe: A History of Modern Cosmology* (Oxford: Clarendon Press, 1965).
Oppenheimer, J. R., *Science and the Common Understanding* (New York: Simon and Schuster, 1953).
———, *The Open Mind* (New York: Simon and Schuster, 1955).
Ostwalt, W., *Natural Philosophy,* trans. T. Seltzer (New York: Holt, 1910).
Pannekoek, A., *A History of Astronomy* (New York: Interscience, 1961).
Partington, J. R., *A Short History of Chemistry* (New York: Harper Torchbooks, 1960).
Pauli, W., ed., *Niels Bohr and the Development of Physics* (New York: Pergamon, 1955).
Pauli, W., *Aufsätze und Vorträge über Physik und Erkenntnistheorie* (Braunschweig: Vieweg, 1961).
Planck, M., *Where is Science Going?,* trans. J. Murphy (New York: Norton, 1932).
———, *The Philosophy of Physics,* trans. W. H. Johnston (New York: Norton, 1936).
———, *Scientific Autobiography and Other Papers,* trans. F. Gaynor (New York: Philosophical Library, 1949).
———, *A Survey of Physical Theory* (formerly *A Survey of Physics*), trans. R. Jones and D. H. Williams (New York: Dover, 1960).
Pupin, M., *The New Reformation: From Physical to Spiritual Realities* (New York: Scribner, 1927).
Rabi, I. I., *My Life and Times as a Physicist* (Claremont, Calif.: Claremont College Press, 1960).
Raman, C. V., *The New Physics: Talks on Aspects of Science* (New York: Philosophical Library, 1951).
Rey, A., *La Théorie de la Physique chez les Physiciens Contemporains* (Paris: Alcan, 1907).
Ronchi, V., *Optics: The Science of Vision,* trans. E. Rosen (New York: New York University Press, 1957).
———, *Storia del Cannocchiale* (Vatican City: Pontifical Academy of Sciences, 1964).
———, *Sui Fondamenti dell'Acustica e dell'Ottica* (Florence: Olschki, 1967).
———, *The Nature of Light,* trans. V. Barocas (Cambridge, Mass.: Harvard University Press, 1970).
Rozenthal, S., ed., *Niels Bohr: His Life and Work as Seen by His Friends and Colleagues* (Amsterdam: North-Holland, 1967).
Schilpp, P. A., ed., *Albert Einstein, Philosopher-Scientist* (New York: Harper Torchbooks, 1959).
Schrödinger, E., *Science and Humanism: Physics in Our Time* (London: Cambridge University Press, 1951).
———, *Nature and the Greeks* (London: Cambridge University Press, 1954).
———, *What is Life? and Other Scientific Essays* (New York: Doubleday Anchor Books, 1956).
———, *Science, Theory, and Man* (formerly *Science and the Human Temperament*) (New York: Dover, 1957).
———, *My View of the World,* trans. C. Hastings (London: Cambridge University Press, 1964).

————, *What is Life? The Physical Aspect of the Living Cell* and *Mind and Matter* (London: Cambridge University Press, 1967).

Sciama, D. W., *The Unity of the Universe* (New York: Doubleday Anchor Books, 1961).

Shapley, H., *Of Stars and Men* (New York: Washington Square Press, 1960).

————, *Beyond the Observatory* (New York: Scribner, 1967).

Smith, C. S., *A History of Metallography: The Development of Ideas on the Structure of Metals before 1890* (Chicago: University of Chicago Press, 1960).

Stallo, J. B., *The Concepts and Theories of Modern Physics* (Cambridge, Mass.: Harvard University Press, 1960).

Stratton, J. A., *Science and the Educated Man* (Cambridge, Mass.: MIT Press, 1966).

Strauss, A. L., and Rainwater, L., *The Professional Scientist: A Study of American Chemists* (Chicago: Aldine, 1962).

Taylor, L. W., *Physics, The Pioneer Science,* 2 vols. (New York: Dover, 1959).

Thomson, J. J., *Recollections and Reflections* (New York: Macmillan, 1937).

Truesdell, C. A., *Essays in the History of Mechanics* (New York: Springer, 1968).

van Melsen, A. G., *From Atomos to Atom: The History of the Concept Atom* (New York: Harper Torchbooks, 1960).

Weisskopf, V. F., *Knowledge and Wonder: The Natural World as Man Knows It* (New York: Doubleday, 1962).

Weizsäcker, C. F. von, *The World View of Physics,* trans. M. Grene (Chicago: University of Chicago Press, 1952).

————, *The Relevance of Science: Creation and Cosmogony* (New York: Collins, 1964).

Weyl, H., *Symmetry* (Princeton, N. J.: Princeton University Press, 1952).

————, *Philosophy of Mathematics and Natural Science,* trans. O. Helmer (New York: Atheneum, 1963).

Whitrow, G. J., *The Structure and Evolution of the Universe: An Introduction to Cosmology* (London: Hutchison, 1959).

Whittaker, E., *From Euclid to Eddington: A Study of Conceptions of the External World* (New York: Dover, 1958).

————, *A History of the Theories of Aether and Electricity,* 2 vols. (New York: Harper Torchbooks, 1960).

Wiener, N., *God and Golem, Inc.: A Comment on Certain Points where Cybernetics Impinges on Religion* (Cambridge, Mass.: MIT Press, 1966).

————, *The Human Use of Human Beings: Cybernetics and Society* (New York: Avon Books, 1967).

Wigner, E. P., *Symmetries and Reflections: Scientific Essays* (Bloomington: Indiana University Press, 1967).

Williams, L. P., *The Origins of Field Theory* (New York: Random House, 1966).

III. Biological sciences

Barlow, N., ed., *The Autobiography of Charles Darwin* (*with original omissions restored*) (New York: Norton, 1969).

Beckner, M., *The Biological Way of Thought* (New York: Columbia University Press, 1959).

Bertalanffy, L. von, *Problems of Life: An Evaluation of Modern Biological and Scientific Thought* (New York: Harper Torchbooks, 1960).

————, *Modern Theories of Development: An Introduction to Theoretical Biology,* trans. J. H. Woodger (New York: Harper Torchbooks, 1962).

Bonner, J. T., *Morphogenesis: An Essay on Development* (New York: Atheneum, 1963).

————, *The Ideas of Biology* (New York: Harper Torchbooks, 1964).

Carrel, A., *Man, The Unknown* (New York: Harper & Row, 1935).

Crick, F., *Of Molecules and Men* (Seattle: University of Washington Press, 1966).

Darwin, C., *The Origin of Species by means of Natural Selection or the Preservation of Favoured Races in the Struggle for Life* (New York: Collier, 1962).

de Beer, G., *Charles Darwin: A Scientific Biography* (New York: Doubleday Anchor Books, 1965).

Dobell, C., *Antony van Leeuwenhoek and His "Little Animals"* (New York: Russell and Russell, 1958).

Dobzhansky, T., *The Biological Basis of Human Freedom* (New York: Columbia University Press, 1956).

———, *Mankind Evolving: The Evolution of the Human Species* (New Haven: Yale University Press, 1962).

———, *The Biology of Ultimate Concerns* (Cleveland: World, 1967).

Dubos, R., *The Dreams of Reason: Science and Utopias* (New York: Columbia University Press, 1961).

———, *Man Adapting* (New Haven: Yale University Press, 1965).

———, *The Torch of Life: Continuity in Living Experience* (New York: Simon and Schuster, 1970).

Eiseley, L., *Darwin's Century: Evolution and the Men Who Discovered It* (New York: Doubleday Anchor Books, 1961).

Elsasser, W. W., *Atom and Organism: A New Approach to Theoretical Biology* (Princeton, N. J.: Princeton University Press, 1966).

Ghiselin, M. T., *The Triumph of the Darwinian Method* (Berkeley: University of California Press, 1969).

Greene, J. C., *The Death of Adam: Evolution and Its Impact on Western Thought* (Ames, Ia.: Iowa State University Press, 1959).

Haldane, J. S., *The Sciences and Philosophy* (New York: Doubleday, 1930).

———, *The Philosophy of a Biologist* (Oxford: Clarendon Press, 1935).

Hooykaas, R., *The Principle of Uniformity in Geology, Biology and Theology* (formerly *Natural Law and Divine Miracle*) (Leiden, Netherlands: Brill, 1963).

Jonas, H., *The Phenomenon of Life: Toward a Philosophical Biology* (New York: Dell, 1968).

Lorenz, K., *Evolution and Modification of Behavior* (Chicago: University of Chicago Press, 1965).

Medawar, P. B., *The Uniqueness of the Individual* (New York: Basic Books, 1957).

———, *The Future of Man* (London: Methuen, 1960).

———, *The Art of the Soluble* (London: Methuen, 1967).

Mendelsohn, E., Shapere, D., and Allen, G. E., eds., "Conference on Explanation in Biology: Historical, Philosophical and Scientific Aspects" in *Journal of the History of Biology,* **2** (1969), pp. 1-281.

Monod, J., *Le Hasard et la Nécessité: Essai sur la Philosophie Naturelle de la Biologie Moderne* (Paris: Seuil, 1970).

Moody, P. A., *Introduction to Evolution* (New York: Harper & Row, 1962).

Needham, J., *Order and Life* (Cambridge, Mass.: MIT Press, 1968).

Ramòn y Cajal, S. "Rules and Counsels for the Scientific Investigator" in Craigie, E. H., and Gibson, W. C., *The World of Ramòn y Cajal with Selections from His Nonscientific Writings* (Springfield, Ill.: Charles C. Thomas, 1968), pp. 186-198.

Richet, C., *Natural History of a Savant,* trans. O. Lodge (London: Dent, 1927).

Roberts, C., *The Scientific Conscience: Reflections on the Modern Biologist and Humanism* (New York: Braziller, 1967).

Schierbeek, A., *Measuring the Invisible World: The Life and Works of Antony van Leeuwenhoek* (New York: Abelard-Schuman, 1959).

Sherrington, C., *Man on His Nature* (New York: Doubleday Anchor Boooks, 1955).

Simpson, G. G., *This View of Life: The World of an Evolutionist* (New York: Harcourt, 1964).

———, *The Meaning of Evolution: A Study of the History of Life and of Its Significance for Man* (New Haven: Yale University Press, 1967).

———, *Biology and Man* (New York: Harcourt, 1969).

Singer, C., *A History of Biology: A General Introduction to the Study of Living Things* (New York: Abelard-Schuman, 1950).

Sirks, M. J., and Zirkle, C., *The Evolution of Biology* (New York: Ronald, 1964).

Thompson, D'Arcy W., *On Growth and Form*, J. T. Bonner, ed. (London: Cambridge University Press, 1961).

Thorpe, W. H., *Biology and the Nature of Man* (London: Oxford University Press, 1962).

———, *Science, Man, and Morals* (Ithaca, N. Y.: Cornell University Press, 1966).

Waddington, C. H., *The Nature of Life* (London: G. Allen, 1961).

———, *The Ethical Animal* (Chicago: University of Chicago Press, 1967).

Watson, J. D., *The Double Helix: A Personal Account of the Discovery of the Structure of DNA* (New York: Atheneum, 1968).

Wickler, W., *Antworten der Verhaltensforschung* (Munich: Kösel, 1970).

Young, J. Z., *Doubt and Certainty in Science: A Biologist's Reflections on the Brain* (New York: Oxford, 1960).

IV. Psycho-socio-anthropological sciences

Allport, G. W., *Becoming: Basic Considerations for a Psychology of Personality* (New Haven: Yale University Press, 1955).

———, *Personality and Social Encounter: Selected Essays* (Boston: Beacon, 1964).

———, *The Person in Psychology: Selected Essays* (Boston: Beacon, 1968).

Aron, R., *Main Currents in Sociological Thought*, trans. R. Howard and H. Weaver, 2 vols. (New York: Basic Books, 1965-1967).

Berger, P. L., *Invitation to Sociology: A Humanist Perspective* (New York: Doubleday Anchor Books, 1963).

Berger, P. L., and Luckmann, T., *The Social Construction of Reality: A Treatise in the Sociology of Knowledge* (New York: Doubleday Anchor Books, 1967).

Bertalanffy, L. von, *Robots, Men and Minds: Psychology in the Modern World* (New York: Braziller, 1967).

Bidney, D., *Theoretical Anthropology* (New York: Schocken Books, 1967).

Boring, E. G., *A History of Experimental Psychology* (New York: Appleton, 1950).

———, *History, Psychology and Science: Selected Papers* (New York: Wiley, 1963).

Childe, V. G., *Man Makes Himself* (London: Watts, 1941).

———, *What Happened in History* (Baltimore: Penguin, 1964).

Daniel, G., *The Idea of Prehistory* (London: Watts, 1962).

Durkheim, E., *The Rules of Sociological Method*, trans. S. A. Solovay and J. H. Mueller (New York: Free Press, 1964).

Eliade, M., *Cosmos and History: The Myth of the Eternal Return*, trans. W. R. Trask (New York: Harper Torchbooks, 1959).

———, *The Sacred and the Profane: The Nature of Religion*, trans. W. R. Trask (New York: Harper Torchbooks, 1961).

———, *Patterns in Comparative Religion*, trans. S. Sheed (New York: Meridian, 1963).

———, *Myth and Reality*, trans. W. R. Trask (New York: Harper Torchbooks, 1963).

———, *Myths, Dreams and Mysteries: The Encounter between Contemporary Faiths and Archaic Realities*, trans. P. Mairet (New York: Harper Torchbooks, 1967).

Erikson, E. H., *Identity and the Life Cycle: Selected Papers* (New York: International Universities, 1959).

———, *Insight and Responsibility: Lectures on the Ethical Implications of Psychoanalytic Insight* (London: Faber, 1964).

———, *Identity: Youth and Crisis* (New York: Norton, 1968).

Evans-Pritchard, E. E., *Theories of Primitive Religion* (Oxford: Clarendon Press, 1965).

Frankl, V. E., *Man's Search for Meaning: An Introduction to Logotherapy,* trans. I. Lasch (New York: Washington Square Press, 1963).

Freud, S., *The Future of an Illusion,* trans. W. D. Robson-Scott and J. Strachey (Garden City, N. Y.: Doubleday Anchor Books, 1964).

———, *The Origin and Development of Psychoanalysis* (Chicago: Gateway Books, 1965).

Fromm, E., *Psychoanalysis and Religion* (New Haven: Yale University Press, 1950).

———, *Man for Himself: An Inquiry into the Psychology of Ethics* (Greenwich, Conn.: Fawcett, 1968).

———, *The Revolution of Hope: Toward a Humanized Technology* (New York: Bantam Books, 1968).

Harris, M., *The Rise of Anthropological Theory: A History of Theories of Culture* (New York: Crowell, 1968).

Herskovits, M. J., *Cultural Dynamics* (New York: Knopf, 1964).

Homans, G. C., *The Nature of Social Science* (New York: Harcourt, 1967).

Kroeber, A. L., and Kluckhohn, C., *Culture: A Critical Review of Concepts and Definitions* (New York: Vintage Books).

Lerner, D., ed., *The Human Meaning of the Social Sciences* (New York: Meridian, 1959).

Lévi-Strauss, C., *The Scope of Anthropology,* trans. S. O. Paul and R. A. Paul (London: J. Cape, 1967).

———, *Structural Anthropology,* trans. C. Jacobson and B. G. Schoepf (New York: Doubleday Anchor Books, 1967).

Lienhardt, G., *Social Anthropology* (New York: Oxford, 1966).

Malinowski, B. *Magic, Science and Religion and Other Essays* (New York: Doubleday Anchor Books, 1954).

———, *A Scientific Theory of Culture and Other Essays* (New York: Oxford, 1960).

Merton, R. K., *Social Theory and Social Structure* (New York: Free Press, 1957).

———, *On Theoretical Sociology: Five Essays, Old and New* (New York: Free Press, 1967).

Metzger, W., *Psychologie: Die Entwicklung ihrer Grundannahmen seit der Einführung des Experiments* (Darmstadt: Steinkopff, 1963).

Misiak, H., *The Philosophical Roots of Scientific Psychology* (New York: Fordham University Press, 1961).

Misiak, H., and Sexton, V. S., *History of Psychology: An Overview* (New York: Grune & Stratton, 1966).

Montagu, A., *The Humanization of Man* (Cleveland: World, 1962).

Myrdal G., *Value in Social Theory: A Selection of Essays on Methodology,* P. Streeten, ed. (New York: Harper & Row, 1958).

Piaget, J., *Introduction à l'Epistémologie Génétique,* 3 vols. (Paris: Presses Universitaires de France, 1950).

———, *The Psychology of Intelligence,* trans. M. Piercy and D. E. Berlyne (Totawa, N. J.: Littlefield Adams, 1966).

Radin, P., *Primitive Religion: Its Nature and Origin* (New York: Dover, 1957).

———, *Primitive Man as Philosopher* (New York: Dover, 1957).

Rickman, H. P., *Understanding and the Human Studies* (London: Heinemann, 1967).

Robinson, J., *Economic Philosophy* (New York: Doubleday Anchor Books, 1964).

Sachs, H., *Freud: Master and Friend* (Cambridge, Mass.: Harvard University Press, 1944).

Skinner, B. F., *Science and the Human Behavior* (New York: Free Press, 1965).

Smith, F. V., *Explanation of Human Behavior* (London: Constable, 1960).

Sorokin, P. A., *Sociological Theories of Today* (New York: Harper & Row, 1966).

Taylor, O. H., *A History of Economic Thought: Social Ideals and Economic Theories from Quesnay to Keynes* (New York: McGraw-Hill, 1960).

Timasheff, N. S., *Sociological Theory: Its Nature and Growth* (New York: Random House, 1967).

Weber, M., *The Methodology of the Social Sciences*, trans. E. A. Shils and H. A. Finch (New York: Free Press, 1949).

———, *General Economic History*, trans. F. H. Knight (New York: Collier, 1961).

———, *The Theory of Social and Economic Organization*, trans. A. M. Henderson and T. Parsons (New York: Free Press, 1964).

Wolff, K. H., ed., *Emile Durkheim et alii: Essays on Sociology and Philosophy* (New York: Harper Torchbooks, 1964).

V. SOCIETY AND SCIENCE

Bernal, J. D., *The Social Function of Science* (Cambridge, Mass.: MIT Press, 1967).

Boyko, H., ed., *Science and the Future of Mankind* (Bloomington: Indiana University Press, 1961).

Commoner, B., *Science and Survival* (New York: Viking, 1967).

———, *The Closing Circle: Nature, Man and Technology* (New York: Knopf, 1971).

Ellul, J., *The Technological Society*, trans. J. Wilkinson (New York: Vintage, 1967).

Forbes, R. J., *Man, The Maker: A History of Technology and Engineering* (New York: Abelard-Schuman, 1958).

———, *The Conquest of Nature: Technology and Its Consequences* (New York: Mentor Books, 1969).

Glass, B., *Science and Liberal Education* (Baton Rouge: Louisiana State University Press, 1959).

———, *Science and Ethical Values* (Chapel Hill: University of North Carolina Press, 1965).

Goldsmith, M., and Mackay, A., eds., *Society and Science* (New York: Simon and Schuster, 1964).

Hall, E. W., *Modern Science and Human Values: A Study in the History of Ideas* (Princeton, N. J.: Van Nostrand, 1956).

Hayek, F. A., *The Counter-Revolution of Science: Studies on the Abuse of Reason* (New York: Free Press, 1964).

Hill, A. V., *The Ethical Dilemma of Science and Other Writings* (New York: Rockefeller University Press, 1960).

Holton, G., ed., *Science and Culture: A Study of Cohesive and Disjunctive Forces* (Boston: Beacon, 1965).

Huxley, T. H., *Science and Education* (New York: Philosophical Library, 1964).

Klaw, S., *The New Brahmins: Scientific Life in America* (New York: Morrow, 1968).

Klemm, F., *A History of Western Technology*, trans. D. W. Singer (Cambridge, Mass.: MIT Press, 1964).

Köhler, W., *The Place of Value in a World of Facts* (New York: Mentor Books, 1966).

Lapp, R. E., *The New Priesthood: The Scientific Elite and the Uses of Power* (New York: Harper & Row, 1965).

Lilley, S., *Men, Machines and History: The Story of Tools and Machines in Relation to Social Progress* (London: Lawrence and Wishart, 1965).

Marcuse, H., *One-Dimensional Man: Studies in the Ideology of Advanced Industrial Society* (Boston: Beacon, 1964).

Margenau, H., *Ethics and Science* (Princeton, N. J.: Van Nostrand, 1964).

Matson, F. W., *The Broken Image: Man, Science and Society* (New York: Braziller, 1964).

Mumford, L., *Technics and Civilization* (New York: Harcourt, 1963).

Pearson, K. *The Ethic of Freethought: A Selection of Essays and Lectures* (London: T. Fisher Unwin, 1888).

Rabinowitch, E., *The Dawn of a New Age: Reflections on Science and Human Affairs* (Chicago: University of Chicago Press, 1963).

Rotblat, J., *Science and World Affairs: History of the Pugwash Conferences* (London: Dawsons, 1962).

Russell, B., *The Impact of Science on Society* (New York: Simon and Schuster, 1953).

Salomon, A., *The Tyranny of Progress: Reflections on the Origins of Sociology* (New York: Noonday, 1955).

Schon, D. A., *Invention and the Evolution of Ideas* (formerly *Displacement of Concepts*) (London: Social Science Paperbacks, 1967).

Siu, R. G. H., *The Tao of Science: An Essay on Western Knowledge and Eastern Wisdom* (Cambridge, Mass.: MIT Press, 1957).

Skinner, B. F., *Beyond Freedom and Dignity* (New York: Knopf, 1971).

Snow, C. P., *The Two Cultures: And a Second Look* (New York: Mentor Books, 1963).

Sorokin, P. A., *The Crisis of Our Age: The Social and Cultural Outlook* (New York: Dutton, 1957).

Tiselius A. and Nilsson, S., eds., *The Place of Value in a World of Facts* (New York: Wiley Interscience, 1970).

Toffler, A., *Future Shock* (New York: Random House, 1970).

Tullock, G., *The Organization of Inquiry* (Durham, N. C.: Duke University Press, 1966).

van Melsen, A. G., *Science and Technology* (Pittsburgh: Duquesne University Press, 1961).

——, *Physical Science and Ethics: A Reflection on the Relationship between Nature and Morality* (Pittsburgh: Duquesne University Press, 1967).

Ward, L., ed., *Ethics and the Social Sciences* (Notre Dame, Ind.: University of Notre Dame Press, 1959).

Weinberg, A. M., *Reflections on Big Science* (Cambridge, Mass.: MIT Press, 1967).

White, L., Jr., *Machina ex Deo: Essays in the Dynamism of Western Culture* (Cambridge, Mass.: MIT Press, 1968).

Editors of *International Science and Technology, The Way of the Scientist: Interviews from the World of Science and Technology* (New York: Simon and Schuster, 1966).

INDEX OF NAMES

INDEX OF SUBJECTS

ABSOLUTE, The
object of religion 116, rejected by anthropocentric humanism 2, identified with nature 122, ultimate source of meaning 298f, 326f, attained through dialogue 436; *see also* God, Meaning, Ultimates

experienced by scientists
as implying developmentalness 300f, immanency 298f, transrationality 301-304, vindication of man's dignity 304-307; *see also* Agnosticism, Religion: experience

ontological problem
bearing of science 319f, consistent quest for meaning 320f, 327f, analogical approach 323f, indicativeness of solution 321f, reflectiveness 321, 326f; *see also* Agnosticism, Analogical Language, Atheism, Deism, Materialism, Pantheism

ACTIVITY
ethical consequence of science 347, 354, refusal of passivity 348; *see also* Science and Ethics: contributions

ACTUALISM
consequence of operationalism 250; *see also* Empiricism, Science and Epistemology: inadequateness

ADAPTATION OF THOUGHT
Mach's doctrine 101f, critized by scientists 103-105; *see also* Machian View

ADMIRATION
accompanies discovery psychologically 98f, reflectively 100, rewarding character 100; *see also* Wonder

ADVENTURISM
arrogant self-assurance 148f, overemphasis on originality 148, pursuit as goal 149, results in ivory-tower syndrome 150, pragmatization of science 150, intellectual emptiness 150; *see also* Originality, Science and Ethics: inadequateness, Scientists: humanistic difficulties

AGNOSTICISM
definition 167f, connection with science 168f, questions certainties 169f, results in depression 171f, desperation 171-173, meaninglessness 171-173, stupefaction 173; *see also* Absolute, Science and Epistemology: inadequateness, Scientists: humanistic difficulties

ANALOGICAL LANGUAGE
in science: *see* Model
in ontology and religion: meaningful 323f, unavoidable 324-326; *see also* Absolute: ontological problem

ANTHROPOCENTRISM
alleged foundations: scientific 370, humanistic 371; doctrinal principles: man-centeredness 371f, polemical attitude 371f; opposed by scientists as unfounded 110f,

469

373f, unrealistic 374f, dehumanizing 375f; *see also* Humanism, Self-Actualization

APPROXIMATIONS, Successive
criterion of scientific progress 227f; *see also* Progress

APRIORISM
opposed by scientists 243f; *see also* Epistemology: inadequate alone, Scientists and Philosophy

ARROGANCE, and Science
different from dignity 60; due to reductionism 251f, emotional insecurity 148f, technicalism 152-154, 251; *see also* Dignity, Reductionism, Scientists: humanistic difficulties

ASTONISHMENT
at orderliness of nature 102f, possibility of science 104f, progressivity and universality of science 103f; *see also* Wonder

ATHEISM
definitions: agnostic 315, assertive 315; essential feature 436; foundations: emotional 315f, scientistic 315f; implies dogmatism 316-318, undermines science 317; *see also* Absolute: ontological problem

AWE
definition 106; accompanies major discoveries: experimental 106f, theoretical 107f; compelling character 108, ethical consequence of science 353f; *see also* Religion: experience, Wonder

AXIOMATIC DEDUCTIVISM
typical of prescientific mind, *see* Essentialism; implicit in classical mechanics 198-200, patterned on geometry 198, rejected by practic-

ing scientists 199f, 241-243; *see also* Mathematics, Rationalism

BACONIAN VIEW, of Science
definition 31, inadequate 32, rejected by scientists 33, 54f; *see also* Fact, Method

BEAUTY
of nature 92-94, of theory 91, sought for by science 92f; *see also* Cosmos: experience

BECOMING
definition 278; problem, *see* Change: ontological problem

CERTAINTY, Scientific
preexperimental 64f, experiential 65f, unshakable 63f; *see also* Science: experience, Simplicity

CHANGE
cognition supplied by science: original 269f, ambiguous 270
ontological problem: questions about changeability: extrinsic 278f, intrinsic 279f, causes of 280, reality of 279; *see also* Ontology: problems

COMMON SENSE
Aristotelian 194, equivocal 392f; relation with science: continuity 394f, tension 194f; *see also* Culture and Science: tension, Naturalness, Prescientific Outlook

COMMUNION
consequence of dialogue 436, religious experience 123; scientific, with nature 143; feature of creativity 419f; *see also* Humanization, Openness, Responsiveness

COMMUNITY, Scientific
definition 69; conservative 52, results from research 66f, leader-

412f; *see also* Culture and Science: tension

CULTURE

definition 420; holistic: contributions by science 420f, philosophy 421f

and SCIENCE: tension

produced by scientific revolution 400f, man's sense of fear 402, homelessness in the cosmos 401-402; shock: initial 401f, ongoing 402f; self-confrontation of man 404. *See also* Crisis, Galilean Controversy, Identity, Man, Scientists: humanistic difficulties

two cultures

psychological basis 13f, ambiguous expression 420
See also Dialogue, Humanism, Humanization, Ideal, Wisdom

DEISM

definition 313, historical evolution 313f, destroys meaningfulness of reality 314; *see also* Absolute: ontological problem

DETERMINISTIC VIEW, of Man

aims at dissolving man 285, scientistic ideal 282f, suppresses freedom 283, morality 284f, meaning 285f, "nothing-but" fallacy 287f; *see also* Man: ontological problem, Mechanism, Reductionism

DEVELOPMENTALNESS

scientists' experience of the absolute 300f; of knowledge: *see* Dynamism, Progress

DIALOGUE

communication 430f, attention 431, volition 432, responsiveness 432f, openness 433

humanizing results

personalization 434, universal humanization 434f, integration in totality of human community 435, of cosmos 435, attainment of unconditional meaning 435f, religious authenticity 436
See also Communion, Openness, Responsiveness, Wisdom

DIGNITY, of Man

inspired by science as result of mental creativity 57-60, man's cosmic awareness 396f; expressed as man's responsibility toward truth 60f, autonomy of science relative to religion 249f; coresponsibility for the world: *see* Coresponsibility; problematic 381
underlies objection to personal God 304f, rejection of religious instinctivism 306f
opposed by instinctivism 382f, naturalness 414f
See also Arrogance, Courage, Creativity, Humanization, Self-Discovery

DISCOVERY

factual: definition 71; difficult 71, important 71f, original 71-73
explanatory: definition 74; great 74f, rewarding 38, 76
surprising: *see* Admiration

chance discovery

does not contradict activity of research 27

simultaneous discoveries

compatible with personal originality 66f

DOGMATISM

scientistic 250f, aprioristic 241f; *see also* Scientism

DYNAMISM

of Knowledge

characteristic of science 200f, not revolutionary 201, progressive 201f; experiential origin 201f; sapiential feature 429f

problematic: cognitive break due to science 235, continuity and tension between science and ordinary knowledge 236, 391f, 401-404, knowledge as a living activity 236f; *see also* Progress: epistemological acceptation

of Nature

compatible with orderliness 82, biological evolution 82f, quantum theory 84; expansive character 84f, historical 85f

problematic: *see* Change

ECONOMY, Mental

Mach's principle 214f, criticized by scientists 219; *see also* Machian View

EFFECTIVENESS

ethical consequence of science 348, examples 348f; *see also*: Science and Ethics: contributions

EMPIRICISM

structure: *see* Conventionalism, Instinctivism, Operationalism

criticized by scientists as dehumanizing science 219f, ignoring creativity 218f, sterilizing science 220f

See also Machian View, Positivism, Pragmatism, Scientism

EPISTEMOLOGY

definition 11; *see also* Philosophy

inadequate alone

see Apriorism, Essentialism, Historicism, Occultism, Phenomenism, Positivism, Pragmatism, Rationalism, Saving of Appearances, Subjectivism, Systematism

problems

see Concept, Dynamism, Law, Objectivity

and HUMANISM

see Humanism: epistemological perspective

and SCIENCE

see Science and Epistemology

ESSENTIALISM

typical of prescientific mind 190, 197, opposed by science 190f, 200, 202; *see also* Epistemology: inadequate alone, Prescientific Outlook

ETHICAL HUMANISM

definition 370; *see also* Anthropocentrism

ETHICAL INTELLECTUALISM

definition 357, history 357f, inadequacies 358-360; *see also* Science and Ethics: inadequateness

ETHICS

definition 11, 379, typically human 366; *see also* Philosophy

problems

see Man, Value

and Humanism

see Humanism: ethical perspective

and Science

see Science and Ethics

ETHOS OF SCIENCE

definition 138, features 138, 346; *see also* Purity

EXPERIENCE

definition 6f, different from experiment 7, 183f; as methodological approach to science 9f, essential to scientific theory 39-41;

31

needs complement by reflection 254f

public character: fosters cognitive dynamism 202f, leads to split with philosophers 243-245; *see also* Consensibility, Publicity

of reality
see Cosmos, Meaning, Openness, Religion, Responsiveness, Science
of self
see Creativity, Dignity, Freedom, Self, Value

EXPERIMENT
different from experience 7, from observation 42, 183, test of hypothesis 48, 187

creative (= **Experimentation**)
develops theoretical idea 42f, questions nature and interprets answer 43f, produces pure event 43f; fosters dynamism of knowledge 225f

dependent
submits to interpersonal test 51f, rigor of precision 49f, specialization 51
See also Observation, Theory

FACT
conditioned by theory 32, 35f, insufficient to make science 32; *see also* Hypothesis

FALLACIES
of misplaced concreteness 257, naturalistic: *see* Naturalism, "nothing-but" 287f, *see also* Reductionism

FALLIBILITY, of Man
stressed by science 409f, *see also* Self-Confrontation

FRAGMENTATION, of Science
feature of its development 405f; *see also* Scientists: humanistic difficulties

FREEDOM
inspired by science 354f; opposed by deterministic view of man 282-285, pantheism 309f, naturalism 360-363; problematic: *see* Man: ethical problem

FRIGHT
different from fear 121, 123, may attend scientific discovery 106-108, results from awareness of holiness 123; *see also* Religion: experience, Wonder

GALILEAN CONTROVERSY
not antireligious 291f; sincerity of Galileo's religious motivation 129-131, 294-297; philosophical rejection of occultism 37, 293f, insistence on the dignity of the creative mind 60f, autonomy of science relative to religion 294f; tragic outcome 297, *see also* Science and Culture: tension

GENETICITY
as methodological approach to science 9; *see also* Inductivity

GOD, Personal
experienced by scientists 117-120, implied by religious communion 123, objected to in the name of science 304f; *see also* Absolute, Religion

HISTORICISM
definition 233, results from instinctivism 233f, inimical to scientific progressivity 234f; *see also* Instinctivism, Revolution

HUMANISM
definition: confused issue 1f, danger of dogmatism 2f, ethical 370, heuristic 3, suiting the scientific age 5

INDUCTIVITY

as methodological approach to science 8; *see also* Geneticity

INEXHAUSTIBILITY, of Reality

perceived by great creators in the humanities 111f, science 110f
Newton's attitude: described 109, explained 109f
See also Mystery, Transrationality, Wonder

INSTINCTIVISM

epistemological: definition 213f, grounds 214; implies principle of mental economy 214f, historicism 233f; *see also* Economy, Historicism, Machian View
ethical: dehumanizing features 382f
religious: opposed by scientists 306f
See also Naturalness

INTELLECTION

constitutes cognition 206, different from intuition 205f, rationality 206; indispensable for humanization 417f, but insufficient alone 165, 426f; must be integrated by dialogue (*see*) and wisdom (*see*); *see also* Knowledge, Rationality, Understanding

INTELLECTUALISM

misleading ideal: *see* Adventurism, Ethical Intellectualism, Rationalism, Scientists: humanistic difficulties

INTERCONNECTEDNESS, of Nature

aspiration of research 139, ontological insight of science 268, inspires ethics 343f; *see also* Nature: intelligibility

INTERPERSONALITY

criterion of scientific objectivity 51f, essential to science 202; not impersonal: *see* Impersonality

KNOWLEDGE

characterized by experience 205, intellection 206, openness and service 206f, tacitness 255

ordinary and scientific

continuity of the two 391f, 394f, 418-420; novelty of the latter 392f, its superiority 393f, *see also* Novelty; humanizing synthesis of the two 417; tension: *see* Culture and Science

as value

experienced by scientists as an end in itself 150; see also Purity, Truth
presuppposed by research as ethical good 339; novelty of this outlook 340, not impersonal but committed to objectivity 341f; threatened by adventurism 150, naturalness 414f, technicalism 406f; *see also* Science and Ethics: contributions to
See also Intellection, Rationality, Understanding

LAPLACIAN IDEAL, of Science

description and implications 271-273; *see also* Axiomatic Deductivism, Mechanism, Rationalism

LAW, Scientific

definition 71, manifestation of the absolute 326f, problematic 223; *see also* Epistemology: problems

LOVE

experienced by scientists as passionate dedication 142f, dignified engagement 143, concerned respectfulness 143, personal communion 143f; *see also* Purity, Responsiveness, Reverence

MACHIAN VIEW, of Science

definition 213f; features: *see* Adap-

and intelligibility of nature 124-126, value of research 125-127

religion to science, psychological: trust in the intelligibility of nature 127-129, quest for the absolute

science to religion: awareness of the transcendence of God 129-131, compatibility of revelation (Bible) and scientific research 294f, conquest of anthropomorphism 304f, instinctivism 306f

mutual tension

arduous problem 116, emotion-laden 116, unavoidable 116, 120; must take into account mutual autonomy 131f, 318f, objective interconnection 319f, analogical thinking 323-326; *see also* Galilean Controversy, Absolute: ontological problem

humanizing synthesis

see Communion, Dialogue, Openness, Responsiveness, Responsibility, Reverence, Wisdom
See also Absolute, God, Mystery

RESEARCH, Scientific
quest for understanding 134f, invariants 135, ultimates 129, 139; *see also* Science: experience

RESPECT, for Reality
typical of scientific research 141f, expressed in care for precision 49f, concern for nature as intelligible 143; *see also* Meaning, Reverence

RESPONSIBILITY
definition 344, inspired by religion 124, and science 344-346, non-delegable character 351, opposed by pantheism 309f; *see also* Responsiveness, Science and Ethics: contributions to

RESPONSIVENESS
feature of science: attraction experienced 45-47, personal call 139-141

feature of humanizing dialogue 431-433

feature of lived wisdom 429, 433-435;
See also Humanization: inspired by science; Responsibility

REVERENCE
typical of religion 122-124
presupposed by science 119f, *see also* Respect; inspired by science 353f, *see also* Awe, Mystery

REVOLUTION, Scientific
upheaval produced by the rise of science, owing to its novelty: epistemological 194f, 392, ontological 330, ethical 340; *see also* Galilean Controversy, Knowledge: ordinary and scientific

as immanent in the development of science: asserted by philosophers 233f, denied by scientists 201, 228f; *see also* Dynamism of Knowledge

SAVING OF THE APPEARANCES
typical of the prescientific mind 195, entails occultism 195, 292-294, 296, reduces science to prediction 37

rejected by Galileo 37, incompatible with scientific ontology 265-267, 294f, 331, 333; *see also* Galilean Controversy, Prescientific Outlook

SCIENCE
definition

natural or experimental 3, Galilean 4; formal acceptation 6, methodological 6, experiential 6f; multiple 13, unified 13-15; dif-

and ONTOLOGY

contributions to

awareness of nature's intelligibility as intrinsic 265-267, universal 267f, accessible to mind 86-88; *see also* Cosmos: experience, Nature: intelligibility, Order

inadequateness

see Atheism, Contingentism, Deism, Deterministic View, Materialism, Mechanism, Pantheism

problems

see Absolute: ontological problem, Change: ontological problem, Man: ontological problem

synthesis

dangers in confusion and separation 334f, need for mutual autonomy and cooperation 335, common goal 336f; *see also* Humanism, Meaning, Reflection, Understanding, Wisdom

SCIENCE OF SCIENCE

definition 416, humanistically inadequate 416, 422; *see also* Philosophy of Science

SCIENTISM

definition 249; dissolves man 250f, 285, empties ethics 284f, 377f, pragmatizes and technicalizes science 251, 407f; *see also* Empiricism, Machian View, Positivism

entails: reductionism 249f, actualism 250, subjectivism 250, pragmatism 251, technicalism 251, 407f, mythology 251f, superstition 252

SCIENTISTS

humanistic difficulties

cognitive: frustration about science as deceptive ideal 160f, endless questioning 161, disintegrating knowledge 161, 405f; sense of

failure 162, despair 163; *see also* Agnosticism, Science and Epistemology: inadequateness

environmental: science as source of pressure 424, competition 424, exhaustion and restlessness 424f, confinement 425, pessimism 425, disenchantment 425f, estrangement 426, meaninglessness 426, self-alienation 426

ethical: disgust engendered by science as an instrument of war 164f, of political tyranny 164f, 175f, coldly intellectualistic 165, 176, sadistic exploitation of man 166f, inherently ambiguous 175; *see also* Arrogance, Science and Ethics: inadequateness

psychological: science experienced as a burden because of its demands: concentration 155, loneliness 155f, continual striving for novelty 147f, 156f; and consequences: professional deformations 157f, despondency 158f, total disappointment 160

See also Science and Humanism: inadequateness

and PHILOSOPHY

interested in

serious about metaphysics 330-332, philosophy 238-240; *see also* Metaphysics, Philosophy

opposed to

philosophical irrelevancy 241f, systematism 242, apriorism 242, 244, disregard for experience 243-245; *see also* Empiricism, Machian View, Positivism

SELF

Self-Actualization

claimed as absolute goal of man 371-373, as a consequence of science 373f; declared to be attainable only through self-transcend-